THE FATE OF THE MALE
GERM CELL

ADVANCES IN EXPERIMENTAL MEDICINE AND BIOLOGY

A Continuation Order Plan is available for this series. A continuation order will bring delivery of each new volume immediately upon publication. Volumes are billed only upon actual shipment. For further information please contact the publisher.

THE FATE OF THE MALE GERM CELL

Edited by

Richard Ivell

Institute for Hormone and Fertility Research
University of Hamburg
Hamburg, Germany

and

Adolf-Friedrich Holstein

Anatomical Institute
University of Hamburg
Hamburg, Germany

SPRINGER SCIENCE+BUSINESS MEDIA, LLC

Library of Congress Cataloging-in-Publication Data

The fate of the male germ cell / edited by Richard Ivell and Adolf
-Friedrich Holstein.
        p.   cm. -- (Advances in experimental medicine and biology ; v.
424)
     "Proceedings of a symposium on the fate of the male germ cell,
held December 5-7, 1996, in Hamburg, Germany"--t.p. verso.
   ISBN 978-0-306-45696-1   ISBN 978-1-4615-5913-9 (eBook)
   DOI 10.1007/978-1-4615-5913-9
   1. Spermatozoa--Congresses. 2. Spermatogenesis--Congresses.
3. Fertilization (Biology)--Congresses.   I. Ivell, Richard.
II. Holstein, A. F. (Adolf Friedrich) III. Series.
QP255.F38 1997
612.6'1--dc21.   97-27283
                                                            CIP

Proceedings of a symposium on The Fate of the Male Germ Cell,
held December 5 – 7, 1996, in Hamburg, Germany

ISBN 978-0-306-45696-1

© 1997 Springer Science+Business Media New York
Originally published by Plenum Press, New York in 1997

http://www.plenum.com

10 9 8 7 6 5 4 3 2 1

PREFACE

THE FASCINATION

The male germ cell is the only cell of the human organism that leaves the body when it has achieved its final, highly sophisticated structure and properties. The male germ cell is designed for one purpose only: to reach the female gametes and to fertilize them. The various stages in the development of the male germ cell are characterized by proliferative phases, by the recombination of the maternal and paternal chromosomes, and by the differentiation and development of a specialized transport vehicle, the spermatozoon. Furthermore, the establishment of a special pool of stem cells, the spermatogonia, guarantees the continuity of the sperm-generation process from puberty to old age.

THE FATE OF THE MALE GERM CELL

The destiny of any individual germ cell is determined by a program that we know only in fragments. On the one hand every human male is able to produce many billions of germ cells in his lifetime, yet the chance of any single sperm reaching and fertilizing the female germ cell is exceedingly rare. A fertility disturbance means that somewhere during the complicated playing out of the germ cell program mistakes are made, and the program fails. It is still a fact that more than 50% of men presenting with male factor infertility have to be diagnosed as idiopathic, largely because of our lack of knowledge and consequent lack of appropriate diagnostic tools. This ignorance has led to an extensive use of artificial methods of fertilization, where possibly an appropriate method of therapy could be derived to encourage more natural procreation. The methods of assisted reproduction and *in vitro* fertilization are successful. There is no doubt that in certain circumstances their use is fully justified. But they should not be used by default, where our ignorance simply prevents a more rational therapy. The success of these methods should not preempt and destroy basic research. We need basic research in order to recognize and understand the natural processes of sperm formation and maturation, which lead to the fertilization of an egg cell. And in one special context, we need to understand what goes wrong when a male germ cell becomes a tumor cell.

THE AIM OF THIS BOOK

This book brings together the views and opinions of a group of experts who have made extensive research into fundamental aspects of sperm production and maturation and the destiny of spermatozoa within the female tract. The contents of each chapter were presented as lectures during a symposium on the fate of the male germ cell held in Hamburg in December 1996. Much of the data are the results of targeted program funding by the German Federal Ministry of Education, Research and Technology (BMBF) to scientists in Giessen, Göttingen, and Hamburg and by the German Research Council (DFG) to the Münster-Hamburg Confocal Research Group. The editors particularly wish to thank all contributors and sponsors for their enthusiasm in helping to put together this volume. We should like to think that the reader of this book will share in the results of these endeavors to understand the fate of the male germ cell, where it comes from, how it is modified and specialized, and where it ends up. These results are intended to be a contribution toward improving the health of the male, so that children might be born by healing the father rather than by artificially bringing male and female gametes together in a culture dish.

Richard Ivell
Adolf F. Holstein

ACKNOWLEDGMENTS

The organization of a large symposium and the publication of a scientific volume is only possible with the help of many people. We have been especially fortunate in the generosity shown to us by a number of academic and industrial sponsors who have believed in the importance of basic research in the male to help the massive problem of male infertility. Special thanks are due not only to those bodies, such as the German Ministry of Education, Research and Technology and the German Research Council who have supported our research, but also to the Ferring company, the GEFEF foundation, the Ernst Schering research foundation, the German Society for Endocrinology, SmithKline Beecham Stiftung, the Hamburgische Wissenschaftliche Stiftung, the Alfred Marchionini Stiftung, Boehringer-Mannheim, Molecular Dynamics, and the KEME society.

We should also like to thank our advisory committee, Professors Freimut Leidenberger, Wolfgang Wuttke, Wolf-Bernhard Schill, and Eberhard Nieschlag, for advice and encouragement in the planning and organization of the symposium. Thanks are particularly due to our local team, without whom neither the symposium in Hamburg nor this volume would have been possible: Marion Böge, Gerhard Boje, Bernhard Biester, Rüdiger Färber, Ursula Fischer, Carmen Kretzschmar, Marianne Lück, Ingeborg Harkensee, Almuth Salewski, Elke Schäfer, Meike Schwartz, Sabine Schwartz, Nikola Urukalovic, and Slaven Urukalovic.

Finally, we are very grateful to Professor Heinrich Schulte for his unquestioning support, especially during the early stages where risks needed to be taken and the possibility of sponsorship seemed a distant hope, and, last but not least, to Petra Behring, who with extreme endurance and patience has taken on much of the practical organization, as well as the editorial and secretarial work, during the conference and in the preparation of this volume.

CONTENTS

Section C: The Role of the Testicular Accessory Cells

Section E: Post-Testicular Sperm Maturation

Section F: Control of the Male and Female Tracts

Section G: Gamete Interaction and Fertilization

THE FATE OF THE MALE
GERM CELL

HUMAN REPRODUCTION

The Missing Parts of the Puzzle

B. P. Setchell

Department of Animal Science
University of Adelaide
Waite Campus, Glen Osmond, SA 5064
Australia

I have been given a very challenging but interesting title. To cover this topic properly would require much more time and space than I have been allocated, so I intend to concentrate on only some aspects, particularly those where interesting and important questions remain which are not being currently addressed by the powerful modern techniques, which have increased our knowledge so much in certain aspects of reproductive biology over the last few years.

1. TESTIS SIZE

At first sight, this would not even appear to be a problem, but sperm production is limited by testis size. Sperm production per unit testis weight appears to be reasonable constant across many species, with the human probably less efficient in this regard (see Sharpe, 1994). Testis size is in general related to body weight (Kenagy & Trombulak, 1986, Moller, 1988, 1989), but within the primates, it is influenced by mating behaviour, with monogamous species, or those in which a dominant male has exclusive access to the females in the troup, having smaller testes, while in those species where the females mate when receptive with a number of males, the testes are in general larger (Harcourt et al, 1981). Many other groups of mammals have testes which are much larger as a fraction of body weight than humans; in some rodents and marsupials, the two testes comprise around 5% of body weight (Kenagy & Trombulak, 1986). If a 80kg man had testes of this size, each testis would weigh about 2 kg. There are two closely related species of Australian rodents, *Notomys alexis* and *Pseudomys australis*, which differ by a factor of 2 in body weight, but by a factor of 100 in testis weight (Breed 1982). We know very little about what determines final testis size, except that it appears to be limited by the numbers of Sertoli cells, which do not divide after puberty. The numbers of Sertoli cells, and conse-

The Fate of the Male Germ Cell, edited by Ivell and Holstein
Plenum Press, New York, 1997

quently final testis size can be increased by making young animals hypothyroid (Bunick et al, 1994), and in sheep, testis size is correlated with ovulation rate in related females (Land & Carr,1975). FSH is known to increase Sertoli cell divisions in fetal rats (Orth, 1984) and treatment of genetically hypogonadal mice with FSH immediately after birth increases Sertoli cell numbers, but not to values seen in normal animals (Singh & Handelsman, 1995). It is also interesting that mice lacking p27^{Kip1}, a cyclin dependent kinase inhibitor, have testes which are 80% larger than control, while body weight is only about 20% greater (Nakayama et al, 1996, but cf Fero et al, 1996). A better knowledge of what normally determines testis size may enable us more logically to attempt to treat men whose sperm production is abnormally low.

2. DESCENT OF THE TESTIS

It has been known since the 18th century, or probably even before then, that the human testis originates in the abdominal cavity in the fetus, and subsequently migrates into the scrotum. The process involves a transient structure known as the gubernaculum, and there have been some important recent discoveries concerning the mechanism for this migration (Wensing & Colenbrander, 1986, Hutson et al, 1990), particularly the involvement of the genitofemoral nerve and calcitonin gene related peptide (Hutson et al, 1996). Furthermore, in contrast to other organs such as the kidney which move during development , the testis retains its original vascular connection, which becomes elongated during descent, and in many species the testicular artery becomes even longer by forming coils in the spermatic cord (Setchell, 1970, 1978). However, the real question is not how the testis migrates, but why. Descent of the testes into a scrotum does not occur even in all mammals (Carrick & Setchell, 1977, Williams & Hutson, 1991) and none of the other classes of vertebrates have extra-abdominal testes. For the individual mammal with scrotal testes, it poses considerable risks in having the testes outside the abdomen, where they are much more vulnerable to traumatic damage, and furthermore, testis function is seriously compromised if for whatever reason the testes fail to descend properly, or if they are exposed to temperatures no higher than those inside the abdominal cavity. There must be some advantage in having the testes at a lower temperature or with an elongated vascular supply, but what that advantage is, we do not know. It has been suggested that there would be lower rate of mutation at the lower temperature (Ehrenberg et al, 1957), or even that the scrotum is a form of ornamentation to attract females (Portman, 1952) or that it arose in those mammals which leap about a lot (Chance, 1996), but none of these suggestions has been tested experimentally, indeed it would be dificult to do so, and the question remains unanswered.It is interesting that in dolphins, in which the testes remain inside the abdominal cavity, venous blood from the dorsal fin and flukes runs into the abdominal cavity, and then between the arteries carrying blood to the testes (Rommel et al, 1992); this anatomical arrangement appears to cool the testes and adjacent colon (Rommel et al, 1994, Pabst et al, 1995). An analogous system is found in Phocid seals, in which the testes lie outside the body wall, but underneath the blubber; in these animals an inguinal venous plexus carrying blood from the flippers overlies the median aspect of the testis, and cools it (Blix et al, 1983, Rommel et al, 1995).

Regardless of the reason for testicular descent, the scrotal testis is susceptible to damage if its temperature is raised to body temperature for as little as 48 h continuously, or exposed intermittently for as little as 8h per 24 for several weeks (Mieusset et al, 1991). Similarly, as little as 20 minutes exposure of a rat's testes to a temperature of 43 C, which is no hotter than most people's hot bath, causes almost complete elimination of the mid-pachytene spermatocytes and many young spermatids (Collins & Lacy, 1969), and this exposure or 41 C for 1 h leads to a temporary fall to about half in testis weight and a period when virtually no spermatozoa leave the testis (Setchell & Waites, 1972, Galil & Setchell 1988). The temperatures involved are not high enough to damage other cells, and the reason for this extraordinary susceptibility to heat is still unknown.

3. ENDOTHELIAL AND PERIVASCULAR CELLS

The endothelial cells of the capillaries in the rat testis are unusual among those of endocrine tissues in that they are unfenestrated (see Setchell et al, 1994), although some capillaries in the human testis do have fenestrations (Ergun et al, 1996). The endothelium of the testicular blood vessels in the rat is also unusual in having a high level of alkaline phosphatase (Kormano 1967) and gamma glutamyl transpeptidase (Niemi and Setchell, 1983, Holash et al, 1993); the latter enzyme is often associated with a high level of amino acid transport, and the endothelial cells of the small veins appear to have a saturable transport system for leucine (Bustamante and Setchell, 1982). The endothelial cells in the human testis also express P-glycoprotein as detected by a mouse monoclonal antibody against the human multidrug-resistance gene product (Cordon-Cardo et al, 1989, Holash et al, 1993). Inhibition of P- glycoprotein by verapamil blocks the transport of vincristine by mouse brain endothelial cells (Tatsuda et al, 1992), and disruption of the mouse mdr1a P-glycoprotein gene leads to a deficiency in the blood-brain barrier to ivermectin and vinblastine, and there is also greater exclusion of these two drugs from the testes of mdr1a-/-mice than controls (Schinkel et al, 1994). As cortisol and aldosterone are also transported by human P-glycoprotein (Ueda et al, 1992), we have studied the effect of its inhibition by verapamil on the distribution of testosterone between interstitial extracellular fluid and venous blood plasma, but could not detect any significant effect (Setchell and Zupp, unpublished observations). Microvessels from the brain and testis also show a high level of immunostaining for erythroid-type glucose transporter, also known as GLUT-1, which is abundant in tissues with barrier functions (Harik et al, 1990, Holash et al, 1993). This transporter is inhibited by 2,3,7,8-tetrachlorodibenzo-p-dioxin, which also causes a fall in blood testosterone levels in rats (Moore et al, 1985, 1991, Kleeman et al, 1990), but no studies have yet been undertaken on the effect of this inhibitor on the distribution of glucose or testosterone in the various compartments of the testis. Other glucose transport sytems, GLUT-3 and GLUT-5 are found in the cells inside the seminiferous tubules (Burant & Davidson, 1994). It has also been known for many years that the testicular endothelial cells are particularly susceptible to the effects of cadmium salts (see Setchell et al, 1994), but no convincing explanation has been given for this sensitivity.

Another interesting feature of the testicular vasculature, though probably originating in the perivascular cells rather than the endothelial cells, is vasomotion, that is rhythmic variation in blood flow, with a periodicity of about 10 per minute (Damber et al, 1982, 1986). In contrast to the situation in other tissues, vasomotion in the rat testis can easily be observed in anaesthetized animals, and contrary to earlier reports, it is present in the testes of a number of species of mammals, including mice, rams and several marsupials (Collin et al, 1996). Furthermore, when the testes of rats were heated, vasomotion became less apparent and finally disappeared when the temperature reached a value between 37 and 40 C, depending on the individual animal, reappearing as the testis was subsequently cooled, and becoming much more pronounced, although with a very slow frequency as the temperature of the testis fell below normal (Setchell et al, 1995).

4. FUNCTIONS OF TESTOSTERONE IN THE TESTIS

The question of how testosterone functions in the testis, as opposed to its androgenic activity elsewhere in the body is made more difficult by the various values which are quoted for the levels of this hormone in the organ. Many authors quote the concentration in the whole tissue, and while Leydig cells do not store large amounts of the hormone, they do contain considerably more than the fluids which bathe them and the cells inside the seminiferous tubules. Values for whole tissue concentrations are therefore at best misleading. A large number of studies have used so-called testicular interstitial fluid collected post mortem from rat testes. Testosterone concentrations in fluid collected in this way are considerably higher than in testicular venous blood from the same animals, implying that there is a large gradient of this lipid soluble hormone across the endothelial cells, a most unlikely situation. When interstitial extracellular fluid is collected in such a way as to avoid leakage of hormone from the Leydig cells, the concentrations in testicular venous blood plasma and interstitial extracellular fluid are similar (Maddocks and Setchell, 1988). The validity of this approach was established by examining rats in which the secretory function of the Leydig cells was suppressed by the administration of testosterone by subcutaneous silastic capsules. In these animals, in contrast to controls with empty implants, the testosterone concentrations in fluid collected post mortem was similar to venous blood, and to fluid collected by a physiological technique from anaesthetized animals; in the animals with larger implants which produce high circulating levels of testosterone, the concentrations in blood plasma, both peripheral and testicular venous were actually higher than the interstitial fluid collected by either technique, an observation which has not been satisfactorily explained. Similarly, in rats from which rete testis fluid, seminiferous tubular fluid and venous blood plasma were collected and analysed for testosterone, the concentrations were similar in control animals, but in those injected with human chorionic gonadotrophin (hCG) or with testosterone, the levels in the venous blood plasma were higher than in the tubular or rete fluids (Setchell, 1980). This was taken to indicate a saturable transport system moving testosterone into the tubules, but such a system could not be demonstrated in isolated tubules, so the true explanation is still awaited. The situation in the human testis is even less clear, but by analogy with the rat, testicular venous blood, collected from a vein on the surface of the testis should give the best indication of hormone levels in the tissue.

Another confusing factor is the variable occurence of androgen binding proteins in various species. Humans have a high concentration of sex hormone binding globulin (SHBG), which binds testosterone as well as oestrogens in blood plasma, and the Sertoli cells secrete an androgen binding protein (ABP) into the tubular fluid; rams are similar except that the concentrations of SHBG in plasma are lower. In rats, there is a very low concentration of SHBG in blood plasma, but a high concentration of ABP in the tubular fluid, where it appears to play an important role in transporting androgens to the epididymis. In pigs, there is neither SHBG nor ABP, but there are very large numbers of Leydig cells in the testis, up to 35% of the entire tissue (Setchell et al, 1994). It is difficult to postulate a function for these carrier proteins which can be reconciled with all these observations.

In many other tissues, testosterone is converted to dihydrotestosterone (DHT) which has a higher affinity for the androgen receptor. However, in the testis, testosterone, rather than

DHT is the principal androgen in nuclei from the seminiferous tubules of adult rats. Nevertheless, inhibition of 5a-reductase, the enzyme forming DHT from testosterone impairs the testosterone-dependent restoration of spermiogenesis in adult rats (O'Donnell et al, 1996), suggesting that the role of the two steroids in spermatogenesis needs reappraisal.

Even more puzzling is the observation that in mice lacking the transcription factor NGFI-A (also known as Egr-1, zif/268 and Krox-24) in which there is a deficiency in LH-ß, Leydig cells are markedly atrophic, but testis weights were normal, as were testosterone levels and seminal vesicle weights, and the males were fertile, in contrast to similarly deficient females (Lee et al, 1996).

Recently, the Leydig cells have also been shown to produce various peptide hormones (see Setchell et al, 1994), oxytocin (see Ivell et al, Chapter 47, this volume) and an insulin-like peptide (Adham et al, 1996). The functions of these peptides is still unclear, and it is curious that no oxytocin could be detected in testicular lymph from ram, bull or boar (Setchell and Bicknell, unpublished observations).

5. SPERMATOGENESIS

One important aspect of this topic to be established in recent years is that each Sertoli cell can only deal with a definite number of germ cells. The exact number depends on the species, with the human at the lower end of the range, but this does mean that the number of Sertoli cells imposes an upper limit to the numbers of germ cells which a testis can produce per unit time (see Sharpe, 1994).

Two major problems still remain. The first concerns the stem cells A_S or A_0 spermatogonia, which give rise to the differentiating spermatogonia and thence the spermatocytes, spermatids and eventually spermatozoa. There is still no general agreement about the way these cells are maintained in the adult. Most accept that they are a slowly multiplying population of cells, which maintain their numbers by divisions not associated with other events in the cycle and feed into the synchronised development of the germ cells as A_1-spermatogonia at a specific stage of the spermatogenic cycle (Huckins, 1971, Oakberg, 1971, de Rooij, 1982, de Rooij and Janssen, 1987). However, there is still some support for the idea that A_4 spermatogonia can give rise either to differentiating intermediate spermatogonia or revert to A_1 spermatogonia, and that the stem cells (A_0-spermatogonia) only divide if cell numbers in the testis are reduced (Dym et al, 1995). A recent review (Meistrich and van Beck, 1993) concluded that the identity of the stem cells in the testis still remains a mystery. It is most important that this long-standing controversy is settled, and furthermore that a way is found to enumerate the stem cells in the testis, distinguishing them from the differentiating spermatogonia. The only technique currently available is the time-consuming spermatogenic colony assay (Withers et al, 1974), which is only of use in estimating the stem cells when their numbers are severely reduced, for example after irradiation. However, it is interesting that using this technique, it has been possible to show that local heating of the testis, conventionally thought not to affect the stem cells, leads to a significant reduction in their numbers, and furthermore, to potentiate the effects of irradiation (Reid et al, 1981).

In some recent studies, it has been shown that A-spermatogonia in mice (Yoshinaga et al, 1991) and rats (Dym et al, 1995) express c-kit receptor, for which the ligand is stem cell factor (also known as steel factor) which is produced by the Sertoli cells (Rossi et al, 1993). Furthermore, Yoshinaga et al (1991) concluded that A_1 to A_4 spermatogonia are c-kit dependent, whereas A_0 cells are not, suggesting that they are not stem cells, and the c-kit is in the phosphorylated form in A_1 to A_4 spermatogonia treated with the soluble form of the ligand (Dym et al, 1995). In the human testis, c-kit was found in early spermato-genic cells, and also Leydig cells, whereas its ligand, stem cell factor, could be demon-strated in Sertoli cells and Leydig cells (Sandlow et al, 1996).

Spermatogonial numbers in the testes of mice and hamsters are reduced by paren-teral or intratesticular injection of inhibin preparations (van Dissel-Emiliani et al, 1989) and spermatogonial proliferation in co-cultures with Sertoli cells is stimulated by activin (Mather et al, 1990, 1992). However, none of the studies on paracrine effects in the testis have taken into account recent evidence from Xenopus embryos that activin, which is an important morphogen in this system, can at a low concentration stimulate the expression of Xbra gene, whereas with higher concentrations of the morphogen, Xbra is not ex-pressed, but the gene Xgsc is instead (Gurdon et al, 1995). Obviously, the concentration of the paracrine factor can determine which genes it causes to be expressed, and this needs to be taken into account when considering how factors might act, particularly lipophobic fac-tors which would be restricted by the blood-testis barrier

The other major problem in spermatogenesis is how the spermatogonia, which di-vide mitotically like somatic cells elsewhere in the body, are switched to become prelep-totene spermatocytes, and following a final period of DNA synthesis, enter the meiotic prophase. In the locust, *Schistocerca gregaria*, treatment with actinomycin D causes po-tential spermatocytes to fail to enter meiosis, and instead, the expected meiotic sequence is replace by an additional mitotic division (Jain and Singh, 1967). As actinomycin D blocks RNA synthesis, this would suggest that expression of a particular gene or genes at the critical stage is necessary for the mitosis-meiosis transition. Attempts to replicate this experiment in mammals were frustrated by the exclusion of actinomycin D from the tes-tes, to a similar extent to the brain (Ro and Busch, 1965), suggesting that the endothelial stage of the blood-testis barriers (Ploen and Setchell, 1992) is involved, possibly through the multidrug resistance gene product, P-glycoprotein, which is present in these cells (see section 3). A similar exclusion of vinblastine and its metabolite O^4-deacetylvinblastine from brain and testis (van Tellingen et al, 1993) may be due to the same multidrug resis-tance gene product. Alternative approaches should be considered to this vital problem. A search has been undertaken for genes expressed at a time in spermatogenesis when they could have a role in meiosis (e.g. Don et al, 1994), and several genes have been identified by knockout procedures to cause either the formation of abnormal spermatids and sperma-tozoa (PMS2 defective mice, Baker et al, 1995) or arrest of the spermatocytes at pachytene, with no production of spermatids (MLH1 deficient mice, Edelmann et al, 1996). However, all these genes appear to be acting at too late a stage to be the trigger for entry into meiosis, although it is interesting that arrest or cell death at pachytene is a fea-ture of a number of treatments damaging spermatogenesis (see Setchell, 1978), and most attempts to maintain spermatogenesis in mammalian systems in vitro are plagued with ar-rest of development at pachytene (see Dirami et al, 1996). Cells which have already passed this critical point in vivo are then capable of continuing through the meiotic divi-sions in vitro (Parvinen et al, 1983). However, there are several old papers in the literature reporting successful development of male meiosis in vitro using human material (Matte and Sasaki, 1971, Ghatnekar et al, 1974), and the question probably merits reinvestigation.

6. EPIDIDYMAL FUNCTIONS

It has been clear for some years that during their passage through the epididymis, many aspects of sperm function and structure are altered (Bedford, 1979). However, with the advent of intracytoplasmic sperm injection, and the successful achievement of off-spring following electrofusion of mouse oocytes with round spermatids (Ogura et al, 1994), it is now not clear whether these changes are necessary for succesful fertilization and syngamy, or whether they are necessary only for the delivery of the sperm to the correct place at the appropriate time.

Another aspect of epididymal physiology which is still not clear is how the testis regulates the various activities of the different parts of the epididymis. Clearly, testosterone carried from the testis to the epididymis in the blood is important for the control of many activities. However, some functions, particularly those of the initial segment and the efferent ducts seem to depend more on factors carried there in the luminal fluid secreted by the testis though the rete testis (Jones et al, 1988, 1989). There is also some evidence for a third channel of communication, namely via lymph from the testis through lymphatic vessels which run across the surface of the epididymis. When radioactive albumin was injected into one testis of rats, much more radioactivity could be found in the ipsilteral than in the contralateral epididymis. Similar accumulation of radioactivity in one epididymis was obtained when radioactive albumin was infused into an actual lymphatic vessel on the surface near the caudal pole of the testis; this vessel emptied into a large lymphatic vessel which runs between the ductus deferens and the body of the epididymis (Setchell, 1986). There is some evidence supporting the importance of this route of communication from studies in rams in which the epididymis was surgically separated from the testis. In these animals, there were changes, albeit rather variable, in the motility characteristics of ejaculated spermatozoa collected over several months following the operation, and the survival of the embryos produced by these rams in control ewes was also reduced (Quintana Casares and Setchell, unpublished observations). Finally, there is the question of oestrogen effects on the epididymis. We showed some years ago that in ram lambs treated with oestradiol as a growth promotant, there was significant enlargement of the epididymis (Papachristoforou et al, 1985). Recently, evidence has been obtained from mice in which the oestrogen receptor gene was deleted, that there is failure of fluid resorption in the ductuli efferentes, leading to accumulation of fluid in the rete testis and seminiferous tubules, and ultimately to degeneration of the spermatogenic tissue (Lubahn et al, 1996), similar to that seen after efferent duct ligation. If oestrogens are involved in the control of epididymal function, it is perhaps relevant that in the stallion (Setchell and Cox, 1982) and boar (Setchell et al, 1983), there are considerable concentrations of the conjugated oestrogen, oestrone sulphate, in testicular lymph. It has also been shown that the ram testis secretes significant amounts of oestradiol, both into blood and lymph (Setchell et al, 1991).

7. SPERM QUALITY AND NUMBERS

It is often said that human sperm exhibit a wide range of abnormalities, even in good quality samples of semen, and that the proportion of abnormal forms is much greater than in any of the domestic animals. This does appear to be the case, particularly with regard to spaces in the condensed chromatin of the head of the sperm, but recent evidence suggests that in some rodents, there are large vacuoles in the chromatin of the sperm head (Breed et al, 1988, Breed, 1993), so the human may not be as unusual in this regard as we thought.

Considerable interest has been shown in recent reports (Carlsen et al, 1992, Irvine, 1994, van Waeleghem et al, 1994, Auger et al, 1995, Auger & Jouannet, 1995) that the number of spermatozoa in normal human semen has fallen appreciably over the last 50 years, especially as this fall was not apparent in two other studies from Europe (Vierula et al, 1995, Bujan et al, 1996) nor in two studies from USA (Paulsen et al, 1996, Fisch et al, 1996). Furthermore, criticisms have been levelled at some of the studies which drew on data from a number of centres (Farrow, 1994, Sherins, 1995). It was also noticed that the decrease appeared to be between the data collected before 1960 and those obtained after 1970, with no trend in the later years (Olsen et al, 1995), but this has been challenged in later studies (Auger et al, 1995, Irvine et al, 1996).It was also suggested that the apparent decline was a result of a changed reference range of "normal" in the human population (Bromwich et al, 1994) and that human sperm counts are not normally distributed (Berman et al, 1996). There is also some even more puzzling evidence of large geograhical variations in "normal" sperm concentrations (Fisch & Goluboff, 1996). If this decline over the years is real, and if it is due to some change in the environment, one would expect to see a similar fall in the sperm numbers in the semen of domestic animals. Semen has been collected from these animals for artificial insemination since the turn of the century (Iwanoff, 1907), and sperm counts have been done since the early 1930's, using haematocytometers as with human semen, or turbidimetric techniques calibrated against haematocytometers. I have collected 137 studies from the literature involving bulls, 76 for boars and 129 for rams (Setchell, 1997). All involved adult animals, from which semen was collected regularly but at a frequency which would not be likely to cause a fall in sperm counts, and were usually control animals in nutritional or breeding experiments. The references were obtained systematically from Animal Breeding Abstracts, and where possible the original articles consulted to obtain mean values for each study; where the original reference was not easily obtainable, values were taken from the abstract. The bull data showed absolutely no correlation of sperm count with year of publication ($r^2 = 0.000$), with the boars there was a slight but not significant positive correlation ($r^2 = 0.041$) and with the sheep, there was a slight, but significant rise in sperm counts with time ($r^2 = 0.124$ for sperm counts and 0.126 for total sperm per ejaculate; not all authors gave both values). Therefore, it would appear that if the fall in human sperm counts is real, then it must be due to something which is specifically affecting our species.

One suggested cause of the fall in humans is exposure to oestrogens (Sharpe and Skakkebaek, 1993). It is therefore probably relevant that in the 1940's and 1950's in Australia, large numbers of sheep were exposed to pastures dominated by subterranean or red clovers, which contained large amounts of phytoestrogens, genistein, formenonetin and other related compounds. These pastures caused serious infertility in the ewes, with uterine prolapse and dystokia (Bennett et al, 1946, Schinckel, 1948, Moule et al, 1963, Adams, 1981), enlargement of the bulbourethral (Cowper's) glands in wethers, but apparently had no effect on the fertility of the rams (Bennett, 1946). The only paper apparently published on semen from animals grazing these pastures (Meyer, 1970) does not give actual sperm concentrations, but the colour and consistency of the semen (which in this species are well correlated with sperm number), and sperm motility were unaffected, even though the rams had been exposed to so much oestrogen that they were lactating. In another study, rams were grazed from 3 months to 12 months of age on red clover dominated pasture which was equivalent to 15mg stilboestrol per day intramuscularly in its effect on teat length in wethers also grazing the same pasture, a dose large enough to interfere with spermatogenesis (Moule and Mattner, 1961). Testis size, sperm numbers in the epididymis and seminal vesicle weight in the rams were no different from similar animals grazed on a non-oestrogenic grass pasture (George and Turnbull, 1966). However, the administration of genistein, one of the active compounds in subterranean clover, to male mice did impair their fertility (East, 1955). Different oestrogens may have a different effect, and there may also be species differences in the response in semen quality, but these findings do cast some doubt on the possibility of environmental oestrogens being of major significance in male fertility.

8. PATERNAL EFFECTS ON FERTILIZATION AND EMBRYONIC DEVELOPMENT

It has been known for many years that exposure of male animals to hot environments or local heating of their testes leads to infertility due to azoospermia or severe oligospermia (see Setchell 1978). However, it is less generally accepted that as well as frank infertility, these treatments can also lead to effects on fertilization rates even when adequate numbers of spermatozoa are present, and the embryos which do result have a reduced chance of survival to term. This has been demonstrated in mice exposed to a hot environment (34.5°C and 65% humidity for 24 h, Bellve, 1972, 1973 or 32°C and 65% humidity for 24 h, Burfening et al, 1970), in boars (34.5°C for 8h and 31°C for 16 h each day for 90 days, Wettemann et al, 1976), in rams whose scrota were placed in insulating bags which raised subcutaneous scrotal temperature by about 2°C for 16h per day for 11 days (Mieusset et al, 1992) and in rats whose testes were heated by immersion in a water bath at 43°C for 30 min (Setchell et al, 1988). The heated rats when mated to normal females produced smaller than normal litters for up to 14 weeks after a single exposure (Zupp & Setchell, 1988). Part of this effect may have been due to the production of chromosomal non-dysjunction during spermatogenesis (Garriott & Chisman, 1980, Waldbieser & Chisman, 1986, van Zelst et al, 1995) leading to the formation of chromosomally abnormal embryos, whose chances of survival are reduced. Another possibility is that following heating of the testes, sperm are produced which can inititate development of the egg, but do not contribute their genes to the embryo. Embryos with no paternal genes obtained by nuclear tranfer show poor trophoblast development, with some retardation of embryo growth (Surani et al, 1984).

In rams whose scrota were insulated for as little as 8h per day so as to produce a rise in testicular temperature of about 2 C (the normal abdominal-scrotal gradient in this species is about 6 C), sperm selected for good motility by swim-up procedures were less capable than control sperm of fertilising ova in vitro, even when comparable numbers were present, and the embryos which did arise developed less rapidly in culture up to blastocyst stage (Ekpe et al, 1992, 1993, Setchell, 1994). Also, when the testes of male mice were heated once to 42 C for about half an hour, the embryos they sired in normal females were smaller until about 11.5 days of pregnancy, although thereafter they were normal in size (Ekpe et al, 1994, Setchell et al, 1997). This effect did not appear to be due to later mating by the heated males. Furthermore, the pattern of growth retardation did not correspond to that seen in mice deficient in the gene for IGF2, which were normal in size until day 13.5, and then showed retarded growth up to birth (Baker et al, 1993). This gene was considered to be a possibility, since only the paternal copy is expressed (De Chiara et al, 1990, 1991) and it is expressed in early embryos (Lee et al, 1990). The effect of heating the males is probably more similar to that described by Snow & Tam (1979) and Tam & Snow (1981), in embryos from pregnant female mice treated with mitomycin. Embryo size was reduced at 7.5 days to less that 10 percent of normal, but the embryos were almost normal in size by 13.5 days, presumably due to compensatory growth. Bellve (1973) has shown that embryos sired by heat-exposed males show developmental retardation or arrest at the morula stage.

There is a clear need for more investigations in this area, especially as it has recently been shown, that contrary to long-held belief, the male does contribute some mitochondrial DNA to the embryo in mice (Gyllensten et al, 1991).

9. CONCLUSION

I hope that in this paper I have provided a few examples of gaps in our understanding of male reproduction in the human, which are not being addressed by current techniques. While it is important to exploit new techniques to the full, and to follow up leads which the experiments with these techniques reveal, it is also important sometimes to stand back from what we are doing, and to look at the gaps in our knowledge, and how a better understanding of these problems may help us in our practice of andrology.

10. REFERENCES

Adams NR 1981 A changed responsiveness to oestrogen in ewes with clover disease. Journal of Reproduction and Fertility Supplement 30: 223–230.

Adham IM, Burkhardt-Gottges E & Engel W 1996 The Leydig insulin-like peptide. Journal of Animal Breeding and Genetics 113: 229–235.

Auger J, Kunstmann JM, Czyglik F & Jouannet P 1995 Decline in semen quality among fertile men in Paris during the past 20 years. New England Journal of Medicine 332: 281–285.

Baker, J, Liu JP, Robertson EJ & Efstratiadis A 1993 Role of insulin-like growth factors in embryonic and post-natal growth. Cell 75: 73–82.

Baker SM, Bronner CE, Zhang L, Plug AW, Robatzek M, Warren G, Elliott EA, Yu J, Ashley T, Arnheim N, Flavell RA & Liskay RM 1995 Male mice defective in the DNA mismatch repair gene PMS2 exhibit abnormal chromosome synapsis in meiosis. Cell 82: 309–319.

Bedford JM 1979 Evolution of sperm maturation and sperm storage functions of the epididymis. In The Spermatozoon, pp 7–21, Eds DW Fawcett & JM Bedford, Urban & Schwarzenberg, Baltimore.

Bellve AR 1972 Viability and survival of mouse embryos following parental exposure to high temperatures. Journal of Reproduction and Fertility 30: 71–81.

Bellve AR 1973 Development of mouse embryos with abnormalities induced by parental heat stress. Journal of Reproduction and Fertility 35: 393–403.

Bennetts HW 1946 Metaplasia in the sex organs of castrated male sheep maintained on early subterranean clover pastures. Australian Veterinary Journal 22: 70–78.

Bennetts HW, Underwood EJ & Shier FL 1946 A specific breeding problem of sheep on subterranean clover pastures in Western Australia. Australian Veterinary Journal 22: 2–12.

Berman NG, Wang C & Paulsen CA 1996 Methodological issues in the analysis of human sperm concentration data. Journal of Andrology 17: 68–73.

Blix AS, Fay FH & Ronald K. 1983 On testicular cooling in Phocid seals. Polar Research 1 n.s.: 231–233.

Breed WG 1982 Morphological variation in the testes and accessory sex organs of Australian rodents in the genera Pseudomys and Notomys. Journal of Reproduction and Fertility 66: 607–613.

Breed WG 1993 Novel organization of the spermatozoa in two species of murid rodents from Southern Asia. Journal of Reproduction and Fertility 99: 149–158.

Breed WG, Cox GA, Leigh CM & Hawkins P 1988 Sperm head structure of a murid rodent from Southern Africa: The red veld rat Aethomys chrysophilus. Gamete Research 19: 191–202.

Bromwich P, Cohen J, Stewart I & Walker A 1994 Decline in sperm counts: an artefact of changed reference range of "normal". British Medical Journal 309: 19–22.

Bujan L, Mansat, Pontonnier F & Mieusset R 1996 Time series analysis of sperm concentration in fertile men in Toulouse, France between 1977 and 1992. British Medical Journal 312: 471–472.

Bunick D, Kirby J, Hess RA, & Cooke PS 1994 Developmental expression of testis messenger ribonucleic acids in the rat following propylthiouracil-induced neonatal hypothyroidism. Biology of Reproduction 51: 706–713.

Burant CF & Davidson NO 1994 GLUT3 glucose transporter isoform in rat testis: localization, effect of diabetes mellitus, and comparison to human testis. American Journal of Physiology 267: R1488–1495.

Burfening PJ, Elliott DS, Eisen EJ & Ulberg LC 1970 Survival of embryos resulting from spermatozoa produced by mice exposed to elevated ambient temperature. Journal of Animal Science 30: 578–582.

Bustamante JC & Setchell BP 1982 Kinetics of the facilitated diffusion of leucine into the perfused testis of the rat. Journal of Physiology 324: 49–50P.

Carlsen E, Giwercman A, Keiding N & Skakkebaek N 1992 Evidence for decreasing quality of semen during the past 50 years British Medical Journal 305: 609–612.

Carrick FN & Setchell BP 1977 The evolution of the scrotum. In Reproduction and Evolution, pp165–170, Eds JH Calaby & CH Tyndale-Biscoe. Australian Academy of Science, Canberra.

Chance MRA 1996 Reason for externalization of the testis of mammals. Journal of Zoology 239: 691–695.

Collin O, Zupp JL & Setchell BP 1997 Testicular vasomotion in different mammals. Journal of Reproduction and Fertility (in press).

Collins PM & Lacy D 1969 Studies on the structure and function of the mammalian testis. II. Cytological and histochemical observations on the testis of the rat after a single exposure to heat applied for different lengths of time. Proceedings of the Royal Society of London, series B 172: 17–38.

Cordon-Cardo C, O'Brien JP, Casals D, Rittman-Grauer L, Biedler JL, Melamed MR & Bertino JR 1989 Multidrug-resistance gene (P-glycoprotein) is expressed by endothelial cells at blood-brain barrier sites. Proceedings of the National Academy of Science USA 86: 695–698.

Damber JE, Lindahl O, Selstam G & Tenland T 1982 Testicualr blood flow measured with a laser Doppler flowmeter: acute effects of catecholamines. Acta Physiologica Scandinavica 115: 209–215.

Damber JE, Bergh A, Fagrell B, Lindahl O & Rooth P 1986 Testicular circulation in the rat studied by video-photometric capillaroscopy, fluorescence microscopy and laser Doppler flowmetry. Acta Physiologica Scandinavica 126: 371–376.

De Chiara TM, Efstratiadis A & Robertson EJ 1990 A growth-deficiency phenotype in heterozygous mice carrying an insulin-like growth factor gene disrupted by targeting. Nature 345: 78–80.

De Chiara TM, Robertson EJ & Efstratiadis A 1991 Parental imprinting of the mouse insulin-like growth factor II gene. Cell 64: 849–859.

De Rooij DG 1982 Proliferation and differentiation of undifferentiated spermatogonia in the mammalian testis. In Stem Cells, pp 89–117, Ed CS Potten, Churchill Livingston, Edinburgh.

De Rooij DG & Janssen JM 1987 The regulation of the density of spermatogonia in the seminiferous epithelium of the Chinese hamster. I Undifferentiated spermatogonia. Anatomical Record 217: 124–130.

De Rooij DG, van Dissel-Emiliani FMF & van Pelt AMM 1994 Regulation of spermatogonial proliferation. Annals of the New York Academy of Science 140–153.

Dirami G, Ravindranath N, Jia MC & Dym M 1996 Isolation and culture of immature rat type A spermatogonial stem cells. In Signal Transduction in Testicular Cells, pp 142–165, Eds V Hansson, FO Levy & K Tasken, Springer, Berlin.

Don J, Winer MA & Wolgemuth DJ 1994 Developmentally regulated expression during gametogenesis of the murine gene meg1 suggests a role in meiosis. Molecular Reproduction and Development 38: 16–23.

Dym M, Jia MC, Dirami G, Proce JM, Rabin SJ, Mocchetti I & Ravindranath N 1995 Expression of c-kit receptor and its autophosphorylation in immature rat type A spermatogonia. Biology of Reproduction 52: 8–19.

East J 1955 The effect of genistein on the fertility of mice. Journal of Endocrinology 13: 94–100.

Edelmann W, Cohen PE, Kane M, Lau K, Morrow B, Bennett S, Umar A, Kunkel T, Catttoretti G, Chaganti R, Pollard JW, Kolodner R & Kucherlapati R 1996 Meiotic arrest in MLH1-deficient mice. Cell 85: 1125–1134.

Ekpe G, Seamark RF, Sowerbutts SF & Setchell BP 1992 Effect of intermittent scrotal insulation on fertilising ability of ram spermatozoa and the development of the embryos to blastocysts in vitro. Proceedings of the Australian Society for Reproductive Biology 24: 51.

Ekpe G, Zupp JL, Seamark RF & Setchell BP 1993 Fertilising ability of spermatozoa from rams subjected to intermittent scrotal insulation and development of the resultant embryos in vitro. Proceedings of the Australian Society for Reproductive Biology 25: 88.

Ekpe G, Zupp, JL & Setchell BP 1994 Retarded embryo development in normal female mice mated to males whose testes had been heated. Miniposters 8th European Workshop on Molecular and Cellular Endocrinology of the Testis p 110.

Ehrenberg L, von Ehrenstein G & Hedgram A 1957 Gonad temperature and spontaneous mutation rate in man. Nature 180: 1433–1434.

Ergün S, Davidoff M & Holstein AF 1996 Capillaries in the lamina propria of human seminiferous tubules are partly fenestrated. Cell and Tissue Research 286: 93–102.

Farrow S 1994 Falling sperm quality: fact or fiction? British Medical Journal 309: 1–2.

Fero ML, Rivkin M, Tasch M, Porter P, Carow CE, Firpo E, Polyak K, Tsai L-H, Broudy V, Perlmutter RM, Kaushansky K & Roberts JM 1996 A syndrome of multiorgan hyperplasia with feature of gigantism, tumorigenesis, and female sterility in p27^{kip1}-deficient mice. Cell 85: 733–744.

Fisch H & Goluboff ET 1996 Geographic variations in sperm counts: a potential cause of bias in studies of semen quality Fertility and Sterility 65: 1044–1046.

Fisch H, Goluboff ET, Olson JH, Feldshuh J, Broder SJ & Barad DH 1996 Semen analysis in 1,283 men from the United States over a 25-year period: no decline in quality. Fertility and Sterility 65: 1009–1014.

Galil KAA & Setchell BP 1988 Effects of local heating of the testis on testicular blood flow and testosterone secretion in the rat. International Journal of Andrology 11: 73–85.

Garriott ML & Chrisman CL 1980 Hyperthermia induced dissociation of the X-Y bivalent in mice. Environmental Mutagenesis 2: 465–471.

George JM & Turnbull KE 1966 The effect of red clover pasture on the reproductive tract of ram lambs. Australian Journal of Agricultural Research 17: 919–922.

Ghatnekar R, Lima-de-Faria A, Rubin S & Menander K 1974 Development of human male meiosis in vitro. Hereditas 78: 265–272.

Gurdon JB, Mitchell A & Mahony D 1995 Direct and continuous assessment of cells of their position in a morphogen gradient. Nature 376: 520–521.

Gyllensten U, Wharton D, Josefsson A & Wilson AC 1991 Paternal inheritance of mitochondrial DNA in mice. Nature 352: 255–257.

Harik SI, Kalaria RN, Andersson L, Lundahl P & Perry G 1990 Immunocytochemical localization of the erythroid glucose transporter: abundance in tissues with barrier functions. Journal of Neuroscience 10: 3862–3872.

Harcourt AH, Harvey PH, Larson SG & Short RV 1981 Testis weight, body weight and breeding system in primates. Nature 293: 55–57.

Holash JA, Harik SI, Perry G & Stewart PA 1993 Barrier properties of testis microvessels. Proceedings of the National Academy of Science USA 90: 11069–11073.

Huckins C 1971 The spermatogonial stem cell population in adult rats. I. Their morphology, proliferation and maturation. American Journal of Anatomy 169: 533–558.

Hutson JM, Williams MPL, Fallat ME & Attah A 1990 Testicular descent: new insights into its hormonal control. Oxford Reviews of Reproductive Biology 12: 1–56.

Hutson JM, Terada M, Zhou B & Williams MPL 1996 Normal testicular descent and the aetiology of cryptorchidism. Advances in Anatomy and Embryology 132: 1–56.

Irvine S, Cawood E, Richardson D, MacDonald E & Aitken J 1996 Evidence of deteriorating semen quality in the United Kingdom: birth cohort study in 577 men in Scotland over 11 years. British Medical Journal 312: 467–471.

Iwanoff E 1907 De la fecondation artificielle chez les mammiferes. Archives des Sciences Biologiques, St Petersbourg 12: 377–511.

Jain HK & Singh U 1967 Actinomycin D induced chromosome breakage and suppression of meiosis in the locust Schistocerca gregaria. Chromosoma 21: 463–471.

Jones RC, Stone GM, Hinds LA Setchell BP 1988 Distribution of 5a-reductase in the epididymis of the tammar wallaby (Macropus eugenii) and dependence of the epididymis on systemic testosterone and luminal fluids from the testis. Journal of Reproduction and Fertility 83: 779–783.

Jones RC, Walsh AL, Setchell BP & Clulow J 1989 Growth factor activity in luminal fluids from the male reproductive tract of the ram, rat, tammar wallaby (Macropus eugenii) and Japanese quail (Coturnix coturnix japonica). Journal of Reproduction and Fertility 86: 513–516.

Kenagy GJ & Trombulak SC 1986 Size and function of mammalian testes in relation to body size. Journal of Mammalogy 67: 1–22.

Kleemann JM, MooreRW & Peterson RE 1990 Inhibition of testicular steroidogenesis in 2,3,7,8-tetrachlorodibenzo-p-dioxin-treated rats: evidence that the key lesion occurs prior to or during pregnenolone formation. Toxicology and Applied Pharmocology 106: 112–125.

Kormano M 1967 Dye permeability and alkaline phosphatase activity of testicular capillaries in the post-natal rat. Histochemie 9: 327–338.

Land RB & Carr WR 1975 Testis growth and plasma LH concentration following hemicastration and its relation with female prolificacy in sheep. Journal of Reproduction and Fertility 45: 495–501.

Lee JE, Pintar J & Efstratiadis A 1990 Pattern of insulin-like growth factor II gene expression during early mouse embryogenesis. Development 110: 151–159.

Lee SL, Sadovsky Y, Swirnoff AH, Polish JA, Goda P, Gavrilina G & Milbrandt J 1996 Luteinizing hormone deficiency and female infertility in mice lacking the transcription factor NGFI-A (Egr-1) Science 273: 1219–1221.

Lubahn DB, Taylor JA, Seo K, Bunick D & Hess RA 1996 Estradiol receptor-minus mice have abnormal seminiferous tubules, rete testes and efferent ductules. Abstracts 10th International Congress of Endocrinology, San Francisco, California.

Maddocks S & Setchell BP 1988 The physiology of the endocrine testis. Oxford Reviews of Reproductive Biology 10: 53–123.

Mather JP, Attie KM, Woodruff TK, Rice GC & Phillips DM 1990 Activin stimulates spermatogonial proliferation in germ-Sertoli cell cocultures from immature rat testis. Endocrinology 127: 3206–3214.

Mather JP, Woodruff TK & Krummen LA 1992 Paracrine regulation of reproductive function by inhibin and activin. Proceedings of the Society for Experimental Biology and Medicine 201: 1–15.

Matte R & Sasaki M 1971 Autoradiographic evidence of human male germ-cell differentiation in vitro Cytologia 36: 298–303.

Meistrich ML & van Beek MEAB 1993 Spermatogonial stem cells. In Cell and Molecular Biology of the Testis, pp 266–295, Eds C Desjardins & LL Ewing, Oxford University Press, Oxford.

Meyer EP 1970 Lactation in rams grazing subterranean clover. Australian Veterinary Journal 46: 305–307.

Mieusset R, Quintana Casares PI, Sanchez-Partida LG, Sowerbutts SF, Zupp JL & Setchell BP 1991 The effects of moderate heating of the testes and epididymides of rams by scrotal insulation on body temperature, respiratory rate, spermatozoa output and motility and on fertility and embryonic survival in ewes inseminated with frozen semen. Annals of the New York Academy of Science 637: 445–458.

Mieusset R, Quintana Casares PI, Sanchez-Partida LG, Sowerbutts SF, Zupp JL & Setchell BP 1991 The effects of heating the testes and epididymides of rams by scrotal insulation on fertility and embryonic mortality in ewes inseminated with frozen semen. Journal of Reproduction and Fertility 94: 337–343.

Moller AP 1988 Ejaculate quality, testes size and sperm competition in primates. Journal of Human Evolution 17: 479–488.

Moller AP 1989 Ejaculate quality, testes size and sperm production in mammals. Functional Ecology 3: 91–96.

Moore RW, Potter CR, Theobald HM, Robinson JA & Peterson RE 1985 Androgenic deficiency in male rats treated with 2,3,7,8-tetrachlorodibenzo-p-dioxin. Toxicology and Applied Pharmacology 79: 99–111.

Moore RW, Jefcoate CR & Peterson RE 1991 2,3,7,8-Tetrachlorodibenzo-p-dioxin inhibits steroidogenesis in the rat testis by inhibiting the mobilization of cholesterol to cytochrome $P450_{scc}$. Toxicology and Applied Pharmacology 109: 85–97.

Moule GR & Mattner PE 1961 Seminal degeneration induced in Merino rams by the administration of stilboestrol dipropionate. Nature 192: 364–365.

Moule GR, Braden AWH & Lamond DR 1963 The significance of oestrogens in pasture plants in relation to animal production. Animal Breeding Abstracts 31: 139–157.

Nakayama K, Ishida N, Shirane M, Inomata A, Inoue T, Shishoda N, Horii I, Loh DY & Nakayama K 1996 Mice lacking p27[Kip1] display increased body size, multiple organ hyperplasia, retinal dysplasia, and pituitary tumors. Cell 85; 707–720.

Niemi M & Setchell BP 1986 Gamma glutamyl transpeptidase in the vasculature of the rat testis. Biology of Reproduction 35: 385–391.

Oakberg EF 1971 Spermatogonial stem-cell renewal in the mouse. American Journal of Anatomy 169: 515–532.

O'Donnell L, Stanton PG, Wreford NG, Robertsons DM & McLachlan RI 1996 Inhibition of 5a-reductase activity impairs the testosterone-dependent restoration of spermatogenesis in adult rats. Endocrinology 137: 2703–2710.

Ogura A, Matsuda J & Yanagamachi R 1994 Birth of normal young after electrofusion of mouse oocytes with round spermatids. Proceedings of the National Academy of Science USA 91: 7450–7462.

Olsen GW, Bodner KM, Ramlow JM, Ross CE & Lipshultz LI 1995 Have sperm counts been reduced 50 percent in 50 years? A statistical model revisited. Fertility and Sterility 63: 887–893.

Orth JM 1984 The role of follicle-stimulating hormone in controling Sertoli cell proliferation in testes of fetal rats. Endocrinology 115: 1248–1255.

Pabst DA, Rommel SA, McLellan WA, Williams TM & Rowles TK 1995 Thermoregulation of the intra-abdominal testes of the bottlenose dolphin (Tursiops truncatus) during exercise. Journal of Experimental Biology 198: 221–226.

Papachristoforou C, D'Occhio MJ. Horsfall D, Tilley W & Setchell BP 1985 Response of the epididymis in ram lambs to oestradiol-17ß. Proceedings of the Australian Society for Reproductive Biology 17: 74.

Parvinen M, Wright WW, Phillips DM, Mather JP. Musto NA & Bardin CW 1983 Spermatogenesis in vitro: completion of meiosis and early spermiogenesis. Endocrinology 112: 1150–1152.

Paulsen CA, Berman NG & Wang C 1996 Data from the greater Seattle area reveals no downward trend in semen quality: Further evidence that deterioration of semen quality is not geographically uniform. Fertility and Sterility 65: 1015–1020.

Ploen L & Setchell BP 1992 Blood-testis barriers revisted. A homage to Lennart Nicander. International Journal of Andrology 15: 1–4.

Portmann A 1952 Animal Forms and Patterns Faber & Faber, London.

Reid BO, Mason KA, Withers HR & West J 1981 Effects of hyperthermia and radiation on mouse testis stem cells. Cancer Research 41: 4453–4457.

Ro TS & Busch H 1965 Concentration of [^{14}C]-actinomycin D in various tissues following intravenous injection. Biochimica Biophysica Acta 108: 317–318.

Rommel SA, Pabst DA, McLellan WA, Mead JG & Potter CW 1992 Anatomical evidence for a countercurrent heat exchanger associated with dolphin testes Anatomical Record 232: 150–156.

Rommel SA, Pabst DA, McLellan WA, Williams TM & Friedl WA 1994 Temperature regulation of the testes of the bottlenose dolphin (Tursiops truncatus): evidence from colonic temperatures. Journal of Comparative Physiology B 164: 130–134.

Rommel SA, Early GA, Matassa KA, Pabst DA & McLellan WA 1995 Venous structure associated with thermoregulation of phocid seals. Anatomical Record 243: 390–402.

Rossi P, Dolci S, Albanesi C, Grimaldi P, Ricca R & Geremia R 1993 Follicle-stimulating hormone induction of steel factor (SLF) mRNA in mouse Sertoli cells and stimulation of DNA synthesis in spermatogonia by soluble SLF. Developmental Biology 155: 68–74.

Sandlow JI, Feng HL, Cohen MB & Sandra A 1996 Expression of c-KIT and its ligand, stem cell factor, in normal and subfertile human testicular tissue. Journal of Andrology 17: 403–408.

Schinckel PG 1948 Infertility in ewes grazing subterranean clover pastures. Australian Veterinary Journal 24: 289–294.

Schinkel AH, Smit JJM, van Tellingen O, Beijnen JH, Wagenaar E, van Deemter E, Mol CAAM, van der Valk MA, Robanus-Maandag EC, te Riele HPJ, Berns AJM & Borst P 1994 Disruption of the mouse mdr1a P-glycoprotein gene leads to a deficiency in the blood-brain barrier and to increased sensitivity to drugs. Cell 77; 491–502.

Setchell BP 1970 Testicular blood supply, lymphatic drainage and secretion of fluid. In The Testis, Volume I, pp 101–239, Eds AD Johnson, NL Vandemark & WR Gomes. Academic Press, New York.

Setchell BP 1978 The Mammalian Testis. Elek Books, London.

Setchell BP 1980 The functional significance of the blood testis barrier. Journal of Andrology 1: 33–10.

Setchell BP 1986 Physiological communications between the testis and epididymis. Proceedings of the International Union of Physiological Sciences 16: 494.

Setchell BP 1994 Possible physiological bases for contraceptive techniques in the male. Human Reproduction 9: 1081–1987.

Setchell BP 1997 Sperm concentrations in semen of farm animals 1932–1995. International Journal of Andrology (in press)

Setchell BP & Cox JE 1982 Secretion of free and conjugated steroids by the horse testis into lymph and blood. Journal of Reproduction and Fertility Supplement 32: 123–127.

Setchell BP & Waites GMH 1972 The effects of local heating on the flow and composition of rete testis fluid in the rat, with some observations on the effects of age and unilateral castration. Journal of Reproduction and Fertility 30: 225–233.

Setchell BP, Laurie MS, Flint APF & Heap RB 1983 Transport of free and conjugated steroids from the boar testis in lymph, venous blood and rete testis fluid. Journal of Endocrinology 96: 127–136.

Setchell BP, D'Occhio MJ, Hall MJ, Laurie MS, Tucker MJ & Zupp JL 1988 Is embryonic mortality increased in normal female rats mated to subfertile males? Journal of Reproduction and Fertility 82: 567–574.

Setchell BP, Locatelli A, Perreau C, Pisselet C, Fontaine I, Kuntz C, Saumande J, Fontaine J & Hochereau-de Reviers MT 1991 The form and function of the Leydig cells in hypophysectomized rams treated with pituitary extract when spermatogenesis is disrupted by heating the testis. Journal of Endocrinology 131: 101–112.

Setchell BP, Maddocks S & Brooks DE 1994 Anatomy, vasculature, innervation and fluids of the male reproductive tract. In Physiology of Reproduction, edn 2, pp 1063–1175. Eds E.Knobil & JD Neill. Raven Press, New York.

Setchell BP, Bergh A, Widmark A & Damber JE 1995 Effect of testicular temperature on vasomotion and blood flow. International Journal of Andrology 18: 120–126.

Setchell BP, Ekpe G, Zupp JL & Surani MAH 1997 Transient retardation in embryo growth in normal female mice made pregnant by males whose testes had been heated. Human Reproduction (in press).

Sharpe RM 1994 Regulation of spermatogenesis. In The Physiology of Reproduction edn 2, pp 1393–1434. Eds E.Knobil & JD Neill. Raven Press, New York.

Sharpe RM & Skakkebaek N 1993 Are oestrogens involved in falling sperm counts and disorders of the male reproductive tract? Lancet 341: 1392.

Singh J & Handelsman DJ 1996 Neonatal administration of FSH increases Sertoli cell numbers and spermatogenesis in gonadotropin-deficient (hpg) mice. Journal of Endocrinology 151: 37–48.

Snow MHL & Tam PPL 1979 Is compensatory growth a complicating factor in mouse teratology? Nature 279: 555–557.

Surani MAH, Barton SC & Norris ML 1984 Development of reconstituted mouse eggs suggests imprinting of the genome during gametogenesis. Nature 308: 548–550.

Tam PPL & Snow MHL 1981 Proliferation and migration of primordial germ cells during compensatory growth in mouse embryos. Journal of Embryology and Experimental Morphology 64: 133–147.

Tatsuta T, Naito M, Oh-hara T, Sugawara I & Tsuruo T 1992 Functional involvement of P-glycoprotein in blood-brain barrier. Journal of Biological Chemistry 267: 20383–20391.

Ueda K, Okamura N, Hirai M, Tanigawara Y, Saeki T, Kioka N, Komani T & Hori R. 1992 Human P-glycoprotein transports cortisol, aldosterone and dexamethasone, but not progesterone. Journal of Biological Chemistry 267: 24248–24252.

van Dissel-Emiliani FMF, Grootenhuis AJ, de Jong FH & de Rooij DG 1989 Inhibin reduces spermatogonial numbers in testes of adult mice and Chinese hamsters Endocrinology 125: 1899–1903.

van Tellingen O, Beijnen JH, Nooijen WJ & Bult A 1993 Tissue disposition, excretion and metabolism of vinblastine in mice as determined by high-performance liquid chromatography. Cancer Chemotherapy and Pharmacology 32: 286–292.

van Waeleghem K, de Clerq N, Vermeulen L, Schoonjans F & Comhaire F 1994 Deterioration of sperm quality in young Belgian men during recent decades. Human Reproduction 9 Suppl 4: 73.

van Zelst SJ, Zupp JL, Hayman DL & Setchell BP 1995 X-Y chromosome dissociation in mice and rats exposed to increased testicular or environmental temperatures. Reproduction, Fertility and Development 7: 1117–1121.

Vierula M, Niemi M, Keiski A, Saaranen M, Saarikoski S & Suominen J 1996 High and unchanged sperm counts of Finnish men. International Journal of Andrology 19: 11–17.

Waldbieser GC & Chrisman CL 1986 X-Y univalency in the testes of hyperthermic mice. I. Concomitant formation of multinucleated giant cells. Gamete Research 15: 153–160.

Wensing CJG & Colenbrander B 1986 Normal and abnormal testicular descent. Oxford Reviews of Reproductive Biology 8: 125–230.

Wettemann RP, Wells ME, Omveldt IT, Pope CE & Turman EJ 1976 Influence of elevated ambient temperature on reproductive performance of boars. Journal of Animal Science 42: 664–669.

Williams MPL & Hutson JM 1991 The history of ideas about testicular descent. Pediatric Surgery International 6: 180–184.

Withers HR, Hunter N, Berkley HT & Reid BO 1974 Radiation survival and regeneration characteristics of spermatogenic stem cells of mouse testis. Radiation Research 57: 88–103.

Yoshinaga K, Nishikawa S, Ogawa M, Hayashi SI, Kunisada T, Fujimoto T & Nishikawa SI 1991 Role of c-kit in mouse spermatogenesis: identification of spermatogonia as a specific site of c-kit expression and function. Development 113: 689—699.

Zupp JL & Setchell BP 1988 Prolonged effect of subfertile males in reducing numbers of fetuses in normal female rats. Proceedings of the Australian Society for Reproductive Biology 20: 70.

2

HUMAN Y CHROMOSOME DELETIONS IN Yq11 AND MALE FERTILITY

P. H. Vogt

Reproduction Genetics in the Institute of Human Genetics
University of Heidelberg
Im Neuenheimer Feld 328, 69120 Heidelberg
Germany

1. SUMMARY

An overview is given about the current knowledge and research activities on the molecular analysis of interstitial deletions in the euchromatic part of the long arm of the human Y chromosome (Yq11). These mutations are associated with the male specific phenotype of azoospermia and severe oligozoospermia. The fact is stressed that only "de novo" microdeletions in Yq11 are of any diagnostic value in the infertility clinic because numerous polymorphic deletion events in Yq11 have also been reported. Three different "de novo" Yq11 microdeletions associated with male infertility are now found repeatedly (31 cases) in more than 700 patients. They strongly support the presence of at least three spermatogenesis loci in Yq11. They have been designated as AZFa, AZFb, and AZFc. Each of them should contain at least one Y gene functional in spermatogenesis and, if mutated, it should induce the same sterile phenotype as the corresponding AZF locus. These genes have not yet been found. However, some candidate genes exist: RBM for AZFb. DAZ and SPGY for AZFc. It is remarkable that all three encode testis specific RNA binding proteins with a similar sequence structure. Their structure and potential relationship is disussed.

2. INTRODUCTION

Y chromosome deletions interfering with male fertility were first proposed 20 years ago (Tiepolo and Zuffardi, 1976). The authors observed terminally deleted Y chromosomes in the karyotype of six sterile males with azoospermia. In all cases, the deletion included the large heterochromatin block in the long Y arm (Yq12) and an undefined amount of the adjacent euchromatic part (Yq11). Since at that time, nobody considered the presence of genetic activity in heterochromatin, it was postulated that genetic Y factor(s) important for male germ cell development are located in Yq11. They were defined as AZoospermia Factor (AZF), as their deletion correlates to the sterile male phenotype "azoospermia", which means that no mature sperm cells were detectable in the patients' seminal fluid. Since then, the presence of AZF in Yq11 was confirmed by numerous studies at the cytogenetic level (Sandberg, 1985) and at the molecular level (Anderson et al., 1988; Bardoni et al., 1991; Vogt, 1992). However, the genetic complexity of AZF could not be revealed by these analyses.

This became first possible after the detection of different interstitial microdeletions in Yq11 not visible in the microscope but detectable by molecular deletion mapping (Ma et al., 1992, Vogt et al., 1992; Vogt et al., 1993). Under the microscope, the metaphase Y chromosome with a microdeletion still displays a normal banding pattern because deletions not larger than 2 Mb of DNA are hidden due to the high packaging density of the Y chromatin fiber at metaphase.

Molecular mapping of microdeletions in Yq11 is relatively easy and therefore was quickly undertaken in numerous laboratories. However, a major difficulty in such research programs is often neglected. Any deletion found in Yq11 by PCR or blotting experiments must be proved to be a "de novo" mutation event, i.e. only present in the Y chromosome of the sterile patient but not in the Y chromosome of his father or fertile brother(s). This analysis is crucial because familiar deletion events of the Y chromosome occurring especially in Yq11 are well known in all human populations (Jobling et al., 1996). They are without any diagnostic value in the infertility clinic and are called polymorphic Y chromosomal deletion events. Polymorphic (or familiar) Y deletion events are used frequently to trace back our evolutionary pedigree (Jobling and Tyler-Smith, 1995). Molecular analysis of small deletion events in Yq11 occurring in patients with idiopathic azoospermia must therefore always include a molecular analysis of the deleted Y DNA loci in the Y chromosome of the patient´s father or fertile brother(s). If the Y chromosome of the father presents the same deletion event we have detected a new polymorphic Y-DNA locus but not a deletion interrupting one of the proposed AZF genes. Analysis of "de novo" microdeletions in Yq11 in Heidelberg now has clearly demonstrated that at least three AZF spermatogenesis loci are present in Yq11: AZFa, AZFb, and AZFc (Vogt et al., 1996).

It is assumed that at least one functional spermatogenesis gene´must be present in each AZF locus. However, the molecular extensions of the three AZF loci in Yq11 is large, ranging between 1–3 million nucleotides (Vogt et al., 1996). Space is clearly sufficient for more than one gene. How many Y genes are present in each AZF locus, and which of them are essential for spermatogenesis is, however, not yet known.

An important step towards the molecular isolation and analysis of all functional AZF genes is the isolation and study of so-called "AZF candidate genes". Each Y gene expressed in testis tissue and located in Yq11 in a position overlapping with one of the AZF loci can be designated as an "AZF candidate gene". Three AZF candidate genes were recently published: the RBM gene family (formerly YRRM, Ma et al., 1993), the DAZ gene cluster (Saxena et al., 1996), and the SPGY locus (Maiwald et al., 1996; Vogt, 1996). Notable is the fact that all three encode testis specific RNA binding proteins suggesting a function for them in posttranscriptional processing of other spermatogenesis genes. However, gene specific mutations causing azoospermia have not yet been found in any of them. Therefore, it should surprise nobody if more "AZF candidate genes" will soon be detected.

It should also be not surprising, if some AZF candidate genes do not encode testis-specific proteins but germ line specific RNA structures. Such structures are expressed, for example by some Y chromosomal spermatogenesis genes in the spermatocyte of Drosophila (Hennig et al., 1987). Besides, there exists strong experimental evidence that the intactness of the chromatin structure in Yq11 is also an important prerequisite for a proper Y spermatogenesis function (AZF chromatin code; Vogt, 1996).

In this paper, I wish to give an overview of our current knowledge and research activities on the structure and function of the proposed AZF loci and their candidate genes in Yq11. With it, I will also point to the borderlines of this knowledge and our still large ignorance about the importance of the different AZF loci in human spermatogenesis.

3. AZF COMPLEXITY INCLUDES AT LEAST THREE SPERMATOGENESIS LOCI: AZFa, AZFb, AZFc

Idiopathic sterile patients with interstitial "de novo" microdeletions in Yq11 were first detected in a study with 13 sterile men suffering from idiopathic azoospermia (Vogt et al., 1992). Two different microdeletions in Yq11 were observed and mapped to two non-overlapping positions in proximal and distal Yq11 (Ma et al., 1992). However, further studies of Yq11 microdeletions associated with the phenotype of male sterility first confirmed only the occurrence of microdeletions in distal Yq11. The most extensive study was performed by Reijo et al. (1995) on 89 sterile men with idiopathic azoospermia. Deletion analysis of a large series of STS loci detected microdeletions in Yq11 in 9 cases. They all overlapped and contained a common interval in distal Yq11, defined then as the AZF region (Reijo et al., 1995). This was in contrast to our own observations suggesting at least two AZF spermatogenesis loci in Yq11 with regard to the divergent pictures of spermatogenic disruption phases in testis tissue sections observed in patients with different Yq11 deletions (Vogt et al., 1993). In order to clarify this situation and to answer the question, whether the genetic complexity of AZF was indeed represented by only one, or by more than one spermatogenesis locus in Yq11, we decided to perform a molecular deletion analysis of small interstitial deletions in Yq11 on a large series of patients (370 individuals) including not only idiopathic sterile men with non-obstructive azoospermia, but also men with severe oligozoospermia (< 2 million sperm per ml ejaculate). This investigation was only possible in a large collaborative screening project in which different infertility clinics participated (Vogt et al., 1996). All sterile individuals were screened for deletions of 76 DNA loci in Yq11. They were mapped to a detailed interval map, subdividing this chromosome region in 25 intervals (D1-D25; Fig. 1). A PCR-multiplex procedure was developed (Henegariu et al., 1994) and used as a rapid screening protocol.

Twelve patients with a "de novo" microdeletion in Yq11 were detected. They were mapped to three different subregions in Yq11. One subregion (Yq11, D20-D22) coincided with the position of the AZF region in distal Yq11 as described by Reijo et al. (1995). The second and third Y region deleted was analysed proximal to it and mapped to proximal (Yq11, D3-D6) and middle Yq11 (Yq11, D13-D16), respectively. Testicular histology of these patients suggests disruption of spermatogenesis at the same phase when the microdeletion occurred in the same Yq11 subregion, but at a different phase when the microdeletion occurred in a different Yq11 subregion (Vogt et al., 1996). Since the three subregions deleted were clearly separated in Yq11, and their associated phases of spermatogenic disruptions were clearly distinguished, it could be supposed that the genetic complexity of AZF, the Y chromosomal azoospermia factor in Yq11 postulated 20 years ago (Tiepolo and Zuffardi, 1976) is represented indeed not by one but by at least three discrete

Y encoded spermatogenesis loci. Each of these loci seems to act at a different phase of this male specific cell differentiation process. As the most severe phenotype after deletion of each of the loci was "azoospermia", we designated them as azoospermia factors: AZFa, AZFb, and AZFc (Vogt et al., 1996). In our ongoing AZF screening program (now 700 patients) the same deletion events have been analysed again in now 31 sterile individuals (Fig. 2).

Most notable is the frequent deletion of AZFc and its association to divergent histological pictures in the patients' testis tissues. Moreover, AZFc patients were not only azoospermic but also oligozoospermic with a variability of sperm numbers (ranging between 0–10 million per ml ejaculate). By interval mapping, the molecular extensions of the deletions in distal Yq11 observed in the Y chromosome of azoospermic, respectively, oligozoospermic men could not be distinguished. In one case, we also detected a familiar AZFc deletion, which suggests that AZFc deletions can be inherited. Again the deletion in the Y chromosome of the father and in the Y chromosome of its azoospermic son could not be distinguished by our extensive panel of screening probes. A paternity test by DNA fingerprinting confirmed their family relationship. Therefore, a large overlap of the deleted Y DNA in father and son must exist. We therefore suppose that deletions of AZFc do not induce azoospermia as the primary effect of mutation, but male subfertility.

It is interesting to note that to date no other and no smaller de novo microdeletions in Yq11 have been analysed in sterile individuals. This suggest that three mutational hotspots must exist in Yq11, always inducing deletions of the same Y-DNA sequence region. This does not exclude specific molecular break- and fusion-points in the Y chromosome of the different AZF individuals, but it strongly suggests that these break- and fusion-points must be clustered in three Y regions (in proximal, middle, and distal Yq11, respectively), and therefore are not resolvable by conventional molecular deletion mapping analysis.

It is safe to assume that to date at least one functional Y spermatogenesis gene must be present in each AZF locus. However, the molecular extensions of the three AZF loci in Yq11 is large, ranging between 1–3 million nucleotides (Vogt et al., 1996). It is obvious that this is sufficient for more than one gene. Therefore, an AZF gene cannot be considered equivalent to an AZF locus unless it has been shown that mutations in the gene cause the same pathological phenotype as observed by mutations of the corresponding AZF locus. Each Y gene expressed in testis tissue and located in Yq11 in a position overlapping with one of the AZF loci is therefore first designated only as an "AZF candidate gene". Three AZF candidate genes were recently published: the RBM gene family (formerly YRRM, Ma et al., 1993) the DAZ gene cluster (Saxena et al., 1996) and the SPGY locus (Maiwald et al., 1996; Vogt, 1996). All three encode testis specific RNA binding proteins with an RNA Recognition Motif (RRM).

$$\longrightarrow$$

Figure 1. Deletion map of Yq11 dividing this Y region in 25 intervals (D1-D25), marked at the right. The map is based on analysis of 76 DNA loci in 26 individuals described with different Yq11 anomalies (Vogt et al., 1996). "+" means that the DNA locus analysed is present. "-" declares its absence. "n" means that this DNA locus was not analysed in this individual. The GDB code of each DNA locus is given at the left when registered in the Genome Data Base. The figure is used with permission of Oxford University Press and published originally in Vogt et al. (1996).

locus	primer/probe																															
DYZ3	Y-97																															D1
DYS268	64a																															
DYS270	sY79																															
DYS271	sY81																															D2
DYS272	sY82																															D3
DYS11	12f3																															
DYS273	sY84																															D4
DYS274	Y6HP35																															D5
DYS148	sY86																															D6
DYS275	sY87																															
DYS246	sY165																															
DYS276	sY88																															
KAL-Y	sY182																															
KAL-Y	sY151																															
DYS136	CRI-S232/D																															
DYS136	CRI-S232/E																															D7
DYS279	sY94																															D8
DYS106	Y-253																															
DYS168	Y-198																															
DYS280	sY95																															
DYS109	Y-221																															D9
DYS243	sY161																															
DYS281	sY97																															D10
STS-P	STS114B																															
DYS198	sY102																															
DYS201	sY105																															
DYS22	M1A																															
DYF43S1	sY109																															D11
DYS108	Y-216a/A																															
DYS110	Y-294																															
	Y6BS64																															
DYF46S1	Y6BS65/B																															D12
DYS205	Y6D14																															
DYS206	Y6BaH34																															
DYS209	sY117																															
	Y6PHc54																															
DYF46S1	Y6BS65/C																															
	FR25-II/B																															
DYS107	Y-202																															D13
DYS215	sY124																															
DYS218	sY127																															D14
DYS7	50f2/E																															
DYS224	sY134																															
DYF27	52d/A																															
DYS8	118e/C																															
DYS231	sY143																															
	LLY22g/A																															
	RBM1/A																															
	RBM1/B2																															D15
	Y-367/A																															
	RBM1/F																															D16
DYS8	118e/D																															
DXYS37	GMGXY10																															D17
	RBM1/C																															
DYS108	Y-216a/B																															
DYF46S1	Y6BS65/D																															
	FR25-II/A																															
	LLY22g/B																															D18
DYS20	p69/6																															D19
DYS237	sY153																															
DYS7	50f2/C																															
	Y-367/B																															
DYS75	GMGY18																															
	RDF8																															
	FR15-II																															
DYS239	Y6HP52																															D20
DYS21	p116/21																															
DYS232	sY147																															D21
DYS249	sY157																															
DYS1	49f																															
DYS65	GMGY5																															
DYF46S1	Y6BS65/E																															
	RBM1/E																															D22
DYS12	GMGY1																															D23
	poxY1																															D24
DYZ2	pHY2.1																															D25

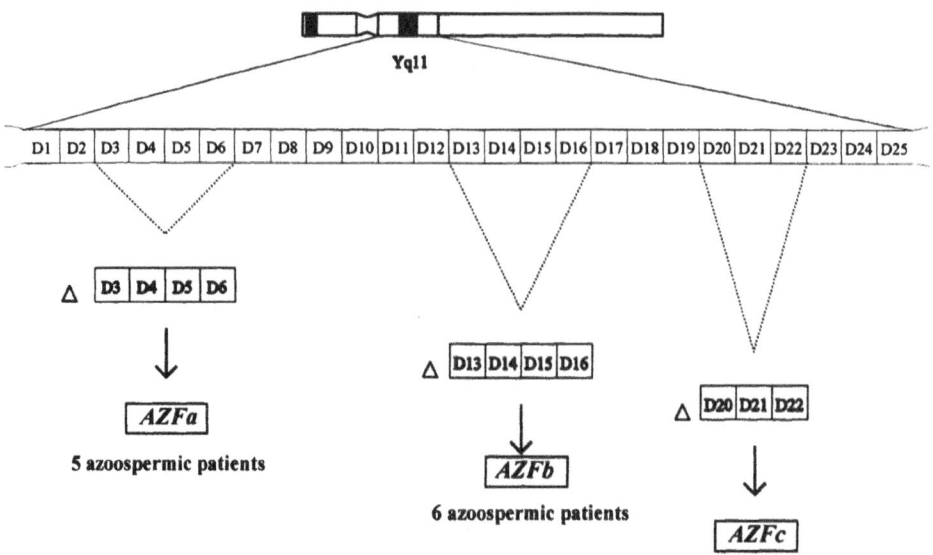

Figure 2. Actual survey of the number of microdeletions occurring "de novo" in 700 consecutive individuals with idiopathic azoospermia and severe oligozoospermia analysed in my laboratory between 1992–1996.

4. CANDIDATE AZF GENES ENCODE TESTIS SPECIFIC RRM-PROTEINS

The "RNA-Recognition Motif" (RRM) is an evolutionary conserved protein motif containing 80–100 amino acids present in a large number of proteins interacting with RNA. It is mostly located in the N-terminal part of the protein. The RRM domain is able to mediate general and specific interactions between a protein and RNA molecules. Two highly conserved peptide boxes (RNP-1 and RNP-2) present in yeast and human RRM proteins help to align their functional sequence structure (Dreyfuss et al., 1993). This suggests that RRM domains have a very ancient origin and that the mode of their interaction with RNAs is highly conserved. The specificity of the RNA-protein binding reaction is thought to be achieved additionally by auxiliary sequence blocks in the C-terminal part of the RRM protein structure (Burd and Dreyfuss, 1994). Most RRM proteins are expressed constitutively. However, cell-type specific RRM proteins exist and are functional in the brain (Robinow et al., 1988), and during sex determination in Drosophila (Amrein et al., 1988; Bell et al., 1988) and as shown here during development of the male germ cell.

4.1. The RBM Gene Family in AZFb

Molecular screening experiments on a cDNA library from adult testis poly(A) RNA with a genomic Y probe deleted in patients with deletion of AZFc (Y-367) succeeded in isolating a gene family expressed in human testis (Ma et al., 1993). It was designated RBM and earlier YRRM due to homology of their putative gene product(s) to the RNA-binding proteins with a RRM domain. From their sequence, RBM genes are closely related to hnRNPG, a protein member of the heterogenous nuclear RNP family known to be associated with nuclear polyadenylated RNA and involved in pre-mRNA metabolism (Dreyfuss et al., 1993).

We could show that most RBM gene copies located in Yq11 were deleted in the Y chromosome of AZFb patients (Vogt et al., 1996). Consequently, the RBM gene family is a candidate gene for expression of AZFb. The functional significance of RBM genes in the region of AZFb was supported by its testicular transcription pattern visible in testis tissue in-situ-hybridization experiments, which shows that RBM genes are transcribed especially in primary spermatocytes (Chandley & Cooke, 1994). Moreover, an antibody raised against a specific part of the RBM protein (repetitive SRGY domain) does detect RBM proteins in spermatogonia and spermatocytes only if the Y region defining AZFb was present (Elliot et al., submitted). However, mutations in exons of RBM genes located in AZFb have not yet been found in azoospermic men. Therefore, it is not yet known whether mutations in the RBM gene structure are able to induce the same sterile phenotype as deletions of AZFb. Moreover, the number of functional RBM genes in the region of AZFb is not known. and it might be that only the deletion of all of them induce the AZFb phenotype (Vogt et al., 1995).

The only RBM gene copy in AZFb isolated as a complete cDNA clone is RBM1 (Ma et al., 1993). Therefore, in the clinic, only PCR analysis of RBM1 is recommended, using the following primer pairs:

RBM1, forward: 5´-ATGCACTTCAGAGATACGG-3´;

RBM1, reverse: 5´-CCTCTCTCCACAAAACCAACA-3´.

Cycling conditions in our RBM1 screening program (Robocycler, 30 cycles) are: 96 °C, 60 sec (melting); 62 °C, 60 sec (annealing); 73 °C, 120 sec (extension). Denaturation is performed for 5 min at 96 °C. The final extension cycle takes 5 min at 73 °C.

However, not all RBM gene copies are located in the Y region of AZFb. Some of them were also mapped on the short arm of the Y chromosome (Yp) or in Yq11 regions not overlapping with any AZF locus (one RBM copy mapped to the region of AZFc is supposed to be a pseudogene; Pauline Yen, personal comm.). The total number of RBM gene copies on the Y chromosome was estimated to range between 20 and 30.

Recently, it has been shown that some RBM gene copies are polymorphic i.e. are deleted also in some fertile male populations (Jobling et al., 1996). One of these gene copies was isolated as a cDNA clone (RBM2; Ma et al., 1993). RBM2 was mapped between AZFb and AZFc to Yq11-interval D18 (Vogt et al., 1996). RBM2 is deleted in the Japanese fertile male population (Nakahori et al., 1994). In our own patient collective and fertile control males (mostly Caucasians), RBM2 deletions were also frequently found. The functional significance of the RBM2 gene is therefore unclear. We assume that RBM2 deletions are most likely not functional and therefore not related to the spermatogenic disruption of the corresponding patient but only a polymorphic deletion event. This illustrates again the importance of demonstrating that each deletion found in Yq11 of sterile men must be proved to be a "de novo" mutation event.

The RBM gene family is conserved on the Y chromosome in different mammals including mouse and marsupials (Elliot et al. 1996; Cooke and Elliot, submitted) and its expression always appears to be specific for male germ cells. This suggests an important function of at least one RBM gene copy (mostly likely located in the region of AZFb) in human spermatogenesis.

4.2. The DAZ Gene Cluster in AZFc

Recently, exon trapping experiments with cosmids selected from the AZFc region succeeded in the isolation of exon probes expressed specifically in adult human testis (Reijo et al., 1995). A corresponding cDNA sequence of 1641 nucleotides was isolated from a human poly-(A) testis cDNA library. It contains the 5′end of a Y gene encoding again an RNA binding protein of the RRM protein family. Although its sequence was clearly different from that of the RBM1 protein, its general sequence structure was similar. Both contain one RRM domain in the N-terminal part of their protein structures and both contain a specific repetitive peptide domain. RBM1 contains 4 tandem copies of a 111 bp exon encoding a repetitive 37aa peptide (SRGY; Ma et al., 1993). The Reijo peptide contains 7 tandem copies of a 72 bp exon encoding a repetitive 24aa peptide (DAZ-repeat; Reijo et al., 1995). The Y gene encoding this novel RNA binding protein was designated as DAZ (Deleted in AZoospermia; Reijo et al., 1995) because the whole sequence was deleted in azoospermic men with deletion of AZFc.

DAZ was first described to be a single copy gene (Reijo et al., 1995). Sequence analysis of genomic cosmid clones then revealed that at least two or three DAZ gene copies exhibiting 99.9 % sequence identity should be clustered in the AZFc region (Saxena et al., 1996). DAZ genes are however only candidate genes for expression of AZFc because gene-specific mutation events in idiopathic azoospermic men have not yet been reported for any DAZ copy. Therefore, it is unclear whether DAZ gene mutations will cause the same pathological phenotype as deletion of AZFc.

This analysis has now become complicated by the fact that deletions of AZFc were shown to be associated with different patterns of testis histology (Vogt et al., 1996). Moreover, patients with AZFc deletion are not only azoospermic but also oligozoospermic with sperm numbers ranging between 1–10 million per ml ejaculate (see above). This implies that deletion of the whole DAZ gene cluster is compatible with spermatogenesis (although reduced in sperm numbers) and that the DAZ genes are not absolutely required for the maturation of a human male germ cell. This conclusion is strengthened by our observation of a familiar deletion of DAZ in rare cases (Vogt et al., 1996). We therefore assume that as for AZFc, the primary mutation effect of deletion of the whole DAZ gene cluster is not azoospermia but male subfertility.

For diagnostic purposes in the clinic, it is proposed to use the following DAZ primers:

forward: 5′-GGAAGCTGCTTGGTAGATAC-3′;

reverse: 5′-TAGGTTTCAGTGTTTGGATTCCG-3′,

and the same cycling conditions as described for the RBM1 gene analysis.

4.3. The SPGY Locus in AZFc

Using a YAC clone (71G11) mapped to the region of AZFc we were able to select a cDNA clone (CT52Y) from a human poly(A) testis library, expressed specifically in testis tissue (Maiwald et al., 1996). Sequence analysis revealed a complete polyadenylated 3′UTR region and a large open reading frame, open at its 5′end i.e. without a 5′UTR region. We designated this gene SPGY1 (SPermatogenesis Gene on the Y) and mapped it toYq11 interval D21 i.e. to the region of AZFc. SPGY1 is therefore a second candidate gene for AZFc. Sequence analysis of SPGY1 revealed a distinct sequence homology to the DAZ genes (Maiwald et al., 1996), however, SPGY1 contained 11 complete tandem copies of the 72bp exon sequence found in the DAZ genes only with 7 copies (Saxena et al., 1996). The consensus sequence of the 72bp exon is the same in DAZ and SPGY genes, however the composition of the 72bp exon repeats in the DAZ genes is different from that in any SPGY gene (Vogt et al., submitted). Gene specific mutations in SPGY1 or any other SPGY copy of idiopathic sterile men have not yet been found. Therefore, like RBM and DAZ genes, also SPGY genes are still only candidate genes of the corresponding AZF locus (AZFc).

For diagnostic purposes of SPGY analysis in the clinic we recommend to use the following SPGY1 primers:

forward: 5´-TTTCACATACAGCCATTAAGTTTAGC-3´;

reverse: 5´-CAATTTTGATAGTCTGAACACAAGC-3´.

PCR cycling condtions are the same as used in the DAZ experiment.

In contrast to the RBM gene family, the position of DAZ and SPGY genes on the Y chromosome is not highly conserved. In the mouse, a DAZ copy was mapped only to chromosome 17 (DAZLA: DAZ like autosomal (Cooke et al., 1996). DNA blot experiments with genomic DNA of different species (zoo-blot) indicated that DAZ and SPGY genes were first transposed to distal Yq11 during primate evolution. We were able to show that in the human a SPGY copy exists on an autosome as well (Shan et al., 1996). It was mapped to the distal part of the short arm of chromosome 3 (3p24) and designated as SPGYLA (SPGY Like Autosomal). A functional interaction between the SPGY locus in distal Yq11 and SPGYLA is most likely because of their sequence homology and testis specific transcription patterns.The autosomal SPGY copy is highly conserved and homologous to the mouse gene Dazla (89% sequence identity) and found with significant sequence homologies also in Drosophila (Shan et al., 1996). The Drosophila gene designated "boule" was described earlier to be homologous to DAZ (Eberhart et al., 1996). A sequence homology of boule to DAZ genes exists just as it does to the Y chromosomal SPGY genes. However, it differs by the absence of the repetitive structure of the 72bp exon sequence and absence of a 130bp exon unit (CTA box) only present in the autosomal SPGY copy SPGYLA and the homologous mouse gene Dazla (Fig. 3).

This divergence results in a protein structure of the DAZ and SPGY proteins divergent from that of SPGYLA and Dazla in the C-terminal region. Loss of boule function in the Drosophila male results in azoospermia. The boule protein seems to be functional during the premeiotic phase of spermatogenesis in the fly (Eberhart et al., 1996). It is therefore assumed that also Dazla in the mouse, or SPGYLA in human fulfill a meiotic function in spermatogenesis.

SPGYLA was isolated not only by us but in parallel also by the group of David Page in Boston (Saxena et al., 1996) and by the group of Pauline Yen in Los Angeles (Yen et al., 1996). Consequently, three different names are now found for the same gene locus in the literature (DAZH: Saxena et al., 1996; Dazla: Yen et al. 1996; SPGYLA: Shan et al., 1996). This unnecessary complexity needs to be resolved in the near future.

Gene mutations for SPGYLA have not yet been found in our patient collective of men with idiopathic azoospermia. Nor do we know the sterile phenotype of mice with a homologous knock-out of the mouse gene Dazla.

SPGYLA-CTAbox homology in Dazla and boule

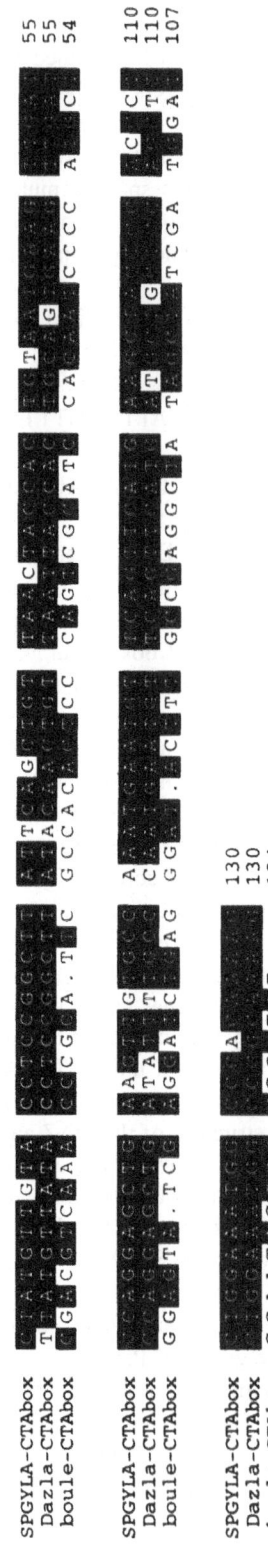

Figure 3. A 130bp sequence unit in SPGYLA (CTA box; sequence position 762–891 in cDNA clone submitted to GenBank database; accession no. U66726) is absent in the homologous SPGY gene copies mapped to the AZFc region in dsital Yq11 but present in the homologous mouse gene Dazla and the homologous Drosophila gene boule (Shan et al., 1996). Protein sequences were aligned by the software tool PRETTYBOX highlighting identical amino acids by black boxes.

5. DISCUSSION

5.1. The Functional Significance of AZFa, AZFb and AZFc in Human Spermatogenesis

Genetic "switch signals" act at specific developmental stages. They initiate downstream specific cascades of molecular events up to the next switch signal and so forth until the specialized fate of each cell is found and manifested. If one assumes that gene activities acting as switch signals are expressed at the most critical steps during any cell differentiation pathway, they should be expected during male germ cell development (1) at the onset of spermatogonial proliferation and differentiation before and at puberty, (2) at the onset of meiosis, (3) at the onset of spermatid development starting with formation of the acrosome and the flagellum as well as the condensation of the nucleus. Our current knowledge about the function of the different AZF loci now mapped to three different sites in the euchromatic part of the long arm of the human Y chromosome (Yq11) suggests that they may encode such genetic switch signals (Vogt, 1996). Deletion of AZFa results in a pure Sertoli-cell-only syndrome, a histological picture compatible with disruption of spermatogenesis at the onset of spermatogonia proliferation. Deletion of AZFb results in maturation arrest of the germ cells at meiosis, a histological picture compatible with disruption of spermatogenesis at the onset of meiosis. Deletion of AZFc results in disruption of spermatogenesis at some postmeiotic germ cell stages. In-situ-hybridization experiments with SPGY1 probes on testis tissue sections indicate their presence in early spermatids during formation of the acrosome (Vogt, 1996).

However, humans are not experimental species. We can only analyse the disruption phase of spermatogenesis in patients with deletion of one AZF locus, after they have consulted an infertility clinic and agreed to take part in an AZF screening experiment. At that time their age is usually between 30–40 years. But the first wave of a complete differentiation process of the male germ cell in human occurs already 15–25 years earlier i.e. at puberty. I assume that this time interval is large enough for the occurrence of secondary apoptotic effects in the patient's testicular tissue which might disturb the pathological testis tissue picture primarily associated with the loss of function of an AZF locus. From the scientific point of view, my experience with the functional research on each AZF locus appears therefore discouraging, because it is rather difficult and perhaps even impossible to get testis tissue samples from sterile men with an identified AZF mutation before or shortly after puberty.

However, from the clinical point of view, my experience is just the opposite. Idiopathic sterile men screened successfully for an AZF mutation in Yq11 were told the reason for their sterility. This helped them a lot in coping with their sterility and can save them from further long and often useless medical treatments. So sterile men with deletion of AZFa have a complete Sertoli-cell-only syndrome and clinical application of a testis biopsy for the aspiration of testicular sperms (TESA) used for in-vitro fertilization by single sperm injection experiments (Silber et al., 1995) can be omitted. The same holds true for sterile men with deletion of AZFb. No postmeiotic germ cell has ever been observed in their testicular biopsies. On the other hand, azoospermic men with deletion of AZFc most likely do have tubules containing mature spermatids, the amount of which might be age-dependent (Vogt et al., 1996).

This would suggest that one should screen in the clinic for possible deletion of AZFa,b,c in idiopathic azoospermic men as soon as possible in order to support a qualified prediction for the success of testicular sperm aspiration experiments in the testis of azoospermic men asking for in-vitro fertilization by Intra-Cytoplasmic-Sperm-Injection (ICSI).

5.2. Mutation Frequency of AZFa, AZFb, AZFc in Idiopathic Sterile Men

The frequency of the observed AZF deletion events in proximal and middle Yq11 (AZFa and AZFb) is the same, ranging below 1% in men with idiopathic azoospermia (Fig. 2). The frequency of deletion events in distal Yq11 (AZFc) seems to be generally higher and is described with a marked broadness ranging between 5 % and 20 % in different laboratories. I speculate that one reason for this diffuse frequency rate of AZFc deletions is the inclusion of not only idiopathic azoospermic and severe oligozoospermic men (0–2 million sperms per ml ejaculate), but also the inclusion of idiopathic oligozoospermic men with higher sperm numbers (2–20 million sperms per ml ejaculate) in some clinical AZF screening programs. This assumption is supported by our analysis of a familiar deletion of AZFc in one patient with azoospermia suggesting subfertility of the father (Vogt et al., 1996). If this holds true, it is expected that the sperm number of AZFc patients at puberty may be at least as high as 20 million per ml ejaculate.

Another explanation for the observed frequency range for AZFc deletions may be the inclusion of polymorphic Yq11 deletion events in some clinical AZF screening programs. This happens easily if, after the detection of a deletion in Yq11 of the patient, the deletion analysis of the father's Y chromosome (or that of the fertile brother) is omitted. Polymorphic deletion events in Yq11 are observed not only in sterile men but in fertile men as well (Jobling et al., 1996). Therefore, it is important to emphasize for any diagnostic AZF screening program in the clinic, that any new Y deletion observed in Yq11 is proved to be a "de novo" mutation event, before it can be given any diagnostic value.

5.3. Candidate Genes for Expression of AZFa, AZFb, AZFc

It is well-known that numerous genes transcribed in the testis have no biological function (Willison & Ashworth, 1987). Therefore, expression of a Y gene in human testis and mapping its position to one of the AZF loci is not sufficient to mark it as an AZF gene but only as a "candidate gene". A candidate gene will become a real AZF gene after it has been shown that mutations in its gene structure (i.e. its exon sequence) are able to cause the same pathological phenotype as observed after deletion of the corresponding AZF locus. Such an AZF gene has not yet been identified. RBM, DAZ, and SPGY genes are only candidate genes for expression of AZFb or AZFc. Gene-specific mutations associated with the phenotype of azoospermia have not yet been found for any of them. From the practical point of view, it is suggested that analysis of each AZF candidate gene isolated should be included in the diagnostic screening schedule of each infertility clinic. However, this should be done not in order to exchange it against the established genomic PCR multiplex system, but in order to improve it by adding the gene specific primer pairs cited above in an additional PCR multiplex experiment.

6. ACKNOWLEDGMENTS

I would like to thank all my friends and colleagues, especially, Angela Edelmann, Peter Hirschmann, Octavian Henegariu, Franklin Kiesewetter, Frank Köhn, Prof. W. Hilscher and Barbara Hilscher, Prof. R.A. Pfeiffer, Prof. E. Nieschlag, and Prof. W.B. Schill whose skills, experience and ideas have contributed significantly to the current knowledge of the human azoospermia factors on the long arm of the human Y chromosome, now designated as AZFa, AZFb, and AZFc. The Bundesministerium für Wissenschaft, Forschung und Technologie (BMBF) is thanked for the grant 01KY 9507/4 without which my research on the genetic complexity of AZF would not have been possible.

7. REFERENCES

Amrein H, Gorman M & Nothiger R 1988 The sex determining gene tra-2 of Drosophila encodes a putative RNA binding protein. Cell 55: 1025–1035.

Andersson M, Page DC, Pettay D, Subrt I, Turleau C, Grouchy J & de la Chapelle A 1988 Y autosome translocations and mosaicism in the aetiology of 45,X maleness: assignment of fertility factor to distal Yq11. Human Genetics 79: 2–7.

Bardoni B, Zuffardi O, Guioli S, Ballabio A, Simi P, Cavalli P, Grimoldi MG, Fraccaro M & Camarino G 1991 A deletion map of the human Yq11 region: implications for the evolution of the Y chromosome and tentative mapping of a locus involved in spermatogenesis. Genomics 11: 443–451.

Bell LR, Maine EM, Schedl P & Cline TW 1988 Sex-lethal, a Drosophila sex determination switch gene, exhibits sex-specific RNA splicing and sequence similarity to RNA binding proteins. Cell 55: 1037–1046.

Bourrouillou G, Dastugue M, Colombies P et al. 1985 Chromosome studies in 952 infertile males with a sperm count below 10 million/ml. Human Genetics 71: 366–367

Burd GB & Dreyfuss G 1994 Conserved structures and diversity of functions of RNA-binding proteins. Science 265: 615–621.

Chandley AC & Cooke HJ 1994 Human male fertility-Y-linked genes and spermatogenesis. Human Molecular Genetics 3: 1449–1452.

Cooke HJ, Lee M, Kerr S & Ruggiu M 1996 A murine homologue of the human DAZ gene is autosomal and expressed only in male and female gonads. Human Molecular Genetics 5: 513–516.

Dreyfuss G, Matunis MJ, Pinol-Roma S & Burd CG 1993 hnRNP proteins and the biogenesis of mRNA. Annual Review of Biochemistry 62: 289–321.

Eberhart CG, Maines JZ & Wasserman SA 1996 Meiotic cell cycle requirement for a fly homologue of human Deleted in Azoospermia. Nature 381: 783–785.

Elliott DJ, Ma K, Kerr SM, Thakrar R, Speed R, Chandley AC and Cooke H 1996 An RBM homologue maps to the mouse Y chromosome and is expressed in germ cells. Human Molecular Genetics 5: 869–874.

Henegariu O, Hirschmann P, Kilian K, Kirsch S, Lengauer C, Maiwald R, Mielke K & Vogt P 1994 Rapid screening of the Y chromosome in idiopathic sterile men, diagnostic for deletions in AZF, a genetic Y factor expressed during spermatogenesis. Andrologia 26: 97–106.

Hennig W, Brand RC, Hackstein J, Huijser P, Kirchhoff C, Kremer H, Lankenau DH & Vogt P 1987 Structure and function of Y chromosomal genes in Drosophila. In: Chromosomes Today, vol. 9, pp 48–58. Eds. Stahl, A., Luciani, J.M. and Vagner-Capodano, A.M.. Chapman & Hall, London.

Jobling MA, Tyler-Smith C 1995 Fathers and sons: the Y chromosome and human evolution. TIG 11: 449–456.

Jobling MA, Samara V, Pandya A, Fretwell N, Bernasconi B, Mitchell RJ, Gerelsaikhan T, Dashnyam B, Sajantily A, Salo PJ, Nakahori Y, Disteche CM, Thangaraj K, Singh L, Crawford MH & Tyler-Smith C 1996 Recurrent duplication and deletion polymorphisms on the long arm of the Y chromosome in normal males. Human Molecular Genetics 5: 1767–1775.

Ma K, Sharkey A, Kirsch S, Vogt P, Keil R, Hargreave TB, McBeath & Chandley AC 1992 Towards the molecular localisation of the AZF locus: mapping of microdeletions in azoospermic men within 14 subintervals of interval 6 of the human Y chromosome. Human Molecular Genetics 1: 29–33.

Ma K, Inglis JD, Sharkey A, Bickmore WA, Hill RE, Prosser EJ, Speed RM, Thomson EJ, Jobling M, Taylor K, Wolfe J, Cooke HJ, Hargreave TB, and Chandley AC 1993 A Y chromosome gene family with RNA-binding protein homology: Candidates for the azoospermia factor AZF controlling human spermatogenesis. Cell 75: 1287–1295.

Maiwald R, Seebacher T, Edelmann A, Hirschmann P, Köhler MR, Kirsch S & Vogt PH 1996 A human Y-chromosomal gene mapping to distal Yq11 is expressed in spermatogenesis. In Lau et al. (eds.), Report of the Second International Workshop on Y Chromosome mapping, Asilomar 1995. Cytogenetics Cell Genetics 73: 33–76.

Nakahori Y, Kobayashi K, Komaki ., Matsushita I & Nakagome Y 1994 A locus of the candidate gene family for azoospermia factor (YRRM2) is polymorphic with a null allele in Japanese males. Human Molecular Genetics 3: 1709.

Reijo R, Lee T-Y, Salo P, Alagappan R, Brown LG, Rosenberg M, Rozen S, Jaffe T, Straus D, Hovatta O, de la Chapelle A, Silber S. & Page DC 1995 Diverse spermatogenic defects in humans caused by Y chromosome deletions encompassing a novel RNA-binding protein gene. Nature Genetics 10: 383–393.

Robinow S, Campos AR, Yao K-M & White K 1988 The elav gene product of Drosophila, required in neurons, has three RNP consensus motifs. Science 242: 1570–1572.

Sandberg AA, 1985 Clinical aspects of Y chromosome abnormalities. In series: Progress and Topics in Cytogenetics, Vol. 6, part B. Alan R. Liss Inc., New York.

Saxena R, Brown SL, Hawkins T, Alagappan RK, Skaltsky H, Reeve MP, Reijo R, Rozen S, Dinulos MB, Disteche CM & Page DC 1996 The DAZ gene cluster on öthe human Y chromosome arose from an autosomal gene that was transposed, repeatedly amplified and pruned. Nature Genetics 14: 292–299.

Shan Z, Hirschmann P, Seebacher T, Edelmann A, Jauch A, Morell J, Urbitsch P, Vogt PH 1996 A SPGY copy homologous to the mouse gene Dazla and the Drosophila gene boule is autosomal and expressed only in the human male gonad. Human Molecular Genetics 5: 2005–2011.

Silber SJ, Van Steirteghem AC, Liu J Nagy Z, Tournaye H & Devroey P 1994 High fertilization and pregnancy rate after intracytoplasmic sperm injection with spermatozoa obtained from testicle biopsy. Human Reproduction 10. 148–152.

Tiepolo L and Zuffardi O 1976 Localization of factors controlling spermatogenesis in the nonfluorescent portion of the human Y chromsome long arm. Human Genetics 34: 119–124.

Vogt P 1992 Y chromosome function in spermatogenesis (review). In: Spermatogenesis-Fertilization-Contraception. Molecular, cellular and endocrine events in male reproduction, vol. 4, pp 226–257. Eds E Nieschlag & UF Habenicht. Springer Verlag, Heidelberg, Berlin, New York.

Vogt P, Chandley AC, Hargreave TB, Keil R, Ma K & Sharkey A 1992 Microdeletions in interval 6 of the Y chromosome of males with idiopathic sterility point to disruption of AZF, a human spermatogenesis gene. Human Genetics 89: 491–496.

Vogt P, Keil R, Kirsch S 1993 The "AZF"-function of the human Y chromosome during spermatogenesis. In: Chromosomes Today, vol.11, pp 227–239 Eds AC Chandley & A Sumner. Chapman & Hall, London.

Vogt PH, Edelmann A, Hirschmann P, Köhler MR 1995 The Azoospermia Factor (AZF) of the Human Y Chromosome in Yq11: Function and Analysis in Spermatogenesis. Reproduction Fertility Development. 7: 685–693.

Vogt PH, 1996 Human Y chromosome function in male germ cell development. Advances in Developmental Biology 4: 193–258.

Vogt PH, Edelmann A, Kirsch S, Henegariu O, Hirschmann P, Kiesewetter F, Köhn F.M, Schill WB, Farah S, Ramos C, Hartmann M, Hartschuh W, Meschede D, Behre H.M, Castel A, Nieschlag E, Weidner W, Gröne HJ, Jung A, Engel W & Haidl G 1996 Human Y chromosome azoospermia factors (AZF) mapped to different subregions in Yq11. Human Molecular Genetics 5: 933–943.

WHO Laboratory Manual 1992 For the examination of human semen and S-cervical mucus interaction. 3rd Edn. Cambridge University Press, Cambridge.

Willison K & Ashworth A 1987 Mammalian spermatogenic gene expression. Trends in Genetics 3: 351–355.

Wolf U 1995 The genetic contribution to the phenotype. Human Genetics 95: 127–148.

Yen PH, Chai NN, Salido EC 1996 The human autosomal gene DAZLA: testis specificity and a candidate for male infertility. Human Molecular Genetics 5: 2013–2017.

Zuffardi O & Tiepolo L 1982 Frequencies and types of chromosome abnormalitites associated with human male infertility. In Genetic control of gamete production and function. pp 261–273. Eds PG Crosignani & BL Rubin. Academic Press, London.

3

FREQUENCY OF Y-CHROMOSOME MICRODELETIONS (Yq11.22–23) IN MEN WITH REDUCED SPERM QUALITY REQUESTING ASSISTED REPRODUCTION

A. Bonhoff,[1] R. Fischer,[1] V. Baukloh,[1] O. G. J. Naether,[1] W. Schulze,[2] and W. Höppner[3]

[1]Endokrinologische Praxisgemeinschaft Hamburg
Lornsenstr.4–6, 22767 Hamburg
Germany
[2]Universitätskrankenhaus Eppendorf
Abt. Andrologie, Martinistr. 52, 20246 Hamburg
Germany
[3]Institute for Hormone and Fertility Research
Grandweg 64, 22529 Hamburg
Germany

Infertility affects about 15% of couples with approximately one half of the cases due to male factors. Today intracytoplasmic sperm injection (ICSI) is the treatment of choice for severe andrological subfertility. According to the pathology involved, spermatozoa from different sources (ejaculated, epididymal, and testicular) are used. Fertilization and pregnancy rates close to those of natural conception are achieved, even in the presence of severely compromised semen parameters. However, severity of spermatogenic impairment and incidence of chromosomal anomalies seem to be positively correlated, as the latter increases where the sperm count is $<10 \times 10^6$ per ml or in azoospermia (Lange et al., 1990). Thus infertility treatment with ICSI is aimed at a male population at risk for genetic disorders.

A genetic region controlling spermatogenesis in the human has been localized at the distal end of the long arm of the Y-chromosome (Yq11), defined later as the azoospermia factor (AZF). Using PCR analysis microdeletions in this region were observed mainly in men with oligo- and azoospermia (Reijo et al., 1995). It is assumed that the deletions disrupt the gene for a fertility factor predicted already 16 years ago. Up to now three different regions (AZFa, b, c) of deletions several kb apart have been mapped in Yq11.22–23 pointing to several spermatogenesis loci. This finding is supported by testicular histology in men with microdeletions which suggests that disruption of spermatogenesis at a specific

Table 1. Genotypes and phenotypes of subfertile patients with Yq structural anomalies

Patient	Age	Sperm parameter	LH mU/ml	FSH mU/ml	Testo ng/ml	Prolactin ng/ml	Testicular histology	Karyotype
A.S.	23	azoospermic	2,2	6,1	3,3	17,8	not done	not done
R.M.	30	azoospermic	4,3	15,6	4,9	8,8	not done	mos 45,X/46,XYq11.2
M.F.	36	cryptozoospermic	5	17,6	4,5	2,6	Sertoli cell only, tubulus atrophy	normal
B.J.	34	cryptozoospermic	3,1	11,2	4,9	7,2	degenerated, spermatogonial depression	not done
J.P.	29	azoospermic	5,7	13,9	5,3	4,9	mixed atrophy, varicocele	normal
normal range for men <40 years			4-15	2-7	4-9	5-14		

phase is correlated with a certain deleted subregion in Yq11. It is proposed that each region might be active during a different phase of male germ cell development (Vogt et al., 1996). Recent studies revealed that more than 10% of men with azoospermia of unknown origin have minute deletions of the Yq_{11}.

From male patients undergoing investigation before IVF/ICSI treatment a total of 235 men with severely impaired spermatogenesis (progressive motile sperm density $<1 \times 10^6$/ml) were analysed for deletions in the Y-chromosome. During clinical examination patients were investigated for bilateral absence of the vas deferens (CBAVD). Testis volume, conventional semen and hormone parameters were also tested.

For molecular analysis genomic DNA was extracted from peripheral leucocytes. A panel of 28 loci was examined using a PCR multiplex procedure for detection of microdeletions in the Y-chromosome: 27 loci screened for deletions in Yq_{11} and locus SRY (Yp) served as intraassay control. The primers are described in detail by Henegariu et al.,1993. Deletions identified were verified by subsequent single primer PCR experiments.

Five patients out of the 235 tested (2.1%) showed interstitial deletions in the Yq11 region, none showed CBAVD. Their data are summarized in Table 1, the raised levels of FSH and LH were indicative of spermatogenic impairment. So far three of the five patients underwent testicular biopsy and histological diagnosis. The individual findings included tubular atrophy, "Sertoli-cell-only" syndrome, spermatogenic depression, and germ-cell maturation arrest. By DNA analysis two patients were found to have larger interruptions in their DNA being proximal in interval 6 (locus DYS205 to DYS1). These deletions span the described region for AZFb and c deletions. In the case of patient R.M. the deletion could also be detected by cytogenetic analysis. Figure 1 shows a representative gel for deletion confirmation and figure 2 the corresponding region mapped on the Y-chromosome. Three patients showed smaller deletions of identical extension size more distally (AZFc type deletion). Karyotyping in two cases of the five under investigation revealed no chromosomal abnormalities, one patient presented with a Turner mosaic.

Findings of interstitial deletions in the Yq_{11} region have led to the postulation that this region contains one or more genes essential for the completion of spermatogenesis. In 235 patients investigated 5 men (2.1%) with severely impaired spermatogenesis were identified as carriers of microdeletions in Yq11.22–23. Azoospermia or cryptozoospermia can be associated with a variety of abnormal testis histologies, ranging from the complete lack of germ cells to meiotic arrest with few or no mature spermatids. We speculate due to

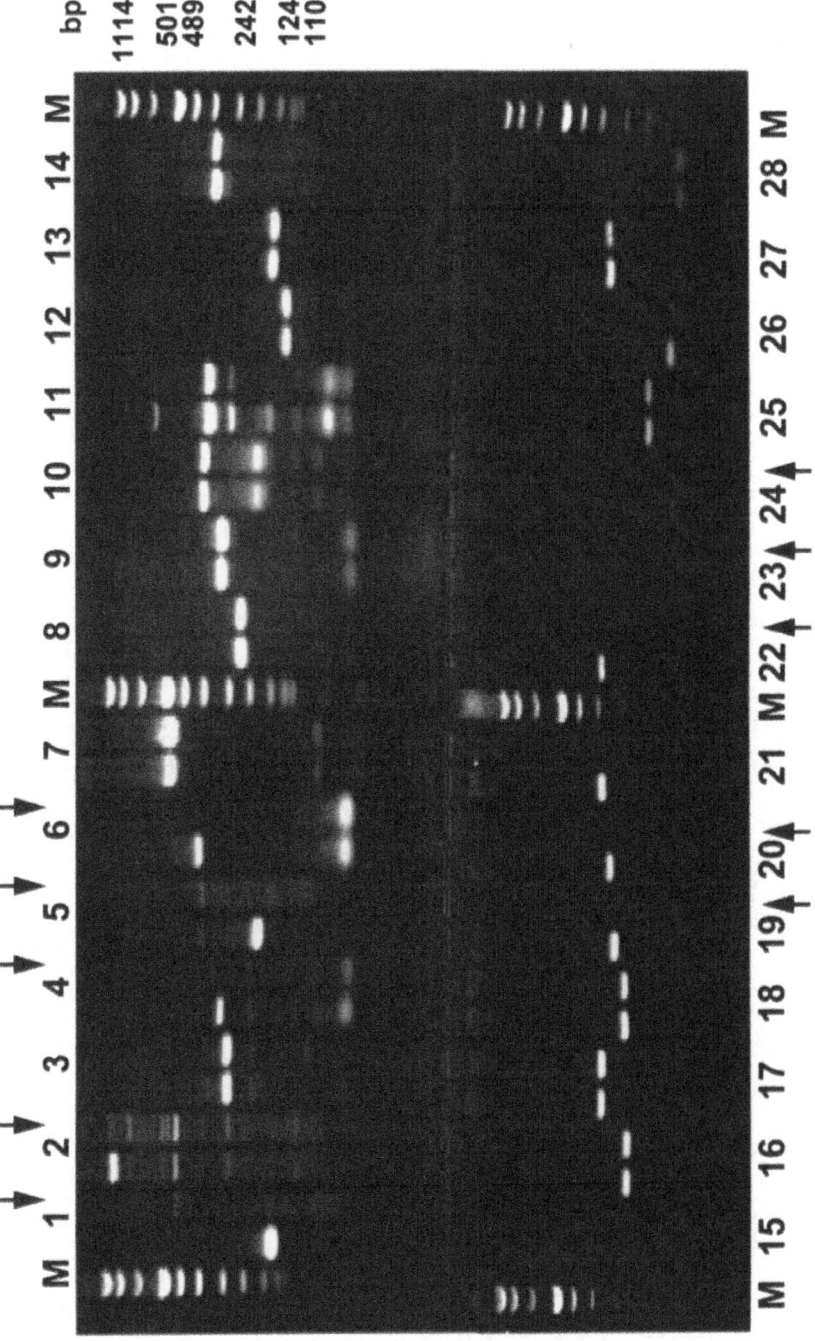

Figure 1. Molecular diagnostic of Yq11 in subfertile men.

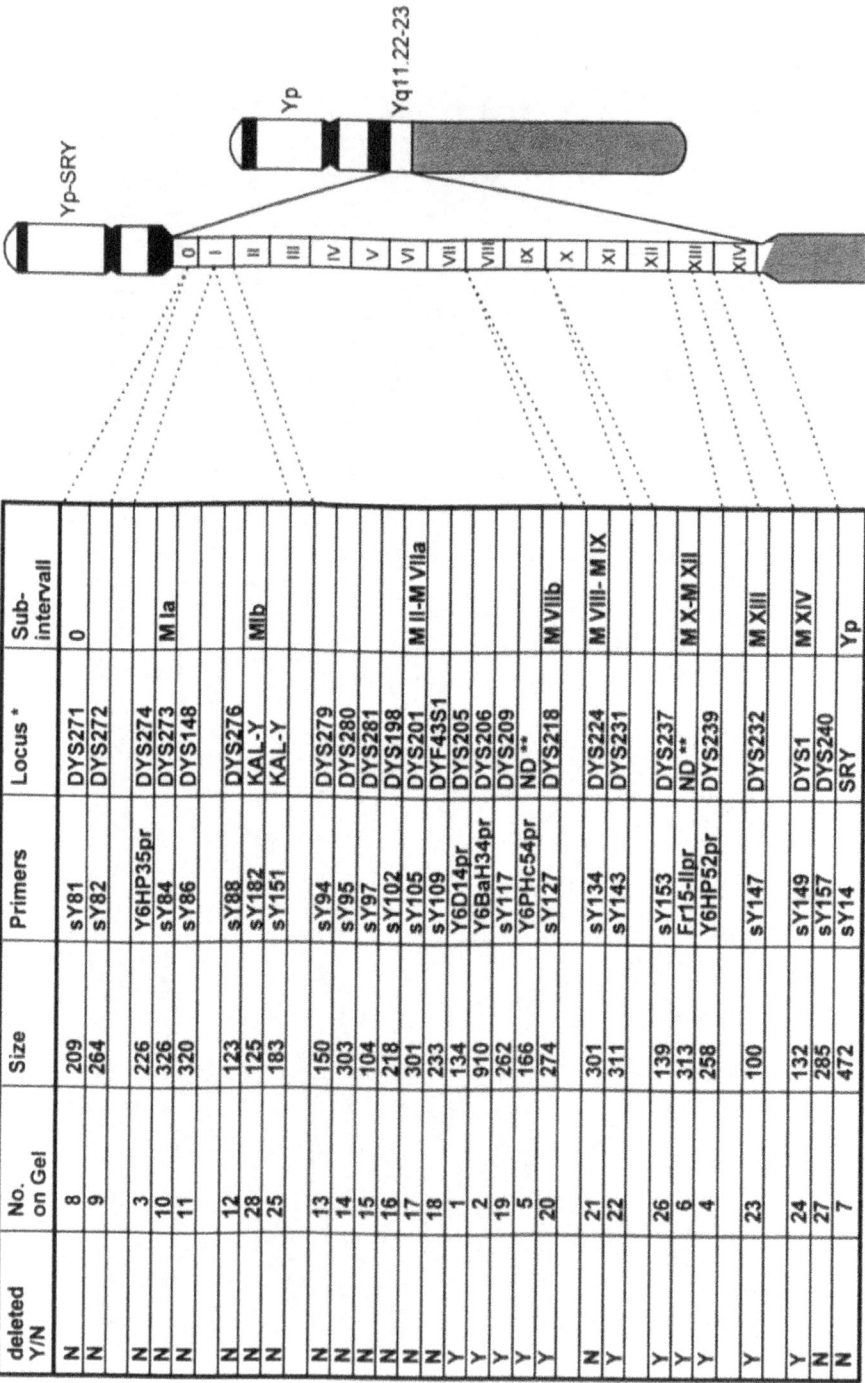

deleted Y/N	No. on Gel	Size	Primers	Locus*	Sub-intervall
N	8	209	sY81	DYS271	0
N	9	264	sY82	DYS272	
N	3	226	Y6HP35pr	DYS274	
N	10	326	sY84	DYS273	M Ia
N	11	320	sY86	DYS148	
N	12	123	sY88	DYS276	
N	28	125	sY182	KAL-Y	M Ib
N	25	183	sY151	KAL-Y	
N	13	150	sY94	DYS279	
N	14	303	sY95	DYS280	
N	15	104	sY97	DYS281	
N	16	218	sY102	DYS198	
N	17	301	sY105	DYS201	M II-M VIIa
N	18	233	sY109	DYF43S1	
Y	1	134	Y6D14pr	DYS205	
Y	2	910	Y6BaH34pr	DYS206	
Y	19	262	sY117	DYS209	
Y	5	166	Y6PHc54pr	ND**	
Y	20	274	sY127	DYS218	M VIIb
N	21	301	sY134	DYS224	M VIII- M IX
Y	22	311	sY143	DYS231	
Y	26	139	sY153	DYS237	
Y	6	313	Fr15-IIpr	ND**	M X-M XII
Y	4	258	Y6HP52pr	DYS239	
Y	23	100	sY147	DYS232	M XIII
Y	24	132	sY149	DYS1	M XIV
N	27	285	sY157	DYS240	
N	7	472	sY14	SRY	Yp

Figure 2. Subintervals of interval 6 on the human Y-chromosome. Patient A.S. *Y-DNA locus as defined in genome data base. **Not yet defined.

the histological findings that the microdeletions observed were *de novo* mutation events. However, analysis of DNA samples of the fathers or fertile brothers have not yet been performed. The majority of our patients revealed no detectable deletions in the Y-chromosome despite severely impaired spermatogenesis. Thus, it can be speculated that in these cases either undetectable point mutations are responsible or alternatively, the AZF locus is not the only crucial gene cluster in spermatogenesis.

REFERENCES

Lange R, Michelmann HW & Engel W 1990 Chromosomale Ursachen der Infertilität beim Mann. Fertilität 6: 17–28.

Ma K et al. 1992 Towards the molecular localisation of the AZF locus: mapping of microdeletions in azoospermic men within 14 subintervals of interval 6 of the human Y chromosome. Human Molecular Genetics 1: 29–33.

Henegariu O et al. 1993 Rapid screening of the Y chromosome in idiopathic sterile men, diagnostic for deletions in AZF, a genetic Y factor expressed during spermatogenesis. Andrologia 25: 97–106.

Reijo R et al. 1995 Diverse spermatogenic defects in humans caused by Y chromosome deletions encompassing a novel RNA-binding proteine gene. Nature Genetics 10: 383–393.

Vogt PH, et al 1996: Human Y chromosome azoospermia factors (AZF) mapped to different subregions in Yq11. Human Molecular Genetics 5: 933–943.

4

HISTONE GENE EXPRESSION AND CHROMATIN STRUCTURE DURING SPERMATOGENESIS

D. Doenecke, B. Drabent, C. Bode, B. Bramlage, K. Franke, K. Gavénis, U. Kosciessa, and Olaf Witt

Abteilung Molekularbiologie
Zentrum Biochemie und Molekulare Zellbiologie
Georg-August-Universität Göttingen
Humboldtallee 23, 37073 Göttingen
Germany

1. SUMMARY

The chromatin of male germ cells is restructured throughout spermatogenesis. Analysis of differential histone protein patterns at specific stages of spermatogenesis may contribute towards an understanding of the changes in chromatin structure and function during this differentiation process. The most striking changes in histone patterns occur at the stage of pachytene spermatocytes when most of the linker H1 histones are replaced by the testis specific subtype H1t. In addition, replacement of core histone subtypes is observed at this stage. These structural changes precede the reorganization of chromatin at haploid stages when histones are replaced first by transition proteins and then by protamines.

2. INTRODUCTION

Chromatin is composed of DNA and proteins at a ratio of about 1:1. The major part of the chromosomal proteins consists of histones. This family of highly conserved, basic proteins is subdivided into 5 major classes, termed H1, H2A, H2B, H3 and H4, ranging from about 220 (H1) to 102 (H4) amino acids in length. H1 is highly enriched in lysine residues, H2A and H2B are moderately lysine-rich and H3 and H4 are defined as arginine-rich histones (for review see Wolffe, 1995). Each of these 5 histone classes, except H4, consists of several subtypes, which differ in their primary structures. This is most evident for the H1 histone group, which in mammals comprises seven subtypes, termed H1.1 to H1.5, H1° and H1t (Lennox and Cohen, 1984; for gene sequences see: Doenecke & Tönjes, 1986; Alonso et al., 1988; Eick et al., 1989; Drabent et al., 1991; Albig et al., 1991; Albig et al., 1993; Drabent et al., 1993; Doenecke et al., 1994; Drabent et al., 1995a; Albig et al., 1997). Similarly, several subtypes have been described within the H2A, H2B and H3 histone classes (Franklin & Zweidler, 1977; Wells & McBride, 1989).

A second criterion for defining different groups of histone proteins is their structural contribution to the basal chromatin structure. At a first level, chromatin is organized as a tandem array of bead-like subunits, termed nucleosomes (for review see Wolffe, 1995). These consist of a core particle composed of eight histones, i.e. two copies of each of H2A, H2B, H3 and H4, and of 146 base pairs of DNA surrounding the histone octamer one and three quarter times. The DNA connecting such core particles is termed linker DNA. It is associated with histone H1 which seals part of the linker DNA to the histone octamer, thus helping to complete a second round of DNA at this subunit particle (Allan et al., 1986; Simpson, 1978). These structural aspects of histone localization within the chain of nucleosomes were the basis for discriminating between linker histones (H1) and core histones (H2A, H2B, H3, H4).

A third possibility to discriminate between different histone proteins or the respective genes is based on their mode of biosynthesis. Most of the histone genes are expressed during the S-phase of the cell cycle in coordination with DNA replication (for review see Osley 1991). The regulation of expression of this group of histone genes, the replication-dependent genes, differs from a second class, the replacement histone genes (Smith et al., 1984). The expression of these histone genes is not blocked by inhibitors of DNA replication (Zlatanova 1980; Pehrson & Cole 1980) and has been described in several terminally differentiated cell systems (Panyim & Chalkley, 1969; for review see Zlatanova & Doenecke, 1994). Thus, S-phase independent synthesis of the replacement histone subtypes H1°, H2A.Z, H2A.X and H3.3 has been observed in several species.

Tissue specific distribution of specific histone subtypes may be taken as a further criterion to discriminate between histone isoforms. The best examples in this respect are the testicular isoforms of histones (Bhatnagar & Bellve 1978; Meistrich et al., 1994; Cole et al., 1986; Wolfe & Grimes, 1997; Drabent et al., 1996; Doenecke et al., 1997). Such testis specific histone subtypes have been described in several mammalian species. Their patterns may reflect the stage specific chromosomal structures and functions during the premeiotic, meiotic and postmeiotic stages of spermatogenesis. Thus, the detailed analysis of histone subtype patterns and of histone gene regulation at different stages of spermatogenesis should contribute to an understanding of chromatin structure and function during male germ cell development.

3. SPERMATOGENESIS AS A MODEL SYSTEM FOR HISTONE GENE EXPRESSION

Cytological and biochemical studies of mammalian testicular cells have shown that the composition and morphological appearance of chromatin changes extensively during the stepwise developmental program which finally yields mature sperm cells. This chromosomal reorganization is particularly evident during the different stages of meiotic prophase which are characterized by drastic changes of chromatin morphology. Later, during haploid stages of germ cell development, chromosomal condensation leads to the tightly packed chromatin of mature sperm cells. Thus, spermatogenesis may be considered as a model system for the analysis of the role of individual types of chromosomal proteins in forming varied chromatin structures at different states of function.

Several methods for the analysis of the structure and composition of chromatin at particular stages of sperm cell development have been used. First, histological techniques with sections obtained from testis at different stages of maturation can be used for an analysis of the distribution of specific chromosomal proteins by immunocytochemical techniques. Secondly, the distribution of transcripts specific for individual genes encoding chromosomal proteins can be studied on such sections. Analysis of testis from different stages of development allows to determine the time of onset of transcription of a specific gene or to detect the stage of the first appearance of a particular protein. As shown in Figure 1, testis from 9 day old mice is still immature, just a small layer of spermatogonia and Sertoli cells is found in the lumen of a testicular tubule. At day 20, several layers of cells have formed, but no mature sperm has developed until this stage. Finally, the tubuli from an adult animal contain several layers of cells including mature spermatozoa. As shown in figure 2, we have used paraffin sections of testis obtained at the same stages in order to define the stages of expression of testis specific protein genes.

Besides this age dependent regulation of the onset of expression of particular genes encoding chromosomal proteins, synchronization of spermatogenesis can be achieved by inducing a vitamin A-deficiency and reversing the resultant block of sperm cell development (Morales & Griswold, 1987). This has been successfully used for the analysis of histone protein patterns at meiotic and postmeiotic stages of rat spermatogenesis (Meistrich et al., 1994).

The biochemical analysis of patterns of chromosomal proteins in particular cell types requires the preparative separation of testicular cells. Several methods have been successfully used for that purpose. Separation by sedimentation velocity on albumin gradients (Romrell et al., 1975) and elutriation centrifugation (Grabske et al., 1975) yield fractions enriched in particular cell types which essentially differ in size. We have used centrifugation elutriation for the assignment of specific histone gene transcripts to particular stages of spermatogenesis (Drabent et al., 1993, 1995).

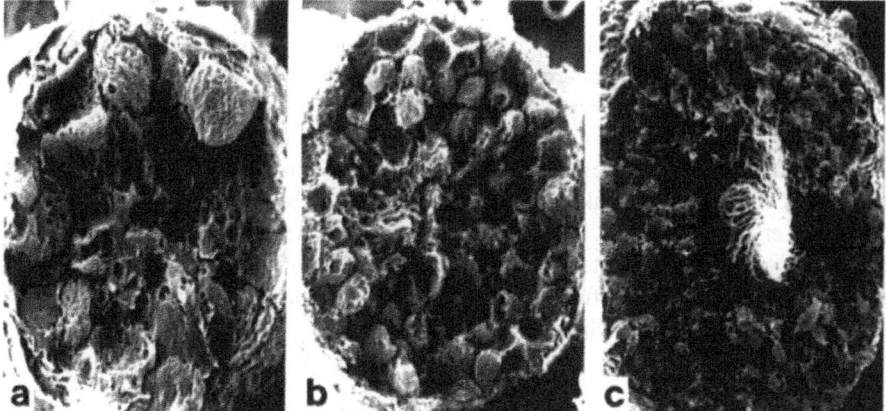

Figure 1. Scanning electron micrograph of testicular tubules from 9- and 20-day-old mice (a, b) and from an adult mouse (c). Testis was fixed in 3% glutaraldehyde, 0.2 M cacodylate buffer, pH 7.4, and after critical point drying and coating with paladium-gold it was subjected to scanning electron microscopy (in collaboration with Prof. G. Steding, Department of Embryology, University of Göttingen). Magnification of tubule from 9 day-old testis is four fold and from 20-day old testis two fold compared with tubule from adult animal.

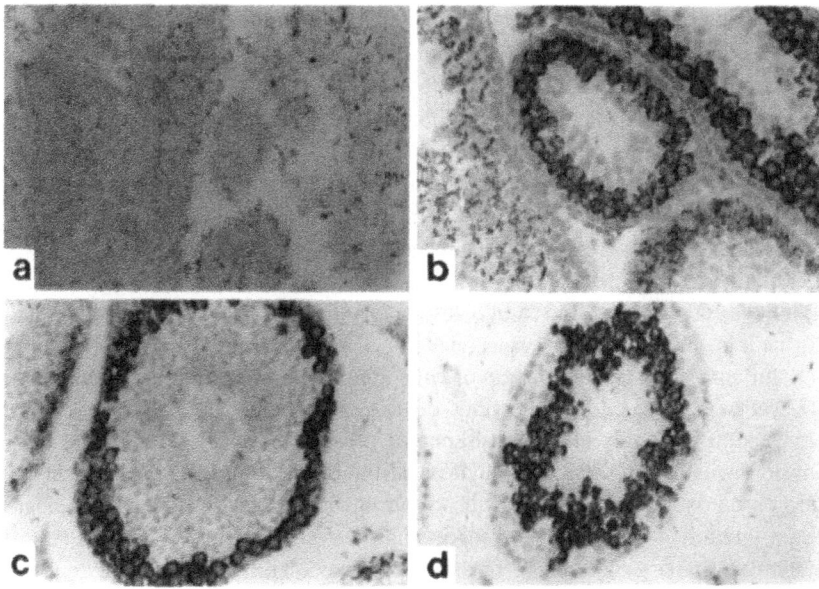

Figure 2. In situ hybridization of mouse H1t and protamine mRNAs. Paraffin sections of testis from 9- and 20-day old (a and b) and adult (c and d) mice were hybridized with non-radioactively labeled antisense RNA prepared from H1t (a-c) and protamine (d) gene probes. Experimental conditions were the same as in Drabent et al. (1996).

As yet, no tumor cell lines derived from a specific stage of germ cell differentiation have been described. Cell lines would be a prerequisite for such expression studies which are based on transfection of promoter and reporter gene constructs aimed at an identification of regulatory DNA motifs and nuclear factors interacting with such sequences. The successful transformation of murine spermatogonia and spermatocytes with the SV40 large T-antigen gene and the gene coding for a temperature sensitive mutant of p53 was an important step towards such studies (Hofmann et al., 1994). These transformed cells, however, appear to have lost part of their stage specific phenotype and thus may not be suitable for an analysis of stage specific chromosomal protein patterns (Wolkowicz et al., 1996).

Further methods for the analysis of chromosomal structures at particular cell stages may include the use of mutant mice showing blocks of spermatogenesis at specific stages. In such mutants, which include testicular feminization (Lyon & Hawkes, 1970), oligotriche (Chubb, 1992), quaking (Bennet et al., 1971; Ebersole et al., 1996) or wobbler (Kaupmann et al., 1992) and CREM-mutant mice (Nantel et al., 1996; Blendy et al., 1996), the analysis of parameters of chromatin composition may contribute to an understanding of the respective dysfunction or, on the other hand, may help to describe the influence of a known defect on chromosomal parameters.

4. STAGE SPECIFIC SYNTHESIS OF CHROMOSOMAL PROTEINS

Stage specific expression of genes encoding chromosomal proteins must be part of a coordinate developmental program which regulates the varied functional, metabolic and structural features of differentiating germ cells. Differential expression of linker and core histone subtype genes occurs during the premeiotic and meiotic stages, whereas transition proteins and thereafter protamines gradually replace histones during the haploid stages of germ cell development.

4.1. Linker Histones

The linker histone family includes the seven H1 subtypes, i.e. H1.1-H1.5, H1° and H1t (for review see Doenecke et al., 1994; for a comparison of H1 nomenclatures, see Lennox & Cohen, 1984, Drabent et al., 1995). The predominant subtype in spermatogonia is the H1.1 isoform which exceeds the H1.2 and H1.4 isoforms several fold (Meistrich et al., 1985). Similarly, the replacement subtype H1° gene is transcribed, though at a very low level, in spermatogonia (García-Iglesias et al., 1993). This expression of the H1° gene at early stages of germ cell development was demonstrated with transgenic mice carrying an H1° gene promoter coupled to a β-galactosidase reporter gene. In that case, expression of β-galactosidase was restricted to a broad layer of early spermatogenic cells (García-Iglesias et al., 1993).

In situ hybridization with a murine H1t antisense RNA probe showed that H1t gene expression is first detected at the pachytene spermatocyte stage (Fig. 2b and 2c). Essentially no H1t transcripts are found in tubules from adult and 20 day old mice in the outer cell layer representing spermatogonia and cells in the early meiotic prophase. In these sections, H1t mRNA is only detectable in pachytene spermatocytes, and no transcripts are found at later stages of germ cell development. This restricted period of expression is also evident in relation to different times after birth, since testis from 9-day-old mice shows no hybridization with the H1t hybridization probe (Fig. 2a), whereas germ cells from 20-day-old and adult mice express the H1t gene. These data agree with H1t expression patterns in the rat (Kremer & Kistler, 1991) and in man (Doenecke et al., 1996). In contrast to the restriction of H1t transcripts to cells of the meiotic prophase, hybridization with a protamine probe occurs in round and elongated spermatids (Fig. 2d) indicating the onset of chromatin condensation which finally results in the tightly packed chromatin structure of the mature sperm head (Oliva & Dixon, 1991).

The duration of the stage specific transcription of genes encoding specific H1 subtypes is not identical to the period of association of the respective histones with the chromatin of spermatogenic cells. In fact, H1.1 exists at a high level at early stages of spermatogenesis; it then persists, but its level decreases upon further development which includes mitotic and meiotic cell divisions (Franke & Doenecke, unpublished; Meistrich et al., 1985; Doenecke et al., 1996). The H1t protein, which is first detectable at the pachytene spermatocyte stage (Drabent et al., 1996), remains associated with the chromatin of the ensuing haploid stage cells before replacement by transition proteins and protamines at late spermiogenesis.

4.2. Core Histones

The histone octamer moiety of the nucleosome core consists of the histones H2A, H2B, H3 and H4. Replacement variants have been described within the H2A and H3 classes. These replication-independent subtypes within the H2A family are H2A.X and H2A.Z (Bonner et al., 1993; Hatch et al., 1988), and H3.3 is the H3 replacement subtype (Wu & Bonner, 1981; Brush et al., 1985; Wells and Kedes, 1985; Albig et al., 1995). Testis-specific subtypes have been described as TH2A (Trostle-Weige et al., 1982), TH2B (Shires et al., 1975; Hwang & Chae 1989) and TH3 (Trostle-Weige et al., 1984) in the rat, but no homologs of these three histone isoforms were found in man or mouse testis chromatin. Recently, we have isolated and characterized a human H3 histone gene (Albig et al., 1996) which is solely expressed in testis (Witt et al., 1996), but the protein encoded by this gene apparently is not a homologue of rat TH3.

The testis specific isoforms of H2A and H2B in the rat, i.e. TH2A and TH2B, are detected first in pachytene spermatocytes and persist until later stages (Kim et al., 1987; Hwang and Chae, 1989). Thus, male germ cells at this stage not only replace the major part of their linker histone moiety by H1t, but also restructure their nucleosomal cores by substituting part of the core histones by testis specific isoforms. It remains to be shown, however, whether homologues of TH2A and TH2B exist in mammals other than rat. The genes encoding rat TH2A and TH2B are located within a major histone gene cluster on chromosome 17 (at 17p12–13; Walter et al., 1996) which corresponds to the human histone gene cluster on chromosome 6 (at 6p21.3, Albig et al., 1993). Within that human histone gene cluster or the corresponding murine gene cluster (Wang et al., 1996), as yet no TH2A or TH2B homologue has been identified.

No gene or sequence data exist on rat TH3 protein, but its amino acid composition (Trostle-Weige et al., 1984) showing four cysteine residues differs from all other H3 subtypes described until now. The protein was found in rat spermatogonia, but was absent at later stages of spermatogenesis. The human H3 variant gene, which is exclusively expressed in testis (Witt et al., 1996), varies from rat TH3 twofold. First, it contains just two cysteine residues in its primary structure, and, secondly, transcripts of this gene are only found in pachytene spermatocytes. Since this stage of expression is the same as for H1t, we proposed the term H3t for this histone and its gene, which is located on chromosome 1 (1q42) outside any histone gene cluster (Albig et al., 1996a; Witt et al., 1996).

The testicular expression of replacement core histone genes has not been analyzed in detail by means of immunocytochemistry or in situ hybridization. H2A.Z and H2A.X are both found in testis (Hatch & Bonner, 1988; Mannironi et al., 1989, 1994; Meistrich et al., 1994). Similarly as the linker histone H1°, these core histone replacement subtypes appear to be mainly synthesized in spermatogonia and persist at a low level during later stages of spermatogenesis.

No definite replacement subtype of H2B has as yet been described. However, a testicular variant of H2B has been found in mouse spermatids (Moss et al., 1989). We have found a cDNA encoding this histone in a cDNA library prepared from RNA of testis from a 42-day-old mouse (Kosciessa & Doenecke, unpublished). This core histone differs from main type H2B at its C-terminus, which is extended by 11, mostly hydrophobic, residues. This variant H2B may be the homologue of an H2B subtype with an alternate C-terminus which is found in chicken spermatids (Challoner et al., 1989). In analogy to these H2B variant transcripts, a testis specific histone mRNA with an alternative 3' end formation has also been found within the H2A class in mouse round spermatids (Moss et al. 1994).

H3.3 is the H3 replacement subtype (Wu & Bonner, 1981). Interestingly, two H3.3 genes termed H3.3A and H3.3B encoding exactly the same amino acid sequence have been described in several vertebrate species (Brush et al., 1985; Wells & Kedes, 1985; Hraba-Renevey and Kress, 1989; Albig et al., 1995). In contrast to main type histone genes, but in analogy to the other replacement histone genes (Kress et al., 1986), both H3.3 mRNAs are polyadenylated (Wells & Kedes, 1985; Hraba-Renevey & Kress, 1989; Albig et al., 1995). Several different H3.3A and H3.3B mRNAs, which differ in relation to the length of their 3' non coding region, have been described. Three different H3.3B mRNAs and two different H3.3A mRNAs were found. We could detect all three H3.3B mRNA species in human testis RNA by RNase protection analysis (Albig et al., 1995).

4.3. Transition Proteins and Protamines

The chromatin of mammalian male germ cells is reorganized during spermiogenesis upon differentiation of spermatids towards mature sperm. During this process of spermiogenesis, the histone complement of early spermatids is first replaced by a spermatid-specific class of basic chromosomal proteins termed transition proteins and then in late spermatids by protamines (Oliva & Dixon, 1991). The genes encoding transition proteins and protamines are transcribed in early spermatids, but the mRNAs are not immediately translated. Instead, they are maintained in a translationally repressed state before being translated at late spermatid stages (Heidaran & Kistler, 1987; Kleene, 1989; for review see Schäfer et al., 1995). This temporal control of translation is mainly based on interactions of regulatory proteins at specific sequences within the 3' untranslated mRNA region as demonstrated by Kwon and Hecht with the mouse protamine 2 mRNA (Kwon & Hecht, 1993). This role of the 3' untranslated part of a protamine mRNA was further proven by Braun et al. (1989) using the growth hormone gene as a reporter for the function of the 3' untranslated region of mouse protamine 2 mRNA in a transgenic mouse approach.

The time course of appearance of transition proteins and protamines was recently studied by protein analysis of microdissected, staged seminiferous tubule segments of the rat (Kistler et al., 1996). It was concluded that transition protein 2 (TP2) is the first chromosomal protein replacing histones in elongated spermatids, followed by TP1 and then by protamine.

5. CHROMATIN STRUCTURE AND FUNCTION

Differentiation of male germ cells from spermatogonia to mature spermatozoa can be described at the chromatin level as a transition from nucleohistones to the nucleoprotamine complex. The replacement of histones by transition protein occurs after meiosis during spermatid differentiation (Oliva & Dixon, 1991; Kistler et al., 1996; Oko et al., 1996). Thus, histones form the major protein component of the chromatin of premeiotic and meiotic cells, and varied patterns of histone subtypes may contribute to structural and functional changes of the nucleoprotein of differentiating germ cells.

5.1. H1 Linker Histone Subtypes

The H1 family of histones exerts its function at several levels. First, it interacts with the nucleosomal DNA and seals two rounds of DNA at the surface of the nucleosomal core histone octamer (Allan et al., 1986; Pruss et al., 1996). It thereby forms an intermediate chromatin particle containing all histones, the chromatosome (Simpson, 1978). Secondly, H1 is a prerequisite for the formation of higher order chromatin structures (Thoma et al., 1979). Thus, H1 apparently is involved in chromatin condensation, and varied H1 patterns may contribute to a modulation of chromatin compaction, e.g. during the different stages of spermatogenesis. Thirdly, H1 histones act as basal repressors of transcription and transcription factors can counteract this H1-mediated repression (for review see Paranjape et al., 1994). Interestingly, co-crystal structure analysis of the liver-specific transcription factor HNF3-γ and of the central domain of the H1-histone subtype H5 has shown a close structural relationship between the DNA-binding domains of these two proteins (Clark et al., 1993, Ramakrishnan et al., 1993). Similar relationships between histone and transcription factor structures have been found in comparing core histone structural motifs and transcription factors or transcription factor activators (see below).

As shown above, the H1 histone pattern changes during spermatogenesis. The H1.1 gene is expressed in spermatogonia and its gene product (termed H1a in the nomenclature of Lennox & Cohen, 1984) predominates until the pachytene stage. The H1.1 gene product is then replaced by H1t. The functional implications of this replacement are still unknown. De Lucia et al. (1994) have compared the condensing capacity of rat H1t and somatic H1 variants by circular dichroism analysis of oligonucleosomes reconstituted with different H1 preparations. The authors concluded that rat H1t had the lowest condensing capacity of all H1 types tested. Similarly, Khadake & Rao (1995) found that rat H1t had the lowest DNA condensing capacity compared with other rat H1 subtypes.

Before functional conclusions on the function of H1t in pachytene and later stages of spermatogenesis can be drawn, detailed investigations on the distribution of H1t in chromatin and on interactions of H1t with other chromatin components have to be done. Moens (1995) has studied the distribution of rat H1t in meiotic prophase chromosomes and has shown that H1t is detected only after completion of synapsis and synaptonemal complex assembly. At that stage, the H1t protein was evenly distributed over the synaptonemal complex and over hetero- and euchromatin. In the same study, it was shown that hyperacetylated H4 histones accumulated in euchromatic regions of these meiotic prophase cells.

The comparatively low condensing capacity of H1t may indicate that a relatively open state is needed to allow interactions of chromosomes with other nuclear proteins, such as non-histone proteins, transcription factors and RNA polymerases or components of the meiotic apparatus. At later stages, a comparatively open chromatin structure may be needed in order to facilitate the replacement of histones by transition proteins at mid-spermiogenesis.

5.2. Core Histones

The core histones H2A, H2B, H3 and H4 have for a long time been treated just as structural components of the nucleosomal core providing a basis for the first level of DNA compaction. In the meantime, detailed analysis of chromatin structures at defined gene and promoter sequences has shown that transcriptional activation disrupts nucleosomal arrays and that transcription factors and transcription factor activating proteins interact with core and linker chromatin components (for review see Steger & Workman, 1996, Burley & Roeder, 1996).

An important aspect of modulating structure and function of nucleosomes is the covalent modification of histones by acetylation. This process stabilizes the binding of transcription factors to chromatin and thus provides a link between chromatin structure and activation of transcription (reviewed in Roth & Allis 1996, Wolffe & Pruss, 1996a). In the context of spermatogenesis, hyperacetylation of H4 histones has been observed in all species that replace histones by protamines (reviewed by Oliva & Dixon, 1991) at a stage before replacement of histones by protamines. Thus, H4 acetylation may help to displace histones from spermatid chromatin in analogy to the role of histone acetylation in gene activation.

Structural analysis of several proteins involved in transcriptional regulation and chromatin structure has revealed strong sequence and structural similarities compared with core histones. These histone-like proteins include transcription-factor regulatory proteins as well as H3-related components of centromere structures (for review see Wolffe and Pruss, 1996b; Burley and Roeder, 1996). As in the case of structural homologies between the H1 subtype H5 and the liver-specific transcription factor HNF3-γ (see above), the data on similarities between core histones and components of the transcriptional machinery may suggest that variations of patterns of these chromatin components contribute to the functional state of chromatin. Thus, replacement of core histones by testis specific isoforms or by general replacement subtypes may modulate the chromatin structure at specific stages of spermatogenesis as part of a coordinated programme regulating chromosomal structures and functions during germ cell differentiation.

6. ACKNOWLEDGMENTS

The work performed in the laboratory of the authors was supported by grants from the Deutsche Forschungsgemeinschaft and from the Bundesministerium für Bildung, Wissenschaft, Forschung und Technologie. The help and support by Drs. G. Steding (Department of Embryology), R. Herken and N. Miosge (Department of Histology) and R. Ringert and G. Zöller (Department of Urology), all at the University of Göttingen, is gratefully acknowledged. The protamine probe used in the experiment shown in Figure 2d was kindly provided by Dr. N.B. Hecht (Boston).

7. REFERENCES

Albig W, Kardalinou E, Drabent B, Zimmer A & Doenecke D 1991 Isolation and characterization of two human H1 histone genes within clusters of core histone genes. Genomics 10: 940–948.

Albig W, Drabent B, Kunz J, Kalff-Suske M, Grzeschik KH & Doenecke D 1993 All known human H1 genes except the H1° gene are clustered on chromosome 6. Genomics 16: 649–654.

Albig W, Bramlage B, Gruber K, Klobeck HG, Kunz J & Doenecke D 1995 The human replacement histone H3.3B gene (H3F3B). Genomics 30: 264–272.

Albig W, Ebentheuer J, Klobeck G, Kunz J & Doenecke D 1996 A solitary human H3 histone gene on chromosome I. Human Genetics 97: 486–491.

Albig W, Meergans T & Doenecke D 1996b Characterization of the H1.5 gene completes the set of human H1 subtype genes. Gene184: 141–148.

Allan J, Mitchell T, Harborne N, Bohm L & Crane-Robinson C 1986 Roles of H1 domains in determining higher order chromatin structure and H1 location. Journal of Molecular Biology 187: 591–601.

Alonso A, Breuer B, Bouterfa H & Doenecke D 1988 Early increase in histone H1° mRNA during differentiation of F9 cells to parietal endoderm. EMBO Journal 7: 3003–3008.

Bennett WI, Gall AM, Southard JL & Sidman RL 1971 Abnormal spermiogenesis in quaking, a myelin-deficient mutant mouse. Biology of Reproduction 5: 30–58.

Bhatnagar YM & Bellve AR 1978 Two-dimensional electrophoretic analysis of major histone species and their variants from somatic and germ-line tissue. Analytical Biochemistry 86: 754–760.

Blendy JA, Kaestner KH, Weinbauer GF, Nieschlag E & Schütz G 1996 Severe impairment of spermatogenesis in mice lacking the CREM gene. Nature 380: 162–165.

Bonner WM, Mannironi C, Orr A, Pilch D & Hatch CL 1993 Histone H2A.X gene transcription is regulated differently than transcription of other replication-linked histone genes. Molecular and Cellular Biology 13: 984–992.

Braun RE, Peschon JJ, Behringer RR, Brinster RL & Palmiter RD 1989 Protamine 3' untranslated sequences regulate temporal translational control and subcellular localization of growth hormone in spermatids of transgenic mice. Genes and Development 3: 793–802.

Brush D, Dodgson JB, Choi OR, Stevens PW & Engel JD 1985 Replacement variant histone genes contain intervening sequences. Molecular and Cellular Biology 5: 1307–1317.

Burley SK & Roeder RG 1996 Biochemistry and structural biology of transcription factor IID (TFIID). Annual Reviews of Biochemistry 65: 769–799.

Challoner PB, Moss SB & Groudine M 1989 Expression of replication-dependent histone genes in avian spermatids involves an alternate pathway of mRNA 3' end formation. Molecular and Cellular Biology 9: 902–913.

Chubb C 1992 Oligotriche and quaking mutations. Phenotypic effects on mouse spermatogenesis and testicular steroidogenesis. Journal of Andrology 13: 312–317.

Clark KL, Halay ED, Lai E and Burley SK 1993 Co-crystal structure of the HNF-3/fork head DNA-recognition motif resembles histone H5. Nature 364: 412–420.

Cole KD, Kandala JC and Kistler WS 1986 Isolation of the gene for the testis specific histone variant H1t. Journal of Biological Chemistry 261: 7178–7183.

De Lucia F, Faraone-Mennella MR, D'Erme M, Quesada P, Caiafa P & Farina B 1994 Histone-induced condensation of rat testis chromatin: testis-specific H1t versus somatic H1 variants. Biochemical and Biophysical Research Communications 198: 32–39.

Doenecke D and Tönjes R 1986 Differential distribution of lysine and arginine residues in the closely related histones H1° and H5. Analysis of a human H1° gene. Journal of Molecular Biology 187: 461–464.

Doenecke D, Albig W, Bouterfa H & Drabent B 1994 Organization and expression of H1 histone and H1 replacement histone genes. Journal of Cellular Biochemistry 54: 423–431.

Doenecke D, Albig W, Bode C, Drabent B, Franke K, Gavénis K & Witt O 1997 Histones: genetic diversity and tissue-specific gene expression. Histochemistry and Cell Biology 107: 1–10.

Drabent B, Kardalinou E and Doenecke D 1991 Structure and expression of the human gene encoding testicular H1 histone (H1t). Gene 103: 263–268.

Drabent B, Bode C & Doenecke D 1993 Structure and expression of the mouse testicular H1 histone gene (H1t). Biochimica Biophysica Acta 1216: 311–313.

Drabent B, Franke K, Bode C, Kosciessa U, Bouterfa H, Hameister H & Doenecke D 1995 Isolation of two murine H1 histone genes and chromosomal mapping of the H1 gene complement. Mammalian Genome 6: 505–511.

Drabent B, Bode C, Bramlage B & Doenecke D 1996 Expression of the mouse testicular histone gene H1t during spermatogenesis. Histochemistry and Cell Biology 106: 247–251.

Ebersole TA, Chen Q, Justice MJ & Artzt K 1996 The quaking gene product necessary in embryogenesis and myelination combines features of RNA binding and signal transduction proteins. Nature Genetics 12: 260–265.

Eick S, Nicolai M, Mumberg D & Doenecke D 1989 Human H1 histones: conserved and varied sequence elements in two H1 subtype genes. European Journal of Cell Biology 49: 110–115.

Franklin SG & Zweidler A 1977 Non-allelic variants of histones 2a, 2b and 3 in mammals. Nature 266: 273–274.

García-Iglesias MJ, Ramirez A, Monzo M, Steuer B, Martinez JM, Jorcano JL & Alonso A 1993 Specific expression in adult mice and post-implantation embryos of a transgene carrying the histone H1° regulatory region. Differentiation 55: 27–35.

Grabske RJ, Lake S, Gledhill BL & Meistrich ML 1975 Centrifugal elutriation: separation of spermatogenic cells on the basis of sedimentation velocity. Journal of Cellular Physiology 86: 177–190.

Hatch CL & Bonner WM 1988 Sequence of cDNAs for mammalian H2A.Z, an evolutionarily diverged but highly conserved basal histone H2A isoprotein species. Nucleic Acids Research 16: 1113–1124.

Heidaran MA & Kistler WS 1987 Transcriptional and translational control of the message for transition protein 1, a major chromosomal protein of mammalian spermatids. Journal of Biological Chemistry 262: 13309–13315.

Hofmann MC, Hess RA, Goldberg E & Millan JL 1994 Immortalized germ cells undergo meiosis in vitro. Proceedings of the National Academy of Sciences USA 91: 5533–5537.

Hraba-Renevey S & Kress M 1989 Expression of a mouse replacement histone H3.3 gene with a highly conserved 3' non coding region during SV40- and polyoma-induced G_o- to S-phase transition. Nucleic Acids Research 17: 2449–2461.

Hwang I & Chae CB 1989 S-phase specific transcription regulatory elements are present in a replication-independent testis-specific H2B histone gene. Molecular and Cellular Biology 9: 1005–1013.

Kaupmann K, Simone-Chazottes D, Guénet JL & Jockusch H 1992 Wobbler, a mutation affecting motoneuron survival and gonadal functions in the mouse, maps to proximal chromosome 11. Genomics 13: 39–43.

Khadake JR & Rao MRS 1995 DNA- and chromatin-condensing properties of rat testes H1a and H1t compared to those of rat liver H1bdec: H1t is a poor condenser of chromatin. Biochemistry 34: 15792–15801.

Kim YJ, Hwang LL, Tres LL, Kierszenbaum AL & Chae CB 1987 Molecular cloning and differential expression of somatic and testis-specific H2B histone genes during rat spermatogenesis. Developmental Biology 124: 23–34.

Kistler WS, Henriksen K, Mali P & Parvinen M 1996 Sequential expression of nucleoproteins during rat spermiogenesis. Experimental Cell Research 225: 374–381.

Kleene KC 1989 Poly(A) shortening accompanies the activation of translation of five mRNAs during spermiogenesis in the mouse. Development 106: 367–373.

Kremer EJ & Kistler WS 1991 Localization of mRNA for testis-specific histone H1t by in situ hybridization. Experimental Cell Research 197: 330–332.

Kress H, Tönjes R & Doenecke D 1986 Butyrate induced accumulation of 2.3 kb polyadenylated H1° histone mRNA in HeLa cells. Nucleic Acids Research 14: 7189–7197.

Kwon YK & Hecht NB 1993 Binding of a phosphoprotein to the 3' untranslated region of the mouse protamine 2 mRNA temporally repress its translation. Molecular and Cellular Biology 13: 6547–6557.

Lennox RW & Cohen LH 1984 The alterations in H1 histone complement during mouse spermatogenesis and their significance for H1 subtype function. Developmental Biology 103: 80–84.

Lyon MF & Hawkes 1970 X-linked gene for testicular feminization in the mouse. Nature 227: 1217–1219.

Mannironi C, Bonner WM & Hatch CL 1989 H2A.X, a histone isoprotein with a conserved C-terminal sequence, is encoded by a novel mRNA with both replication type and poly A 3' processing signals. Nucleic Acids Research 17: 9113–9126.

Mannironi C, Orr A, Hatch C, Pilch D, Ivanova V & Bonner W 1994 The relative expression of human histone H2A genes is similar in different types of proliferating cells. DNA and Cell Biology 13: 161–170.

Meistrich ML, Bucci LR, Trostle-Weige PK & Brock WA 1985 Histone variants in rat spermatogonia and primary spermatocytes. Developmental Biology 112: 230–240.

Meistrich ML, Trostle-Weige PK & Beek MEAB van 1994 Separation of specific stages of spermatids from vitamin A-synchronized rat testes for assessment of nucleoprotein changes during spermatogemnsis. Biology of Reproduction 51: 334–344.

Moens PB 1995 Histones H1 and H4 of surface-spread meiotic chromosomes. Chromosoma 104: 169–174.

Morales C & Griswold MD 1987 Retinol-induced stage synchronization in seminiferous tubules of the rat. Endocrinology 121: 432–434.

Moss SB, Challoner PB & Groudine M 1989 Expression of a novel histone 2B during mouse spermiogenesis. Developmental Biology 133: 83–92.

Moss SB, Ferry RA & Groudine M 1994 An alternative pathway of histone mRNA 3' end formation in mouse round spermatids. Nucleic Acids Research 22: 3160–3166.

Nantel F, Monaco L, Foulkes NS, Masquiller D, LeMeur M, Henriksen K, Dierich A, Parvinen M & Sassone-Corsi P 1996 Spermiogenesis deficience and germ-cell apoptosis in CREM-mutant mice. Nature 380: 159–162.

Oko RJ, Jando V, Wagner CL, Kistler WS & Hermo LS 1996 Chromatin reorganization in rat spermatids using the disappearance of testis-specific histone, H1t, and the appearance of transition proteins TP1 and TP2. Biology of Reproduction 54: 1141–1157.

Oliva R & Dixon GH 1991 Vertebrate protamine genes and the histone to protamine replacement reaction. Progress in Nucleic Acids Research 25–93.

Osley MA 1991 The regulation of histone synthesis in the cell cycle. Annual Review of Biochemistry 60: 827–861.

Panyim S & Chalkley R 1969 High resolution acrylamide gel electrophoresis of histones. Archives of Biochemistry and Biophysics 130: 337–346.

Paranjape SM, Kamakaka RT & Kadonaga JT 1994 Role of chromatin structure in the regulation of transcription by RNA polymerase II. Annual Reviews of Biochemistry 63: 265–297.

Pehrson J & Cole RD 1980 Histone H1° accumulates in growth inhibited cultured cells. Nature 285: 859–862.

Pruss D, Bartholomew B, Persinger J, Hayes J, Arents G, Moudrianakis EN & Wolffe AP 1996 An asymmetric model for the nucleosome: a binding site for linker histone inside the DNA gyres. Science 274: 614–617.

Ramakrishnan V, Finch JT, Graziano V, Lee PL & Sweet RM 1993 Crystal structure of globular domain of histone H5 and its implications for nucleosome binding. Nature 362: 219–223.

Romrell LJ, Bellve AR & Fawcett DW 1975 Separation of mouse spermatogenic cells by sedimentation velocity. Developmental Biolgy 49: 119–131.

Roth SY & Allis CD 1996 Histone acetylation and chromatin assembly: a single escort, multiple dances? Cell 87: 5–8.

Schäfer M, Nayernia K, Engel W & Schäfer U 1995 Translational control in spermatogenesis. Developmental Biology 172: 344–352.

Shires A, Carpenter MP & Chalkley R 1975 New histones found in mammalian testes. Proceedings of the National Academy of Sciences 72: 2714–2718.

Simpson RT 1978 Structure of the chromatosome, a chromatin particle containing 160 base pairs of DNA and all the histones. Biochemistry 17: 5524–5531.

Smith BJ, Harris MR, Sigournay CM, Mayes ELV & Bustin M 1984 A survey of H1°- and H5-like protein structure and distribution in lower and higher eukaryotes. European Journal of Biochemistry 138: 309–317.

Steger DJ & Workman JL 1996 Remodelling chromatin structures for transcription: what happens to the histones? BioEssays 18: 875–884.

Thoma F, Koller T & Klug A 1979 Involvement of histone H1 in the organization of the nucleosome and of salt dependent superstructures of chromatin. Journal of Cell Biology 83: 403–427.

Trostle-Weige PK, Meistrich ML, Brock WA, Nishioka K & Bremer JW 1982 Isolation and characterization of TH2A, a germ-cell specific variant of histone H2A in the rat testis. Journal of Biological Chemistry 257: 5560–5567.

Trostle-Weige PK, Meistrich ML, Brock WA & Nishioka K 1984 Isolation and characterization of TH3, a germ cell specific variant of histone 3 in the rat testis. Journal of Biological Chemistry 259: 8769–8776.

Walter L, Klinga-Levan K, Helou K, Albig W, Drabent B, Doenecke D, Günther E & Levan G 1996 Chromosomal mapping of rat histone genes H1fv (H1°), H1d, H1t, Th2a and Th2b. Cytogenetics and Cell Genetics 75: 136–139.

Wang ZF, Krasikov T, Frey MR, Wang J, Matera AG & Marzluff WF 1996a Characterization of the mouse histone gene cluster on chromosome 13: 45 histone genes in three patches spread over 1 Mb. Genome Research 6: 688–701.

Wells D & Kedes L 1985 Structure of a human histone cDNA: evidence that basally expressed histone genes have intervening sequences and encode polyadenylated mRNAs. Proceedings of the National Academy of Sciences USA 82: 2834–2838.

Wells D & McBride C 1989 A comprehensive compilation and alignment of histones and histone genes. Nucleic Acids Research 17: r311-r346.

Witt O, Albig W & Doenecke D 1996 Testis-specific expression of a novel human H3 histone gene. Experimental Cell Research 229: 301–306.

Wolfe SA & Grimes SR 1993 Histone H1t: a tissue-specific model used to study transcriptional control and nuclear function during cellular differentiation. Journal of Cellular Biochemistry 53: 156–160.

Wolffe AP 1995 Chromatin: Structure and Function. Academic Press, London.

Wolffe AP & Pruss D 1996a Targeting chromatin disruption: transcription regulators that acetylate histones. Cell 84: 817–819.

Wolffe AP & Pruss D 1996b Deviant nucleosomes: the functional specialization of chromatin. Trends in Genetics 12: 58–62.

Wolkowicz MJ, Coonrod SM, Reddi PP, Millan JL, Hofmann MC & Herr JC 1996 Refinement of the differentiated phenotype of the spermatogenic cell line GC-2spd(Ts). Biology of Reproduction 55: 923–932.

Wu RS & Bonner WM 1981 Separation of basal histone synthesis from S-phase histone synthesis in dividing cells. Cell 27: 321–330.

Zlatanova J 1980 Synthesis of histone H1° is not inhibited in hydroxyurea-treated Friend cells. FEBS Letters 112: 199–202.

Zlatanova J & Doenecke D 1994 Histone H1°: a major player in cell differentiation? The FASEB Journal 8: 1260–1268.

5

HISTONE GENE EXPRESSION IN THE HUMAN TESTIS

O. Witt, K. Gavénis, and D. Doenecke

Zentrum Biochemie, Abt. Molekularbiologie
Universität Göttingen
Humboldtallee 23, 37073 Göttingen
Germany

The function of histones has been classically assigned to the formation of DNA-protein-complexes, called nucleosomes. A nucleosomal core consists of a stretch of DNA wrapped around an octamer of core histones (H2A, H2B, H3 and H4). A further condensation into higher order chromatin is achieved by the linker histone H1, which is associated with the nucleosome. Additionally, recent evidence suggest an active participation of histones in the transcription and regulation of genes (Wolfe, 1995). Several histone subtypes have been isolated which can be divided into replication-dependent, replication-independent (replacement) and tissue-specific isoforms. The expression of these subtypes varies with respect to the cell line and tissue investigated (Meergans et al., 1996). During spermatogenesis, the histone fraction of spermatogenic cells is largely replaced by testis specific isoforms in rat and mice (Doenecke and Drabent, 1995).

We have investigated the expression of a newly described, uniquely localized human H3 core histone gene by RNase protection analysis (RPA) (Albig et al., 1996). Total RNA from human cell lines and tissues were subjected to the RPA. Expression of this gene was exclusively found in human testis. Furthermore, we have conducted *in situ* hybridization experiments of sections from human testis in order to identify the pattern of expression during spermatogenesis. mRNA could only be detected in primary spermatocytes of the pachytene stage, as determined by H.E. staining of serial sections. The pattern of expression of the linker histones H1t and H1° was also investigated in the human testis by *in situ* hybridization with specific probes and immunochemistry with specific polyclonal antibodies at the mRNA and protein level. Transcription of the H1t gene was demonstrated in distinct cell populations of spermatogenesis. Serial sections with H.E. identified these cells as pachytene spermatocytes. The corresponding detection of the H1t protein revealed signals in pachytene spermatocytes as well as in the subsequent stage of spermatogenesis, i.e. round spermatids. In contrast, expression of the DNA-replication-independent H1° histone protein was mainly found in Sertoli and myoid cells. Spermatogenic cells showed uniformly weak staining for H1°.

The Fate of the Male Germ Cell, edited by Ivell and Holstein
Plenum Press, New York, 1997

Since the expression of all testis specific histone variants investigated so far is restricted to pachytene spermatocytes, it appears likely that these histone isoforms play an important role in remodeling the chromatin structure during meiosis. This could also contribute to a switch in gene expression required during the haploid stages of spermatogenesis.

ACKNOWLEDGMENTS

this work was supported by a grant from the Bundesministerium für Bildung, Wissenschaft, Forschung und Technologie

REFERENCES

Albig W, Ebentheuer J, Klobeck G, Kunz J, and Doenecke D 1996 Human Genetics 97: 486–491.
Doenecke D & Drabent B 1995 in Advances in Spermatozoal Phylogeny and Taxonomy (Jamieson, B. G. M., Ausio, J., and Justine, J.-L., Ed.). Vol. 166: 525–535. Mémoires du Muséum National d'Histoire Naturelle, Paris.
Meergans T, Albig W, and Doenecke D 1997 submitted for publication.
Wolffe A 1995 Chromatin Structure and Function (London: Academic Press).

6

ENDOCRINE CONTROL OF GERM CELL PROLIFERATION IN THE PRIMATE TESTIS

What Do We Really Know?

G. F. Weinbauer and E. Nieschlag

Institute of Reproductive Medicine of the University
Domagkstrasse 11, 48129 Münster
Germany

1. SUMMARY

The present chapter reviews current knowledge concerning hormonal regulation of gametogenesis in the primate testis. LH/testosterone and FSH are the prime regulators of primate spermatogenesis. Although either hormone is capable of stimulating all phases of the spermatogenic process including the formation of sperm, the combination of both hormones is necessary in most instances to achieve quantitatively normal germ cell numbers. Sertoli cell proliferation can also be induced by either hormone in juvenile monkeys. Evidence for differential effects of testosterone and FSH on gametogenesis, however, is lacking and a synergistic effect is observed when they are combined. Receptors for androgens and FSH occur exclusively on testicular somatic cells and, hence, the trophic effects of these hormones on germ cell numbers are indirect ones. Interestingly, both hormones seem to have a common target, the spermatogonial population but it is unknown how such an indirect albeit highly specific effect is mediated. Whether the trophic hormone action influences germ cell numbers via increased proliferation or decreased cell death or both remains to be seen. There is evidence to suggest that the local androgen requirements for primate spermatogenesis might be comparatively high.

2. INTRODUCTION

It is established beyond any doubt that the pituitary gonadotropic hormones luteinizing hormone (LH) and follicle-stimulating hormone (FSH) are the prime regulators of spermatogenesis in mammals and that the synthesis and secretion of the gonadotropic hormones is governed by the hypothalamic gonadotropin-releasing hormone (GnRH) (Weinbauer and Nieschlag, 1996a for review). It has been considerably more difficult, however,

The Fate of the Male Germ Cell, edited by Ivell and Holstein
Plenum Press, New York, 1997

to define precisely the relative roles of LH (acting via testosterone in the testis) and FSH in the spermatogenic process. There have been several reasons for this difficulty: the complexity of male gametogenesis, the inability to study spermatogenesis under in-vitro conditions and the lack of an experimental paradigm appropriate for relevant in-vivo investigations. For obvious reasons, clarification of hormonal control of spermatogenesis has been more problematic in primates than in readily accessible rodent models. The present chapter intends to provide the reader with an up-to-date account of what is really known about the endocrine control of germ cell proliferation in the primate testis. Our presentation is focussed around data gained from studies in non-human primates, serving as a preclinical model for the human, and reference is made to men whenever relevant data are available. Among nonhuman primates, the rhesus monkey (*Macaca mulatta*) and the cynomolgus monkey (*Macaca fascicularis*) are the species used most regularly.

3. ORGANISATION OF PRIMATE SPERMATOGENESIS

Spermatogonia of type A divide mitotically and become B-type spermatogonia. These spermatogonia enter meiosis and are named spermatocytes therafter. Following the meiotic reduction division, the haploid spermatids evolve and develop into testicular sperm through a series of complex morphological differentiation steps. The morphology of the spermatogenic process has been described in detail in the rhesus monkey on the basis of the acrosome development in spermatids as seen after staining with periodic acid-Schiff's base (Clermont 1972). For macaques, a 12-stage classification system has been devised and this scheme has been adopted for the cynomolgus monkey (Fouquet and Dadoune, 1986; van Alphen et al., 1988). These stages are arranged in a longitudinal manner along the seminiferous tubule, although there is some evidence for a helical arrangement in a small proportion of seminiferous tubules in the cynomolgus monkey testis.

In men, 12 stages of spermatid development were originally described (Clermont and Leblond, 1995) but in terms of the organisation of the entire germinal epithelium a 6-stage system is conventionally used (Clermont, 1963). In human spermatogenesis several spermatogenic stages are present in a seminiferous tubule cross-section. By means of three-dimensional reconstruction studies it has been suggested that a helical arrangement of germ cell development accounts for this observation (Schulze and Rehder, 1984). This concept, however, has been questioned recently (Johnson, 1994). Common to primates is the fact that the type A-spermatogonia can be distinquished morphologically into A-dark and A-pale type spermatogonia. The A-dark type spermatogonia are believed to be the reserve stem cells, whereas the A-pale type spermatogonia are regarded as renewing stem cells providing germ cells for the spermatogenic cycle (Meistrich and van Beek, 1993).

4. EXPERIMENTAL APPROACHES FOR STUDYING THE ENDOCRINE CONTROL OF PRIMATE SPERMATOGENESIS

Hypophysectomy or trans-section of the pituitary stalk have been used to eliminate the influence of gonadotropic hormones in nonhuman primate models. Apart from the fact that the surgical interventions are quite traumatic to the animals, these approaches have the distinct disadvantage that pituitary hormones other than LH and FSH are affected. The discovery of the amino acid sequence of the decapeptide GnRH and the subsequent development of GnRH analogues has been extremely helpful for the study of endocrine control of primate spermatogenesis. Once the biological relevance of each of the ten amino acids had been unravelled, analogues were designed with either agonistic or antagonistic properties. Antagonists of GnRH are particularly suited for that purpose because, unlike GnRH agonists, gonadotropin suppression is effected within a few hours of administering the compound (Weinbauer and Nieschlag, 1996b).

GnRH antagonists have several amino acid substitutions and act via competitive blockade of the pituitary GnRH receptor (Weinbauer et al., 1992). Most importantly, GnRH antagonists selectively suppress LH and FSH secretion (Behre et al. 1994, Weinbauer and Nieschlag, 1993b), leading to a suppression of testicular function that is reversed upon cessation of GnRH antagonist treatment (Behre et al., 1994; Weinbauer et al., 1994). Evidently, GnRH antagonists provide selective and reversible gonadotropinectomy in the primate and, thus, represent an experimental model particularly well suited for the study of primate germ cell proliferation. Other approaches that have been used successfully in primates are (1) immunization against reproductive hormones by administering either the antigen or the respective antiserum and (2) treatment of immature monkeys with reproductive hormones (Weinbauer and Nieschlag, 1993a).

5. ANDROGEN AND GONADOTROPIN CONTROL OF PRIMATE SPERMATOGENESIS

LH acts indirectly on seminiferous tubules via the induction of testosterone production and release in Leydig cells. Follicle-stimulating hormone acts directly in the seminiferous tubules. The relative or specific role of LH/testosterone and FSH in primate and rodent spermatogenesis has received considerable attention and several issues are discussed: (1) what are the relative effects of testosterone and FSH in relation to the spermatogenic status (initiation during puberty, maintenance of ongoing spermatogenesis and reinitiation after suppression of the spermatogenic process) and does either hormone on its own stimulate quantitatively normal numbers of testicular cells, (2) do specific germ cell types respond to the trophic actions of testosterone and FSH and (3) are specific genes activated and/or is the production of testosterone/FSH-specific factors induced by testosterone and FSH? The first two issues will be dealt with in the following paragraphs. In contrast, information with regard to the third issue is still lacking.

5.1. Relative Roles of LH/Testosterone and FSH

A synopsis of our current view concerning the relative roles of the hormones is depicted in Fig. 1. Unlike in the rat, the prepubertal period in primates is characterized by quiescence of the reproductive axis lasting for several years. Administration of high amounts of testosterone to prepubertal monkeys stimulated all phases of the spermatogenic process and spermatozoa were produced (Marshall et al., 1984). In boys, spermatogenesis was complete in the vicinity of Leydig cell tumours producing high amounts of testosterone (Chemes et al., 1982). Whether FSH is able to initiate complete spermatid formation is unknown but this is conceivable since the application of purified human FSH to juvenile monkeys for periods of up to 12 weeks markedly stimulated spermatogonial proliferation (Arslan et al., 1993, Schlatt et al., 1995). Longer administration periods probably would have led to the full development of spermatogenesis. Even with testosterone, however, the number of sperm induced was far from being comparable to adult sperm production and we currently have to assume that both hormones are necessary for the quanti-

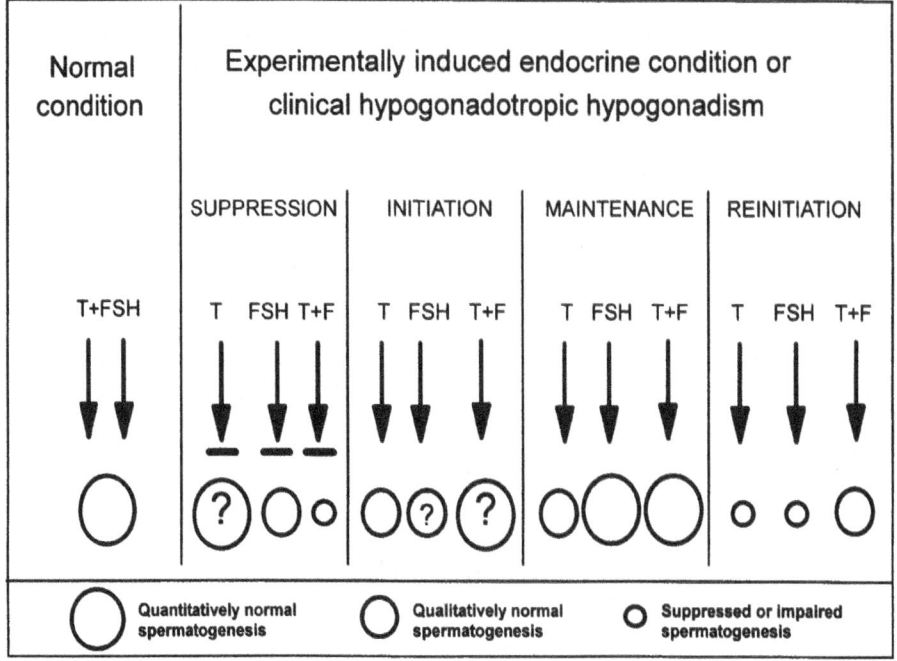

Figure 1. Quantitative and qualitative actions of LH/testosterone (T) and FSH on spermatogenesis in primates. ? = effect is likely but awaits proof; T = testosterone, T+F = testosterone combined with FSH. Inititation refers to the first completion of the spermatogenic process during puberty, maintenance to the requirements of ongoing spermatogenesis and reinitiation to the restart after suppression. Quantititatively normal production of germ cells requires both testosterone and FSH and suppression of spermatogenesis necessitates the inhibition of testosterone and FSH action.

tatively normal induction of germ cell formation. Sertoli cell numbers are also stimulated by either testosterone or FSH to a similar extent (Arslan et al., 1993; Schlatt et al., 1995) and a pronounced increase in the proliferation of Sertoli cells during puberty has been reported (Marshall & Plant, 1996).

With regard to the maintenance of ongoing spermatogenesis, it is well established that exogenous testosterone, depending on the administered dose, prevents the inhibitory effects of gonadotropin withdrawal on spermatogenesis. These effects of testosterone supplementation have been shown in hypophysectomized monkeys and monkeys exposed to GnRH agonists and antagonists (Weinbauer and Nieschlag, 1993a and 1996b). Compelling evidence for the ability of and the need for FSH for germ cell production has been produced from studies in monkeys and men. Immunization against FSH markedly reduced germ cell proliferation (Srinath et al., 1983) and even induced infertility (Moudgal et al., 1992). Moreover, in the GnRH antagonist-treated monkey FSH maintained qualitatively normal spermatogenesis (Weinbauer et al., 1991). FSH-induced stimulation of spermatogenesis is also indicated by the clinical observation that testes are enlarged in patients with pituitary adenomas and hypersecretion of FSH (Heseltine et al., 1989). The most striking piece of clinical evidence in support of the prospermatogenic action of FSH was recently obtained: normal testicular size and complete spermatogenesis were observed in a hypophysectomized patient and was associated with an activating mutation of the FSH receptor (Gromoll et al., 1996; see also chapter by Simoni et al. in this book for more detailed discussion of gonadotropin receptor mutations and gonadal function). This observation provides conclusive evidence that FSH is a highly trophic hormone for the human testis.

Restimulation of suppressed spermatogenesis has been achieved by very high amounts of testosterone in pituitary stalk-sectioned monkeys (Marshall et al., 1983). Doses of human FSH that clearly maintained spermatogenesis in GnRH antagonist-treated monkeys, however, were less effective in reinducing germ cell development (Weinbauer et al. 1991). Clinical experience also indicates that either hormone alone is rather ineffective in restarting male gametogenesis. Best clinical success is achieved by combining preparations exerting FSH and LH activity or by pulsatile administration of GnRH for the purpose of reinducing sperm production (Kliesch et al. 1994). It still remains elusive, however, why the reinitiation of spermatogenesis in primates has different hormonal requirements than the maintenance of the same process and why in most cases sperm fail to return to normality even with the combined hormonal therapy regimen.

5.2. Cell-Specific Actions of Testosterone and FSH in the Primate Testis

At the cellular level, the androgen receptor is expressed in somatic cells of the primate testis: Leydig cells, peritubular cells and Sertoli cells (Ruizeveld de Winter et al., 1991; Saunders et al., 1996) but not in germ cells. The FSH receptor is only expressed in Sertoli cells, both in rats (Kliesch et al., 1992) and in men (Böckers et al., 1994). It is of interest to note here that the FSH receptor, unlike the receptors for LH and androgen, is exlusively expressed in the testis (Dankbar et al., 1995). It follows that the trophic action of testosterone and FSH on gametogenesis must be mediated by somatic cells and for FSH specifically by the Sertoli cell. Given this scenario, it is interesting that both hormones seem to have a distinct and specific action on spermatogenesis. In prepubertal monkeys, either testosterone or FSH stimulated the number of A-pale type spermatogonia (Arslan et al., 1993). Such effect was also seen with hCG in a similar experimental paradigm (Schlatt et al., 1995).

Comparable observations were made in adult monkeys. FSH but not hCG specifically stimulated the proliferation of A-pale type spermatogonia in rhesus and cynomolgus monkeys (van Alphen et al., 1988). When the effects of FSH were studied in the gonadotropin-suppressed monkey, FSH also stimulated the number of this type of spermatogonium (Weinbauer et al., 1991). Since spermatogenesis was studied at two time points, it could be clearly shown that the initial trophic effect of FSH is the increase in A-pale type spermatogonia followed by elevated numbers of spermatocytes, etc. A recent study has suggested that in the presence of testosterone, FSH specifically increases the numbers of B-type spermatogonia (Marshall et al., 1995). Altogether, it is evident that testosterone and FSH as well specifically stimulate spermatogonial numbers in the primate testis. These observations raise interesting questions of how testosterone and FSH exert such a specific effect on germ cells albeit acting indirectly via the somatic Sertoli cells. It would also be important to know whether the effects of these hormones are caused by stimulation of germ cell proliferation or by prevention of germ cell death.

5.3. Testicular Androgens and Primate Spermatogenesis

Generally, testicular concentrations of testosterone far exceed those present in the circulation and, in the rat model, are considerably higher than the amounts needed to support the entire spermatogenic process (Weinbauer and Nieschlag, 1996b). In the rat, approximately 5% of normal testicular androgen levels suffice to stimulate sperm production qualitatively and, in the presence of FSH, as little as 10% of testosterone permit quantitatively normal germ cell yields. The discrepancy between tissue and serum testosterone concentration could be explained by assuming that a high local androgen concentration is required for ensuring pulsatile increases of testosterone upon LH stimulation into the peripheral blood. For primates, the relationship between spermatogenesis and testicular androgens has received only little attention but the available data, although rather scanty, suggest a substantial difference between rats and primates.

In one monkey, 13 weeks after hypophysectomy, testicular testosterone levels were still about 30% of presurgical values but spermatogenesis was severely affected with spermatogonia being the only germ cell type that had survived (Marshall et al., 1986). After a similar duration of GnRH antagonist administration (15 weeks) to 5 monkeys, tissue levels of testosterone remained at a third of baseline but all monkeys were azoospermic (Weinbauer et al., 1988). At face value, these data indicate that primate spermatogenesis requires comparatively higher local androgen concentrations than rat spermatogenesis. Indirect support for this hypothesis comes from the observation that hCG, directly stimulating testosterone production in Leydig cells, is superior to exogenous testosterone for stimulation of spermatogenesis in monkeys (Schlatt et al., 1995) and in men (Schaison et al., 1993). If testosterone can freely diffuse throughout tissue, the testis should be flooded with testosterone. However, it has been suggested that in the rat androgen binding protein (ABP) is able to capture testosterone and to inhibit testosterone action in a Sertoli cell line transfected with an androgen reporter gene (Roberts and Zirkin, 1993). Thus, ABP might function as an intratesticular regulator of androgen action in Sertoli cells. Although the presence of ABP has bene shown in primate Sertoli cells, the relationship of APB to testicular androgens and gametogenesis is unknown. It has been demonstrated, however, that human sex hormone binding globulin, being antigenically related to monkey ABP, is endocytosed by monkey germ cells (Gerard et al., 1991). These observations raise the possibility that ABP delivers androgens into monkey germ cells and that androgens might directly act on germ cells.

6. ACKNOWLEDGMENTS

We are grateful to Susan Nieschlag, MA, for linguistic editing of the manuscript.

7. REFERENCES

Arslan MA, Weinbauer GF, Schlatt S, Shahab M & Nieschlag E 1993 FSH and testosterone, alone or in combination, initiate testicular growth and increase the number of spermatogonia and Sertoli cells in a juvenile non-human primate (*Macaca mulatta*). Journal of Endocrinology 136: 235–243.

Behre HM, Böckers A, Schlingheider & Nieschlag E 1994 Sustained suppression of serum LH, FSH, and testosterone and increase of high-density lipoprotein cholesterol by daily injections of the GnRH antagonist cetrorelix over 8 days in normal men. Clinical Endocrinology 40: 241–248.

Böckers TM, Nieschlag E, Kreutz MR & Bergmann M 1994 Localization of follicle-stimulating hormone (FSH) immunoreactivity and hormone receptor mRNA in testicular tissue from infertile men. Cell and Tissue Research 278: 595–600.

Chemes HE, Pasqualini T, Rivarola MA & Bergada C 1982 Is testosterone involved in the reinitiation of spermatogenesis in humans? A clinicopathological presentation and physiological considerations in four patients with Leydig cell tumours of the testis or secondary Leydig cell hyperplasia. International Journal of Andrology 5: 229–245.

Clermont Y & Leblond CP 1955 Spermiogenesis of man, monkey, ram and other mammals as shown by the "periodic acid-Schiff" technique. 96: 229–253.

Clermont Y 1963 The cycle of the seminiferous epithelium in man. American Journal of Anatomy 112: 35–46.

Clermont Y 1972 Kinetics of spermatogenesis in mammals: Seminiferous epithelium cycle and spermatogonial renewal. Physiological Reviews 52:198–236.

Dankbar B, Brinkworth MH, Schlatt S, Weinbauer GF, Nieschlag E & Gromoll J 1995 Ubiquitous expression of the androgen receptor and testis-specific expression of the FSH receptor in the cynomolgus monkey (Macaca fascicularis) revealed by a ribonuclease protection assay. Journal of Steroid Biochemistry and Molecular Biology 55: 35–41.

Fouquet JP & Dadoune JP 1986 Renewal of spermatogonia in the monkey (Macaca fascilularis). Biology of Reproduction 35: 199–207.

Gerard A, En Nya A, Eggloff M, Domingo M, Degrelle H & Gerard H 1991 Endocytosis of human sex steroid-binding protein in monkey germ cells. Annals of the New York Academy of Sciences 637: 258–276.

Gromoll J, Simoni M & Nieschlag E 1996 An activating mutation of the FSH receptor autonomously sustains spermatogenesis in a hypophysectomized man. Journal of Clinical Endocrinology and Metabolism 81: 1367–1370.

Heseltine D, White MC, Kendall-Taylor P, de Kretser DM & Kelly W 1989 Testicular enlargement and elevated serum inhibin concentrations occur in patients with pituitary macroadenomas secreting follicle-stimulating hormone. Clinical Endocrinology 31: 411–423.

Johnson L 1994 A new approach to study the architectural arrangement of spermatogenic stages revealed little evidence of a partial wave along the length of the human seminiferous tubules. Journal of Andrology 15: 435–441.

Kliesch S, Behre HM & Nieschlag E 1994 High efficacy of gonadotropin or pulsatile gonadotropin-releasing hormone treatment in hypogonadotropic hypogonadal men. European Journal of Endocrinology 131: 347–354.

Kliesch S, Penttilä TL, Gromoll J, Saunders PTK, Nieschlag E & Parvinen M 1992 FSH receptor mRNA is expressed stage-dependently during spermatogenesis. Molecular and Cellular Endocrinology 84: R45-R49.

Marshall GR, Jockenhövel F, Lüdecke D & Nieschlag E 1986 Maintenance of complete but quantitatively reduced spermatogenesis in hypophysectomized monkeys by testosterone alone. Acta Endocrinologica 113: 424–431.

Marshall GR & Plant TM 1996 Puberty occurring either spontaneously or induced precociously in rhesus monkeys (Macaca mulatta) is associated with a marked proliferation of Sertoli cells. Biology of Reproduction 54: 192–1199.

Marshall GR, Wickings EJ & Nieschlag E 1984 Testosterone can initiate spermatogenesis in an immature nonhuman primate, Macaca fascicularis. Endocrinology 114: 2228–2233.

Marshall GR, Wickings EJ, Lüdecke DK & Nieschlag E 1983 Stimulation of spermatogenesis in stalk-sectioned Rhesus monkeys by testosterone alone. Journal of Clinical Endocrinology and Metabolism 57: 152–159.

Marshall GR, Zorub DS & Plant TM 1995 Follicle-stimulating hormone amplifies the population of differentiated spermatogonia in the hypophysectomized testosterone-replaced adult rhesus monkey (Macaca mulatta). Endocrinology 136: 3504–3511.

Meistrich M & van Beek MEAB 1993 Spermatogonial stem cells. In: Cell and Molecular Biology of the Testis, pp 266–295. Eds C Desjardins & LL Ewing. Oxford University Press, Oxford.

Moudgal NR, Ravindranath N, Murthy GS, Dighe RR, Aravindan GR & Martin F 1992 Long-term contraceptive efficacy of vaccine of ovine follicle-stimulating hormone in male bonnet monkeys (Macaca radiata). Journal of Reproduction and Fertility 96: 91–102.

Roberts KP & Zirkin BR 1993 Androgen binding protein inhibition of androgen-dependent transcription explains the high minimal testosterone concentrations required to maintain spermatogenesis in the rat. Endocrine Journal 1: 41–47.

Ruizeveld de Winter JA, Trapman J, Vermey M, Mulder E, Zegers ND & van der Kwast TH 1991 Androgen receptor expression in human tissues: an immunohistochemical study. Journal of Histochemistry and Cytochemistry 39: 927–936.

Saunders PTK, Millar MR, Majdic G, Bremner WJ, McLaren TT, Grigor KM & Sharpe RM 1996 Testicular androgen receptor protein: distribution and control of expression. In Cellular and Molecular Regulation of Testicular Cells, pp 213–229 Ed C Desjardin. Springer -Verlag, New York.

Schaison G, Yaoung J, Pholsena M, Nahoul K & Couzinet B 1993 Failure of combined follicle-stimulating hormone-testosterone admninistration to initiate and/or maintain spermatogenesis in men with hypogonadotropic hypogonadism. Journal of Clinical Endocrinology and Metabolism 77: 1545–1549.

Schlatt S, Arslan M, Weinbauer GF, Behre HM & Nieschlag E 1995 Endocrine control of testicular somatic and premeiotic germ cell development in the immature testis of the primate Macaca mullatta. Eurupean Journal of Endocrinology 133: 235–247.

Schulze W & Rehder U 1984 Organization and morphogenesis of the human seminiferous epithelium. Cell and Tissue Research 237: 395–407.

Srinath BR, Wickings EJ, Witting C & Nieschlag E 1983 Active immunization with follicle-stimulating hormone for fertility control: a 4.5 year study in male rhesus monkeys. Fertility and Sterility 40: 110–117.

van Alphen MAA, van de Kant HJG & de Rooij DG 1988 Follicle-stimulating hormone stimulates spermatogenesis in the adult monkey. Endocrinology 123: 1449–1455.

Weinbauer GF, Behre HM, Fingscheidt U & Nieschlag E 1991 Human follicle-stimulating hormone exerts a stimulatory effect on spermatogenesis, testicular size, and serum inhibin levels in the gonadotropin-releasing hormone antagonist-treated non-human primate (*Macaca fascicularis*). Endocrinology 129: 1831–1839.

Weinbauer GF, Göckeler E & Nieschlag E 1988 Testosterone prevents complete suppression of spermatogenesis in the gonadotropin-releasing hormone antagonist-treated nonhuman primate (Macaca fascicularis). Journal of Clinical Endocrinology and Metabolism 67: 284–290.

Weinbauer GF, Hankel P & Nieschlag E 1992 Exogenous gonadotrophin-releasing hormone (GnRH) stimulates LH secretion in male monkeys (*Macaca fascicularis*) treated chronically with high doses of a GnRH antagonist. Journal of Endocrinology 133 : 439–445.

Weinbauer GF, Limberger A, Behre HM & Nieschlag E 1994 Can testosterone alone maintain the gonadotrophin-releasing hormone antagonist-induced suppression of spermatogenesis in the non-human primate? Journal of Endocrinology 142: 485–495.

Weinbauer GF & Nieschlag E 1993a Hormonal control of spermatogenesis. In: Molecular Biology of the Male Reproductive System, pp 99–142. Ed DM de Kretser. Academic Press, New York.

Weinbauer GF & Nieschlag E 1993b Comparison of the antigonadotropic activity of three GnRH antagonists (Nal-Glu, Antide and Cetrorelix) in a non-human primate model (*Macaca fascicularis*). Andrologia 25: 141–147.

Weinbauer GF & Nieschlag E 1996a Hormonal regulation of reproductive organs. In Comprehensive Human Physiology - from Cellular Mechanisms to Integration, pp 2231–2252. Eds R Greger & U Windhorst. Springer-Verlag, Berlin.

Weinbauer GF & Nieschlag E 1996b The Leydig cell as a target for male contraception. In The Leydig Cell, pp 629–662. Eds AH Payne, MP Hardy & LD Russell. Cache River Press, Vienna.

7

QUANTIFICATION OF SOMATIC AND SPERMATOGENIC CELL PROLIFERATION IN TESTES OF RUMINANTS

H. Roelants and S. Blottner

Institute for Zoo Biology and Wildlife Research
Alfred Kowalkestr. 17, 10315 Berlin
Germany

On account of the high proliferation rates specific to testicular tissue, the production of male germ cells is an important indicator of ecological or anthropogenic factors affecting animal reproduction. Two methods were employed for quantifying the proliferation of post mortem collected testes in cattle and roe deer; furthermore we developed a new method for differentiating between the proliferation of somatic and spermatogenic cells.

The proliferation of testicular cells was quantified by measuring the tissue polypeptide specific antigen (TPS). This marker was determined in homogenised tissue by an ELISA, after selective enrichment of different cell types by density gradient centrifugation. One-parameter flow cytometry was used to monitor the haploid (1C), diploid (2C) and tetraploid (4C) cells, and the percentage of cells in the G2/M phase of mitosis was analysed. Somatic and spermatogenic cells were differentiated by dual-parameter flow cytometry after DNA staining with propidium iodide and labelling of the somatic cells with fluorescein-conjugated vimentin antibody according to our new method, based on Hittmair et al. (1994). The TPS was set in relation to the DNA-content of the cells and to their somatic or spermatogenic type.

High concentrations of TPS were found in homogenised tissue of both species. The TPS values varied with the differing content of spermatogenic and somatic cells in the fractions of the density gradient. TPS was positively correlated with spermatogenic cells (r = 0.434; P< 0.005) and negatively correlated with somatic cells (r = -0.679; P< 0.0001) in roe deer (n=40). This came as a surprise, since TPS is described as an epitope of cytokeratin 18 (Bonfrer et al.,1994), not detected in testicular cells of adults (Steger et al.,1994). The differentiation of germinal and somatic cells demonstrated their different changes of mitotic activity (percentage in G2/M phase) during successive phases of the annual cycle in roe deer (fig. 1). Measuring the amount of spermatogenic tetraploid cells permitted us to calculate the exact ("true") meiotic transformation (ratio of 1C:4C cells), which was significantly higher than the currently used value based on total 4C percentage,

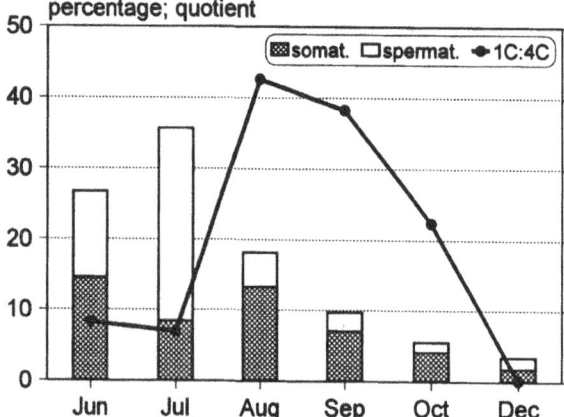

Figure 1. Mitotic activity of somatic and spermatogenic cells (percentage in the G2/M phase) and true meiotic transformation during the annual cycle in roe deer.

and showed an increase from 8.25 to 42.54 from non-breeding (early in June) to breeding season (early in August) (fig. 1).

The results demonstrate the qualification of TPS to quantify the proliferation in the germinal compartment of testes. However, the above mentioned dual-parameter flow cytometry proves to be a still better method to study the changing mitotic and meiotic steps during involution and recrudescence of testes in seasonally breeding ruminants by providing direct relation of proliferative processes to spermatogenic and somatic cells.

REFERENCES

Bonfrer JMG, Groeneveld EM, Korse CM, van Dalen A, Oomen LCJM & Ivanyi D 1994 Monoclonal antibody M3 used in tissue polypeptide-specific antigen assay for the quantification of tissue polypeptide antigen recognizes keratin 18. Tumor Biology 15: 210–222.

Hittmair A, Rogatsch H, Mikuz G & Feichtinger H 1994 Quantification of spermatogenesis by dual-parameter flow cytometry. Fertility and Sterility 61: 746–750.

Steger K, Schimmel M & Wrobel KH 1994 Immunocytochemical demonstration of cytoskeletal proteins in seminiferous tubules of adult rams and bulls. Archives of Histology and Cytology 57: 17–28.

8

THE IMMORTALIZED MOUSE GERM CELL LINES GC-1spg AND GC-2spd AS A MODEL FOR MITOCHONDRIAL DIFFERENTIATION DURING MEIOSIS

A. Meinhardt,[1] H. Renneberg,[1] A. Dersch,[1] G. Wennemuth,[1] J. L. Millán,[2] G. Aumüller,[1] and J. Seitz[1]

[1]Department of Anatomy and Cell Biology
Philipps-University Marburg
Robert-Koch-Str. 6, 35037 Marburg
Germany
[2]Department of Medical Genetics
University of Umea
901 85 Umea, Sweden

INTRODUCTION

It is well recognized that mitochondria of germ cells modify their morphological organization, number and location during spermatogenesis. Mitochondria gradually transform from the usual cristae-rich ("orthodox") type found in the various types of spermatogonia via an intermediate form located in leptotene and zygotene spermatocytes into a so-called "condensed-type" found in pachytene spermatocytes and early spermatids. In pachytene spermatocytes, the mitochondria are grouped together in small clusters. The intramembranous space is greatly enlarged, with the membrane pushed to the periphery leaving almost no cristae. The matrix is condensed to a thin ring-like structure underneath the outer membrane leaving the interior of these mitochondria looking "empty". In early round spermatids, the mitochondria retain the "empty" central regions, but later in spermatogenesis, more cristae become evident. Very little is known about the factors that regulate or trigger the various steps of mitochondrial differentiation during sperm production. Recent studies from our group (Meinhardt et al., 1995; Seitz et al.,1995) indicate the involvement of proteinaceous factor(s), termed paracrine mitochondrial maturation factors (PMMF), which are secreted by Sertoli cells and act in a paracrine fashion on primary spermatocytes. However, the study of factors that induce the differentiation of mitochondria in isolated leptotene and zygotene spermatocytes is hampered by low numbers of this cell type in rat testis. Mitochondrial differentiation is accompanied by the step-wise occurrence of three marker proteins in the matrix, each typical for a distinct organelle form:

heat shock protein 60 (hsp60) for orthodox mitochondria mainly in spermatogonia, LON-protease in leptotene and zygotene spermatocytes and sulfhydryl oxidase (SOx) for the condensed type in pachytene spermatocytes and early round spermatids (Seitz et al, 1995). The immortalized germ cell line GC-1spg has been characterized at a stage between B spermatogonia and an early primary spermatocyte, while the GC-2spd line encompasses germ cells from mid to late meiosis (Hofmann et al, 1992; 1994).

The aim of this study was to investigate if the two immortalized mouse cell lines GC-1spg and GC-2spd express the marker proteins and therefore comprise a suitable model for examining the differentiation of mitochondria in male meiotic germ cells.

MATERIALS AND METHODS

Immunofluorescence: GC-1spg and GC-2spd cells were cultured on glass slides in DMEM supplemented with 10% FCS in an atmosphere of 95% air and 5% CO_2 at either 32°C or 37°C. Cells were fixed for 5 min in icecold methanol. Specimens were incubated with the primary antibodies against sulfhydryl oxidase (SOx; 1:400), heat shock protein 60 (hsp60; 1:100) and subunit IV of cytochrome-c-oxidase (COx-IV; 1:200) for 1h at room temperature and visualized using a secondary antibody labeled with the fluoro-chrome CY_3 (1:100). The cell nuclei were counterstained with DAPI.

SDS-PAGE and Western Blot Analysis

The cell lines were cultured at 37°C or at the permissive temperature of 32°C (GC-2spd) and collected at day 1, day 3 and day 6, respectively. Protein concentrations were determined by the Bradford method. Equal amounts of protein were boiled in sample buff-er and loaded on a 10% SDS-PAGE. The proteins were transferred on a PVDF membrane (Bio-Rad). After blocking the membrane with 5% nonfat dry milk in 10mM Tris-HLC, 0.9% NaCl (pH 7.5) the blots were probed with the polyclonal rabbit antibodies directed against SOx (1:200), hsp60 (1:300) and COx-IV (1:400) overnight at 4°C. Detection was performed using anti-rabbit Ig peroxidase (1:10,000) and the enhanced chemolumines-cence technique (ECL, Amersham).

Polymerase Chain Reaction (PCR)

Total RNA from the GC-1spg and the GC-2spd cell lines was extracted utilizing the RNeasy™-Kit (Qiagen). Contaminating DNA was digested by a subsequent DNase treat-ment. Reverse transcription was performed with MMLV reverse transcriptase (200U) and 0.5µg oligo dT for 60 min at 37°C. PCR was performed using standard procedures with 1 µl of cDNA template, 1.25U of InviTaq-DNA-polymerase and 1µl (50 pM) of each primer (specific for hsp60 and LON-protease). Forty cycles were performed, each consisting of denaturation at 94°C for 30 sec, primer annealing at 55°C (LON) or 65°C (hsp60), and primer extension at 72°C for 1.5 min. PCR products were analyzed on a 1.5% agarose gel in single strength TBE-buffer and stained with ethidium bromide.

RESULTS AND DISCUSSION

The presence of the mitochondrial differentiation markers hsp60, SOx and COx-IV was demonstrated in the GC-1spg and GC-2spd germ cell lines using the immunofluorescence technique. Hsp60 and COx-IV displayed an even distribution pattern of mitochondria throughout the cytoplasm of GC-1 cells, whereas SOx-immunoreactivity in GC-2 cells appeared mainly in clusters around the nuclei. Semi-quantitative Western blotting showed that hsp60-protein was found to be much more strongly synthesized in the GC-1 cell line than in GC-2 germ cells. The opposite was shown for SOx which was found in high amounts in GC-2, but lower in GC-1. This is in good agreement with studies from rat testis that have localized hsp60 mainly to mitochondria in spermatogonia and SOx almost exclusively in the organelles of meiotic germ cells (Meinhardt et al, 1995).

RT-PCR analysis using primers specific for hsp60 resulted in a fragment of the calculated length of 702bp in the GC-1spg, GC-2spd cell line as well as in rat testis. Southern blot analysis confirmed that the amplified fragment is identical with hsp60. In rat testis, LON-protease mRNA expression was shown by RT-PCR analysis (430bp fragment), whereas the amplified products in both mouse cell lines were approx. 200bp larger. PCR on genomic DNA excluded the possibility that the 620bp fragment of LON-protease in GC-1 and GC-2 resulted from a potential contamination with nuclear DNA. LON fragments from rat testis and both cell lines were cloned into the pCR™2.1 vector (Invitrogen) for further analysis.

In conclusion, the immortalized mouse cell lines GC-1spg and GC-2spd seem to provide a suitable model to study the differentiation of mitochondria in germ cells under experimental conditions. Further electron microscopic studies have to reveal whether the influence of paracrine Sertoli cell secreted factors on the change of mitochondrial morphology in this cell lines is comparable to the *in vivo* situation.

REFERENCES

Hofmann MC, Narisawa S, Hess RA & Millán JL. 1992 Immortalization of germ cells and somatic testicular cells using the SV40 Large T Antigen. Experimental Cell Research 201: 417–435.

Hofmann MC, Hess RA, Goldberg E & Millán JL. 1994 Immortalized germ cells undergo meiosis *in vitro*. Proceedings of the National Academy of Sciences 91: 5533–5537.

Meinhardt A, Parvinen M, Bacher M, Aumüller G, Hakovirta H, Yagi A & Seitz 1995 Expression of the mitochondrial heat shock protein 60 in distinct cell types and defined stages of rat seminiferous epithelium. Journal of Biological Reproduction 52: 798–807.

Seitz J, Möbius J, Bergmann M & Meinhardt A. 1995 Mitochondrial differentiation during meiosis. International Journal of Andrology 18 Suppl.2: 7–11.

9

A NOVEL ENDOZEPINE-LIKE PEPTIDE (ELP) IS EXCLUSIVELY EXPRESSED IN MALE GERM CELLS

W. Pusch,[1,2] M. Balvers,[1] N. Hunt,[1] and R. Ivell[1]

[1]Institute for Hormone and Fertility Research at the University of Hamburg
Grandweg 64, 22529 Hamburg
[2]Current address: Institute of Anatomy
University Hospital Hamburg Eppendorf (UKE)
Martinistr. 52, 20246 Hamburg
Germany

Endozepine, which is also referred to diazepam binding inhibitor (DBI) or acyl-CoA binding protein (ACBP), was initially characterized by its ability to displace benzodiazepine pharmaceuticals like diazepam from GABA type A receptors (Costa and Guidotti, 1991). Later an additional peripheral-type benzodiazepine receptor (PBR) was found, which is mainly located in the mitochondrial membrane of various peripheral tissues.

Several functions have been proposed for endozepine, e.g. the stimulation of the cholesterol transport from the outer to the inner mitochondrial membrane (Papadopoulos, 1993) and an action as a cholecystokinin releasing factor (Herzig et al., 1996). However its role in the binding of acyl-CoA esters is best characterized.

Here we report a cDNA isolated from mouse testis coding for a novel endozepine-like peptide (ELP). The ELP mRNA expression was analyzed by northern and in situ hybridisation. An ELP transcript was found in spermatid stages of the maturing germ cells in the testis of sexually mature wildtype mice. However it was absent in the testis of the azoospermic W/W[v] mice, which carry mutations in both alleles of the gene encoding the receptor-tyrosine-kinase *c-kit*. ELP mRNA was not found in any other examined tissue besides the testis.

The part of the ELP cDNA encoding the open reading frame was used to construct expression vectors for the production of ELP protein in bacteria (Pusch et al., 1996).

The recombinant proteins were purified by a nickel-chelate chromatography via an N-terminally introduced hexahistidine peptide and were used to raise polyclonal antibodies in rats. These antibodies identified the predicted ELP protein in protein extracts of mouse testis and epididymis but in no other examined tissue.

Immunohistochemical studies of testis sections showed the ELP antigen in the cytoplasm of late spermatids and in the residual bodies. It was also shown, that the spermatozoa lose their ability to react with the specific ELP antiserum during epididymal transit, probably due to the progressive loss of their cytoplasmic droplet.

We conclude that ELP is an intracytoplasmic peptide exclusively produced in post-meiotic spermatozoa, which may be involved in the energy metabolism of the maturing sperm.

REFERENCES

Costa E & Guidotti A 1991 Diazepam binding inhibotor (DBI): a peptide with multiple biological actions. Life Sciences 49: 325–344.

Hach M, Pedersen S, Börchers T, Hojrup P & Knudsen J 1990 Determination by photoaffinity labelling of the hydrophobic part of the binding site for acyl-CoA esters on acyl-CoA-binding protein from bovine liver. Biochemical Journal 271: 231–236.

Herzig K, Schön I, Tatemoto K, Ohe Y, Li Y & Fölsch U 1996 Diazepam binding inhibitor is a potent cholecystokinin-releasing peptide in the intestine. Proceedings of the National Academy of Science of the USA 93: 7927–7932.

Papadopoulos V 1993 Peripheral-type benzodiazepine/diazepam binding inhibitor receptor: biological role in steroidogenic cell function. Endocrine Reviews 14: 222–240.

Pusch W, Balvers M, Hunt N & Ivell R 1996 A novel endozepine-like peptide (ELP) is exclusively expressed in male germ cells. Molecular and Cellular Endocrinology 122: 69–80.

10

INVESTIGATION ON THE PROLIFERATION OF SPERMATOGONIA IN NORMAL AND PATHOLOGIC HUMAN SEMINIFEROUS EPITHELIUM

K. Steger, I. Aleithe, and M. Bergmann

Institute of Anatomy and Cell Biology
Martin-Luther-University
Große Steinstraße 52, 06097 Halle (Saale)
Germany

We investigated the percentage of S-phase spermatogonia in testis biopsies from azoospermic men with normal histology (control; n=41) and from oligozoospermic men with mixed atrophy (n=222) using antibodies against Ki-67 (MIB-1) and PCNA (PC10).

In the normal seminiferous epithelium MIB-1 reacts with spermatogonia only, PC10 additionally stains primary spermatocytes. 30.6 ± 14.6 % (stage I-III) and 22.6 ± 9.9 % (stage IV-VI) of the spermatogonia were immunopositive for Ki-67, 43.4 ± 15.6 % (stage I-III) and 44.9 ± 10.8 % (stage IV-VI) for PCNA, respectively, indicating no stage-dependent differences.

The staining pattern for the different types of spermatogonia is as follows: Type A pale: 26.1 ± 4.6 % (MIB-1) and 40.9 ± 6.5 % (PC10); Type A dark: 18.2 ± 13.4 % (MIB-1) and 49.8 ± 12.5 % (PC10); Type B: 43.8 ± 15.8 % (MIB-1) and 74.9 ± 27.6 % (PC10).

The low spermatogenic efficiency of oligozoospermic men is associated with histological defects such as hypospermatogenesis (hyp) or arrest of spermatogenesis at the level of spermatids (sda), spermatocytes (sca) or spermatogonia (sga). At present it is unknown whether the proliferation of spermatogonia is associated with these (post)meiotic defects.

There is a significant decrease ($p<0.05$) of Ki-67 staining in hyp, sda, sca and sga and of PCNA staining in sca and sga, respectively.

Spermatogonia immuno-positive for		Control	hyp	sda	sca	sga
anti-Ki-67-antibody	n	32	90	42	87	30
	mv: %	34.6	30.2	24.6	25.1	16.3
	sd: %	9.2	9.3	7.8	8.6	5.9
anti-PCNA-antibody	n	88	148	28	152	45
	mv: %	48.7	48.5	45.8	33.1	19.9
	sd: %	8.8	11.3	13.9	10.9	9.3

n: number of seminiferous tubules; mv: mean value; sd: standard deviation

INTERPRETATION

The different distribution pattern (stained cells, percentage) is caused by the differing half-life of the proteins.

The decrease of spermatogonial proliferation may be due to an impairment of the associated Sertoli cell population as shown in previous studies (Bruning et al., 1993, Bergmann and Kliesch, 1994, Steger et al., 1996).

LITERATURE

Bruning G, Dierichs R, Stümpel C & Bergmann M 1993 Sertoli cell nuclear changes in human testicular biopsies as revealed by three dimensional reconstruction. Andrologia 25: 311–316

Bergmann M & Kliesch S 1994 The distribution pattern of cytokeratin and vimentin immunoreactivity in testicular biopsies of infertile men. Anatomy and Embryology 190: 515–520.

Steger K, Rey R, Kliesch S, Louis F, Schleicher G & Bergmann M 1996 Immunohistochemical detection of immature Sertoli cell markers in testicular tissue of infertile adult men: a preliminary study. International Journal of Andrology 19: 122–128.

11

RAPID METHOD TO DETECT CIS-CELLS

H. Lauke

Institute of Anatomy
Department of Microscopical Anatomy
University of Hamburg (UKE)
Martinistr. 52, 20246 Hamburg
Germany

Seminoma cells, like intratubular tumor cells (carcinoma *in situ*), are characterized by abundant cytoplasmic glycogen. Visualization of glycogen in histologic sections has been shown to be a valuable diagnostic tool for detecting tumor cells in testicular tissue provided that appropriate fixatives were used to preserve the glycogen in the cells. Tumor cells fixed with 5.5 % glutaraldehyde show virtually no loss of their highly soluble content. In combination with other structural characteristics even single tumor cells can be histologically diagnosed (Holstein et al., 1987).

Visualization of glycogen in native cells is possible by application of Lugol's solution (Boeck, 1989). Tumor cells in smear preparations exhibit a brown staining reaction due to cytoplasmic glycogen after being in contact with the aqueous iodine solution. Using testicular tissue adjacent to different germ cell tumors (n=84), the staining results were compared with the final histologic diagnosis. 71 specimen showed intratubular tumor cells (Mumperow et al., 1992). In 48 cases (68%) tumor cells were detected with both methods. From 23 Lugol-negative cases (32%); 18 cases showed only few CIS-tubules and/or severe intertubular fibrosis in the corresponding histological sections. There was no reaction with Lugol's solution in 13 cases that were shown to be free of tumor cells in the corresponding histological sections.

Application of Lugol smear preparations supports the search for early tumor cells if rapid diagnostic results are required. In certain cases prior Lugol staining of isolated tubular fragments considerably reduce sampling errors particularly in regard to cultures of tumor-bearing testicular tubules (Lauke et al., 1991).

REFERENCES

Boeck P 1989 Romeis Mikroskopische Technik, 17. edition, p 392. Urban & Schwarzenberg.
Holstein AF, Schütte B, Becker H & Hartmann M 1987 Morphology of normal and malignant germ cells. International Journal of Andrology 10: 1–18.

Lauke H, Seidl K, Hartmann M & Holstein AF 1991 Carcinoma-in-situ cells in cultured seminiferous tubules. International Journal of Andrology 14: 33–43.

Mumperow E, Lauke H, Holstein AF & Hartmann M 1992 Further practical experiences in the recognition and management of carcinoma *in situ* of the testis. Urologia internationalis 48: 162–166.

12

TESTICULAR TUMOR CELLS PASS THROUGH THE EPIDIDYMAL DUCTS

D. Benson,[1] H. Lauke,[1] and M. Hartmann[2]

[1]Institute of Anatomy, University of Hamburg (UKE)
Martinistr. 52, 20246 Hamburg, Germany
[2]Armed Forces Hospital Hamburg
Department of Urology
Lesserstr. 180, 22049 Hamburg
Germany

The diagnosis of early testicular cancer is based on the histological evaluation of testicular biopsies. Like the germ cells, the tumor cells find their way along the tubular duct system of the male urogenital tract into the ejaculate. But only very few data have been published on the exfoliation of testicular tumor cells into the ejaculate in men suffering from early testicular cancer. The screening of ejaculate to detect early testicular cancer would appear to be a rather unsuccessful method.

In the present investigation the passage of early testicular tumor cells through the epididymal ducts was investigated to elucidate the fate of the tumor cells after leaving the seminiferous tubules.

A systematic light- and electronmicroscopical evaluation of the efferent ductules and the epididymal duct of macroscopically unchanged epididymis was performed in 135 cases of germ cell tumor. In addition in some cases pieces of the rete testis were also investigated in the same way.

Principally, tumor cells of type TC1 and TC2 were found in the lumina of the rete testis, the efferent ductules and the epididymal duct. The percentage was rather low. Only in 2 % were testicular tumor cells found in the lumina of the epididymis. In 32 % of the investigated cases free imature and mature germ cells were found in all sections of the epididymal ducts. In addition, macrophages were seen in the lumina of efferent ductules and epididymal duct.

In contrast to the high number of tumor cells in the seminiferous tubules, the number of tumor cells in the epididymal ducts is very low. Therefore it is understandable that the chance of observing tumor cells in the ejaculate is extremly low. The rare finding of tumor cells in the lumina of the duct system indicates that evaluation of the ejaculate is not a reliable diagnostic parameter for screening patients with an increased risk of testicular cancer.

REFERENCES

Giwercman A, Hopman AHN, Ramaekers FCS & Skakkebæk NE 1990
Carcinoma in situ of the testis. Dectection of malignant germ cells in seminal fluid by means of in situ hybridization. American Journal of Pathology 136: 497–502.

13

AgNOR IN HUMAN LEYDIG CELL TUMORS

M. Mueller[1] and H. Lauke[2]

[1]Armed Forces Hospital Hamburg
Department of Urology
Lesserstr. 180, 22049 Hamburg, Germany
[2]Institute of Anatomy
Department of Microscopical Anatomy
University of Hamburg (UKE)
Martinistr. 52, 20246 Hamburg
Germany

There may be overlap between the microscopic features in benign and rare malignant Leydig cell tumors. Moreover, there is often considerable morphological similarity to normal Leydig cells causing difficulties in predicting which tumors will show a malignant behaviour (Kim et al., 1985).

The AgNOR staining method has been shown to be a valid option for differentiating between benign and malignant cells. Use as a prognostic criterion is documented for carcinomas of the urinary tract (Delahunt et al., 1991) and of the digestive system. The mean AgNOR count per nucleus can also be used as a proliferation marker in various tissues (Egan & Crocker, 1992). Silver ions bind specifically to acidic, non-histone proteins associated with Nucleolar Organizer Regions (NORs). The NORs are genes on the short arms of the D-and G-group chromosomes in the human genome. Development of the nucleolus implicates protein biosynthesis at the NOR sites. The AgNOR staining protocol was recently described (Mueller et al., 1994).

Testicular tissue adjacent to testicular germ cell tumors frequently exhibits increased numbers of Leydig cells. The following cases were selected using paraffin sections and divided into 3 groups based on the assessment of Leydig cell numbers and arrangement: 1. normal number of Leydig cells (n=4), the mean AgNOR counts per nucleus were 8–9 (range 4–15). 2. increased number, cells evenly distributed (n=8), the mean AgNOR counts per nucleus were 7–9 (range 6–14). 3. formation of organized micro-adenomas (n=6), the mean AgNOR counts per nucleus were 8–9 (range 6–17).

Additionally, two Leydig cell tumors were included in the study showing uniform cells histologically similar to normal Leydig cells, occasional Reinke crystals and very rare mitoses. As expected PCNA-reactive cells were also rare in all investigated cases, but corresponding application of the AgNOR method revealed elevated AgNOR counts in Leydig tumor cells. The mean AgNOR counts per nucleus were 12–14,5 (range 10–23).

The Fate of the Male Germ Cell, edited by Ivell and Holstein
Plenum Press, New York, 1997

Compared to the AgNOR values in the investigated cases of Leydig cells adjacent to solid germ cell tumors this possibly indicates an increased proliferation capacity in contrast to the low rate of tumor cells immunoreactive to PCNA and rare mitotic figures found in corresponding tissue sections.

Evaluation of a larger series will prove whether this finding is a unique phenomenon or a characteristic of a subgroup of Leydig cell tumors with a distinct clinical behaviour.

REFERENCES

Delahunt B, Mostofi FK, Sesterhenn IA, Ribas JL & Avallone FA 1991 Nucleolar organizer regions and prognosis in renal cell carcinoma. Journal of Pathology 163: 31–37.

Egan MJ & Crocker J 1992 Nucleolar organizer regions in pathology. British Journal of Cancer 65: 1–7.

Kim I, Young RH & Scully RE 1985 Leydig cell tumors of the testis. A clinicopathological analysis of 40 cases and review of the literature. The American Journal of Surgical Pathology 9(3): 177–192.

Mueller M, Lauke H & Hartmann M 1994 The value of the AgNOR staining method in identifying carcinoma in situ testis. Pathology Research and Practice 190: 429–435.

14

ENDOCRINOLOGICAL DISTURBANCES IN GERM CELL TUMOUR PATIENTS

Comparison of Hormone Levels and Kinetics in Peripheral and Testicular Vein Blood

T. Böhmer,[1] T. Pottek,[1] H. Büttner,[2] and M. Hartmann[1]

[1]Armed Forces Hospital Hamburg
Department of Urology
Lesserstr. 180, 22049 Hamburg, Germany
[2]University of Lübeck
Clinic of Urology
Ratzeburger Allee 160, 23538 Lübeck
Germany

Men with testicular cancer are often not fertile. The disturbances can be seen in the spermiogram and the sex hormone levels. The kinetics of the hormone levels following tumour ablation have never been surveyed, however.

We have investigated the hormone levels of 37 consecutive patients with germ cell tumour. At the time of the primary operation we took testicular vein blood by puncture or venae sectio after preparation of the spermatic cord, and peripheral blood out of the cubital vein. The measurements of cubital vein blood were also performed 24 hours and 4 to 7 days after the operation. All samples were evaluated for the tumour markers ß-HCG, AFP, LDH and PLAP, the gonadotropins LH and FSH, and the sex steroids testosterone, estradiol and prolactin. Significant serial counts and relationships were only shown in this series by ß-HCG, estradiol, testosterone and LH.

We examined 15 patients with a seminoma and 22 patients with a non-seminoma (10 comb. tumour, 7 TC, 5 EC). At the time of the ablation 19 patients were in clinical stage I, 10 in the clinical stage IIa/b, 2 in the clinical stage IIc and 6 in stage III.

In all patients we found distinctly increased levels of estradiol in the testicular vein blood. It is about 30 times higher than in the cubital blood, thus showing that the estradiol is produced in the testis in these patients. Where the estradiol is also increased in the cubital blood there is a strong correlation to the level of testosterone. There are no elevated testosterone levels in patients with normal peripheral estradiol. We found eight patients with increased estradiol and testosterone levels in the cubital blood. All these patients also had high HCG levels and low levels of LH.

The Fate of the Male Germ Cell, edited by Ivell and Holstein
Plenum Press. New York. 1997

75

	E2 (pg/ml)	Test. (µg/ml)	HCG (mU/ml)	LH (mU/ml)
group I (high E2)	89,05	12,2	100,7	0,1
(n=8)	[41,9–182,3]	[6,3–105]	[0–3618]	[0–1,7]
group II (low E2)	16,3	4,8	0,5	2,4
(n=29)	[0,3–34,9]	[2,4–7,9)	[0–53,4]	[0–10,5]

Increases in the estradiol depend on the production of HCG by the tumour. The gonadal-pituitary feedback leads to the suppression of LH-expression. After tumour ablation the high estradiol and testosterone levels return to normal ranges parallel to the decrease of HCG. The LH level is increasing to normal in the same interval.

In patients with a HCG secreting germ cell tumour, this HCG is able to call forth the same effect at the Leydig cells as normal LH, because of the similar structure of the alpha chain. This thus gives rise to a high production rate of testosterone by the Leydig cells, which could then aromatise to estradiol. These high testosterone and estradiol levels are associated with LH suppression, because of the negative gonadal-pituitary feedback mechanism. Our findings show that these disturbances seem to be reversible in those patients with decreased HCG levels after tumour ablation.

15

REINVESTIGATION OF PATIENTS AFTER PRIMARY THERAPY OF TESTICULAR TUMOR

K. H. Schölermann,[1] H. Lauke,[2] H. Huland,[1] and M. Hartmann[3]

[1]Department of Urology
[2]Department of Microscopical Anatomy
University of Hamburg (UKE)
Martinistr. 52, 20246 Hamburg, Germany
[3]Armed Forces Hospital Hamburg
Department of Urology
Lesserstr. 180, 22049 Hamburg
Germany

We studied 263 patients with a germ cell tumor of the testis to evaluate the fertility status and correlation to clinical parameters before initial treatment and after 1 year in the follow-up (mean intervall 397 days +/- 227). Between 1984 and 1995, 760 consecutive patients with germ cell tumor of the testis were treated at the Armed Forces Hospital in Hamburg, Department of Urology (Head of Dep.: M. Hartmann). Patients were followed monthly the first year, every 2 or 3 months in the following years over a minimum of 5 years.

MATERIALS AND METHODS

In this retrospective study we selected n= 263 patients with complete treatment received at our institution. The following criteria had to be fullfilled to attend this protocol:

Complete remission (CR) after therapy, Biopsy of the contralateral testis, histological diagnosis and assessment of spermatogenesis with the semithin section method in all patients. Clinical measures : Patient age, history, primary tumor histology, staging, serum tumor markers alpha-fetoprotein (AFP), human chorionic gonadotropin (beta-HCG); serum testosterone, follicle stimulating hormone (FSH) and luteinizing hormone (LH). In 35 % of patients (92/263) a spermiogram before initial treatment. Follow-up : Testosterone, FSH, LH and a spermiogram in 8 % of patients (22/263). Semen quality scoring (1–5) and Spermatogenesis quality scoring (1–5) in the contralat. biopsy was undertaken.

RESULTS

n = 263 patients, mean age 29 years +/- 7 years. 33 % of patients had a seminoma, 43 % a non-seminoma and 24 % a combined tumor. Stage I was found in 52 % of patients, stage IIa in 13 %, IIb in 21 %, stage IIc in 3 % and stage III in 11 % of patients. 61 % had a normal history (160/263), 20 % a maldescended testis (53/263, 36 ipsilateral, 14 biliteral, 3 contralateral). There was no correlation between fertility status and patient age, tumor histology, stage of disease or a history of maldescended testis. We found normal fertility (score 1–2) in 35 % of the patients (93/263), corresponding to semen quality (mean score 2,31, n = 36), reduced fertility (score 3) in 33 % (87/263), (semen quality : mean score 3,0, n = 30) and severily reduced or annihilated spermatogenesis (score 4–5) 32 % (83/263), (semen quality : mean score 3,77 , n = 26).

Mean levels of testosterone and LH are within the normal range in beta-HCG negative patients. It seems that the hypophysis - testis axis is intact, and elevated level of serum FSH indicate abnormal spermatogenesis. There was no correlation found between fertility status and serum FSH in beta-HCG positive patients (122/263). Obviously beta-HCG has an agonistic, stimulating effect on Leydig-cells, thus elevating testosterone levels and suppressing LH and FSH. 1 year after treatment the hypophysis-testis axis appears to be fully restored.

Fertility status 1 year after initial treatment: No difference is seen in patients with surgical treatment only (72/263) and patients after radiotherapy (77/263). Mean levels of testosterone and LH is within the normal range, FSH was slightly elevated. Following chemotherapy, however, the fertility status of patients (114/263) differs, with a mean FSH level of 23,8 U/L indicating a general infertility status with normal mean testosterone and moderately increased mean LH level.

Assuming a normal pituitary-testis axis after chemotherapy, we see a total loss of spermatogenesis and presume a moderate damage of the Leydigcells, compensated by increased LH levels.

16

CARCINOMA-*IN-SITU* IN TESTES WITH GERM CELL TUMOUR

Comparison of Clinical Parameters with Histological Findings in Testicular Tissue Near to and Distant from the Tumour

T. Pottek,[1] H. Lauke,[2] and M. Hartmann[1]

[1]Armed Forces Hospital Hamburg
Department of Urology
Lesserstr. 180, 22049 Hamburg, Germany
[2]Institute of Anatomy
Department of Microscopic Anatomy
University of Hamburg (UKE)
Martinistr. 52, 20246 Hamburg
Germany

The data from 380 consecutive patients from 1990 to 1995 have been reanalysed. Tumour identity were distributed, as follows, according to the WHO-Classification (MOSTOFI & SOBIN): 169 Seminoma and 211 Non-seminomatous germ cell tumours (67 Embryonal-Carcinoma, 41 Terato-Carcinoma, 3 Chorion-Carcinoma, 3 Yolk-Sac-Tumours, 55 Mixed Tumours with Chorion-Ca., 42 Mixed Tumours with Seminoma). Testicular tissue of all these patients had been examined using the semithin sectioning technique.

The operational procedure of a tumour ablation is standardized: The inguinal incision is followed by the exposure of the spermatic cord and the luxation of the testicle out of the scrotum. The testicle is sectioned through the tumour. Biopsies with a volume of 0,5 ccm are taken out of the parenchyma. One is taken adjacent to the tumour, one distant - if possible near the rete testis.

Samples of testicular tissue adjacent to and distant from solid germ cell tumours were obtained, fixed in phosphate-buffered 5,5% glutaraldehyde and postfixed in osmium tetroxide. After dehydration the tissue was embedded in Epon. Semithin sections (1μm) were cut and finally stained with toluidine blue-pyronine.

380 Reports were retrieved. 303 (80%) patients had CIS in biopsies near to or distant from the tumour. In 53 testes only biopsies close to the tumour could be examined.

CIS was found in 47 (88%) of biopsies. Of 327 testes either tumour-near or -distant biopsies were examined. 262 (80%) of the tumour-near biopsies had CIS and 206 (63 %) of the tumour-distant biopsies had CIS. For 196 patients (60%) both tumour-near and -distant biopsies showed CIS. 65 (20 %) had CIS in tumour-near biopsies but not in those distant from the tumour, whereas only 10 (3%) had CIS in the tumour-distant biopsy only. Of the seminomas, 78 % showed CIS in the tumour-near biopsies and 54 % in the tumour-distant biopsies. Of non-seminomas 86 % showed CIS in the tumour-near biopsies and 67 % in those distant from the tumour. Relations between histology and CIS are statistically not significant (p = 0,436)

There is, however, a statistically significant relationship between tumour size (diameter in cm) and occurence of CIS in the biopsies (p= 0,045). With a cut-off of tumour-diameter ≤ 1 cm the probability of CIS in the biopsies is 0.54. If the tumour has a diameter > 1 cm the probability of CIS in the biopsies is 0.88.

Histological investigations by other authors of whole ablated testis specimens had shown a high incidence (up to 90 %) of CIS adjacent to germ cell tumours. In our material of tumour-near and -distant biopsies the incidence of CIS seems to be lower. This could be interpreted as an indication for non-disseminated distribution of CIS.

Even if mathematical methods are able to show a significant relationship, there is still the question of clinical relevance to be answered: As long as no prospective parameters of high probability are established, every testicle with a germ cell tumour should be regarded as CIS-contaminated. The individual decision for radiation or surveillance after organ-preserving surgery depends on the patients intentions regarding fertility.

To prevent a solid tumour developing in the future, testes which have CIS should be irradiated, especially if the tumour is more than 1 cm in diameter.

17

INTRATESTICULAR SPERM EXTRACTION

Basis for Successful Treatment of Infertility in Men with Ejaculatory Azoospermia

W. Schulze,[1] U. A. Knuth,[2] D. Jezek,[3] D. M. Benson,[4] R. Fischer,[5]
O. G. J. Naether,[5] V. Baukloh,[5] and R. Ivell[6]

[1]Department of Andrology
University of Hamburg (UKE)
Martinistr. 52, 20246 Hamburg, Germany
[2]Gemeinschaftspraxis BKS, Hamburg
Schomburgstr. 120, 22767 Hamburg, Germany
[3]Institute of Histology and Embryology
Zagreb Medical School
Zagreb, Croatia
[4]Institute of Anatomy
University of Hamburg (UKE)
Martinistr. 52, 20246 Hamburg, Germany
[5]Gemeinschaftspraxis Leidenberger, Weise & Partner, IVF Unit
Lornsenstr. 4, 22767 Hamburg, Germany
[6]Institute for Hormone and Fertility Research
University of Hamburg
Grandweg 64, 22529 Hamburg, Germany

According to Wong and Horvath (1987), the causes of male infertility can be separated into three categories: pretesticular, testicular and posttesticular. Disturbances in the hormonal control of the testis are grouped into the first category, for example, hypogonadotropism, steroid hormone excess, hyperprolactinaemia, hypo- or hyperthyroidism, or diabetes mellitus. The testicular category includes, besides idiopathic hypospermatogenesis and maturation arrest, particularly those disturbances which can be attributed to manifest structural defects in the seminiferous tubules and the germinal epithelium. Here one can discriminate between those causes which are developmentally or genetically based (e.g. cryptorchidism or Klinefelter's syndrome) and those which are acquired later in life (e.g. chemotherapy damage, radiation damage, orchitis). The category of posttesticular causes includes both congenital and acquired lesions in the excretory genital ducts, as well as disturbances of the accessory glands.

The Fate of the Male Germ Cell, edited by Ivell and Holstein
Plenum Press, New York, 1997

Whereas for some disturbances in the pre- or posttesticular categories appropriate therapeutic approaches are available (e.g. pulsatile application of GnRH, hCG/hMG treatment, or microsurgical correction such as vasovasostomy, or tubulovasostomy), the possibility of treating disturbances in the testicular category is very limited. Under the terms of evidence-based medicine there is as yet no medication which has proved effective (Vandekerckhove et al., 1993; Nieschlag, 1996 a,b). Even invasive measures, for example the correction of a varicocele, are of debatable effectiveness (Nieschlag et al., 1996). As a consequence treatment was mainly based on the female side to compensate for male impairment (Macleod, 1955). But pregnancy rates were low even after superovulation or IVF. An exception to this has been the introduction of intracytoplasmic sperm injection (ICSI) (Palermo et al., 1992). Even for cases previously considered hopeless presenting with only few isolated sperm in the ejaculate, this technique has achieved high fertilization and pregnancy rates (Van Steirteghem et al., 1993). Furthermore, based on the discovery that even testicular sperm which are not completely mature can be used for ICSI in a comparable manner, it is even possible to offer a treatment to those men who previously had been considered irrevocably infertile (Devroey et al., 1994; Nagy et al., 1995; Silber et al., 1995 a,b; Tucker et al., 1995). These patients show a complete azoospermia in their ejaculate because of a defect in their spermatogenetic tissue. Since we know that there may still be areas with a degree of spermatogenetic activity within the testes of these men, surgical isolation of gametes ("testicular sperm extraction", TESE) is frequently possible.

On the basis of this concept, we have recently developed an optimized treatment protocol which combines the removal of a testicular biopsy for the diagnostic evaluation of the infertility with the possibility of a TESE/ICSI option (Salzbrunn et al., 1996; Fig.1). This involves subdivision of the biopsy sample into at least four portions according to the

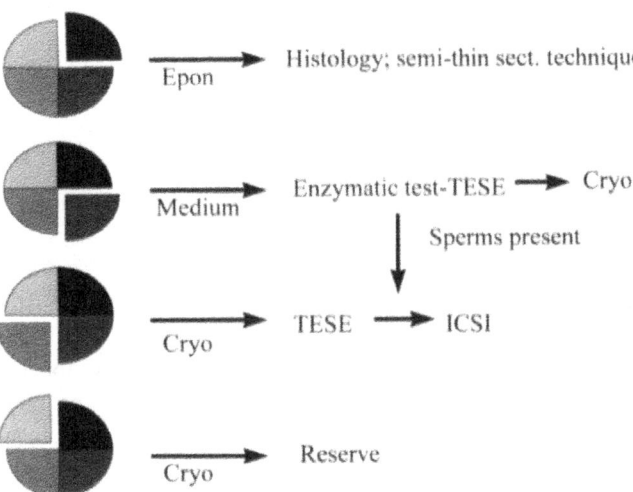

Figure 1. The cryo-TESE/ICSI concept. The testicular biopsy is divided into four portions. One part is fixed in glutaraldehyde and analysed histologically using the semi-thin sectioning technique. A second portion is transferred to IVF medium containing collagenase, type Ia, and subjected to preliminary enzymatic sperm extraction (test-TESE). The remaining portions are cryoconserved. When the preliminary diagnosis indicates the presence of spermatozoa, then the definitive TESE/ICSI treatment can be carried out at a later time-point using the cryopreserved biopsy fragments.

Figure 2. The "sandwich-pattern" for portioning the testicular biopsy. Following an incision in the tunica albuginea, testicular tissue is extruded and cut into three layers. The upper layer is divided into two portions. One portion is analysed histologically, the adjacent piece of tissue is cryopreserved. The test-TESE is performed on the central layer of the biopsy. The lower layer is cryopreserved in toto. Because of the overlapping manner in which the biopsy is apportioned, the diagnostic steps (histology, test-TESE) acquire a high predictive value as to the suitability of the cryopreserved portions for TESE.

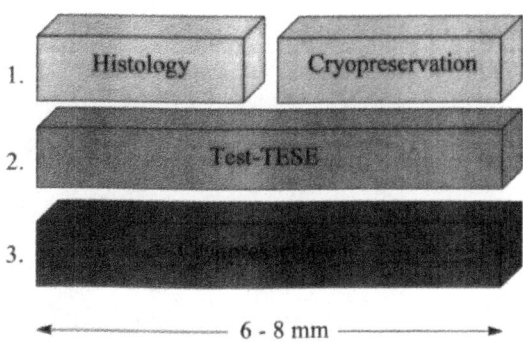

following scheme, which we refer to as the "sandwich pattern" (Fig.2). One of the portions is subjected to a histological examination following semi-thin sectioning, which allows a detailed analysis of the tissue on the basis of cytologic criteria (Holstein and Wulfhekel, 1971). At least two portions are cryopreserved. A further portion is used immediately for a trial extraction of sperm from the testicular tissue after enzymatic digestion ("test-TESE"). Collagenase type Ia is used to digest the tissue in order to isolate individual spermatozoa without damaging them (Figs. 3 and 4). Following the identification of spermatozoa, the remains of this sample can also be cryoconserved. Both the histological analysis as well as the test-TESE provide complementary data on the morphology of the germinal epithelium within the biopsy sample. With this information one can fairly well predict the suitability of the cryopreserved material for TESE. In the event that no haploid gametes are detectable by the described procedures, all further treatment must be abandoned at least in a homologous setting. Thus a negative result would at least prevent the female partner from costly and futile preparations for an IVF procedure. If, however, the preliminary diagnosis indicates the presence of suitable spermatozoa, the definitive TESE/ICSI treatment can be carried out with the cryopreserved tissue portions at a later appointment. The cryopreserved biopsy portions which are not used for the ICSI treatment can be stored for further IVF attempts. If desired they may even be used after a successful IVF/ICSI cycle for the induction of a second pregnancy. A second operation on the man is as a rule not necessary.

In order to gain more information about testicular defects or disturbances which are most likely to give a positive result in TESE, we have investigated biopsy material from 37 azoospermic men with elevated serum FSH concentrations (>9 IU/l). The histological analysis provided the following spectrum of symptoms: 1. Complete tubule sclerosis. 2. Complete germ cell aplasia (Sertoli-cell-only syndrome; SCOS). 3. Focal spermatogenetic activity with the predominating presence of Sertoli-cell-only tubules ("partial SCOS"). 4. Focal spermatogenetic activity with disseminated tubular atrophy. 5. Hypospermatogenesis with occurrence of varying degrees of degenerating seminiferous tubules ("mixed atrophy", according to Sigg and Hedinger (1981)). 6. Severe degeneration of immature germ cells with arrest of the spermatogenetic process at the level of primary spermatocytes or early spermatids. A detailed description and photographic documentation of these pathological situations are to be found, for example, in Holstein et al. (1988), Hedinger and Dhom (1991), and Holstein et al. (1994).

In order to provide a standardized and reproducible basis for judging the TESE findings, we classified the various histological observations according to the scoring system of De Kretser and Holstein (1976), which is a modified version of the "Johnsen score of tes-

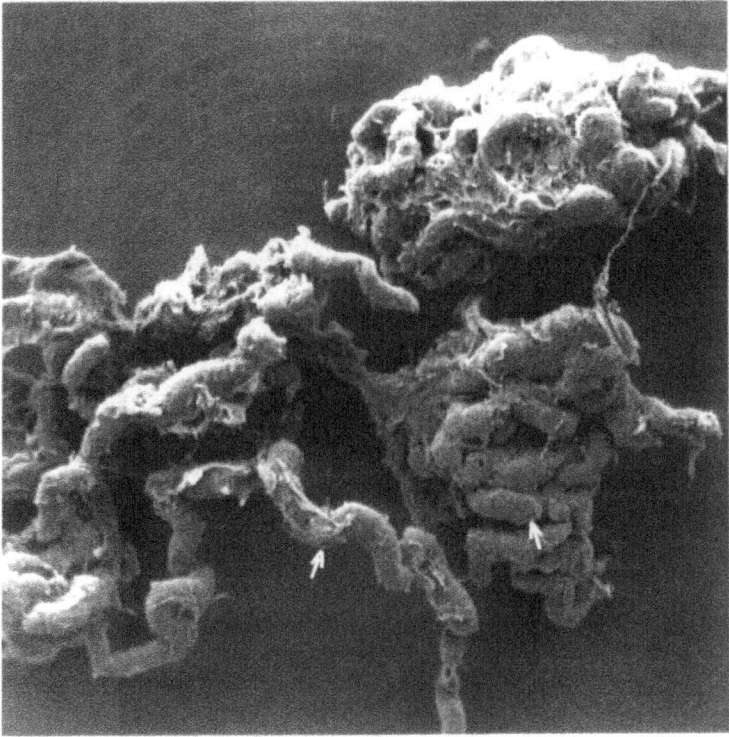

Figure 3. Scanning electron micrograph of a testicular biopsy, which has been incubated for 45 min in IVF medium containing collagenase Ia. In this initial phase of the enzymatic treatment, the tissue already shows a clear dissolution of the instertitium yielding well isolated seminiferous tubules. Small perforations (arrows) arise within the lamina propria of the tubules, through which germ cells can be extruded. With more extensive enzymatic digestion (4–6 hours) there is complete disintegration of the tubule wall.

ticular biopsies" (Johnsen, 1970) (Table 1). The results are listed in Table 2. It is evident that even in the presence of severe parenchymal defects and low scores, below the level of spermatid presence, in some cases it is still possible to to retrieve viable sperm from the testicular tissue. Since results from right and left testicles may differ considerably, testicular biopsies should as a rule be taken bilaterally from azoospermic patients.

The considerable regional variation in the tissue is expressed by the occasional discrepancy between the histological prognosis and the TESE result, particularly in cases of partial SCOS, disseminated tubular atrophy and mixed atrophy. This situation has already been commented on by Tournaye et al. (1996). Our experience shows, however, that it is possible to make a fairly reliable prediction about the general status of the testis, when results of the histological examination and the test-TESE together cover portions of at least 3–5 different testicular lobules. To obtain this amount of tissue it is necessary to make an incision of about 6–8mm length in the tunica albuginea. The probability of obtaining areas with spermatogenetic activity can, in individual cases, be increased if several biopsies are taken from different regions of the testis. However, it should be taken into account that in many men with hypergonadotropic azoospermia, the testicular volume is markedly reduced, such that a single biopsy divided into four portions is probably the limit of what is medically acceptable.

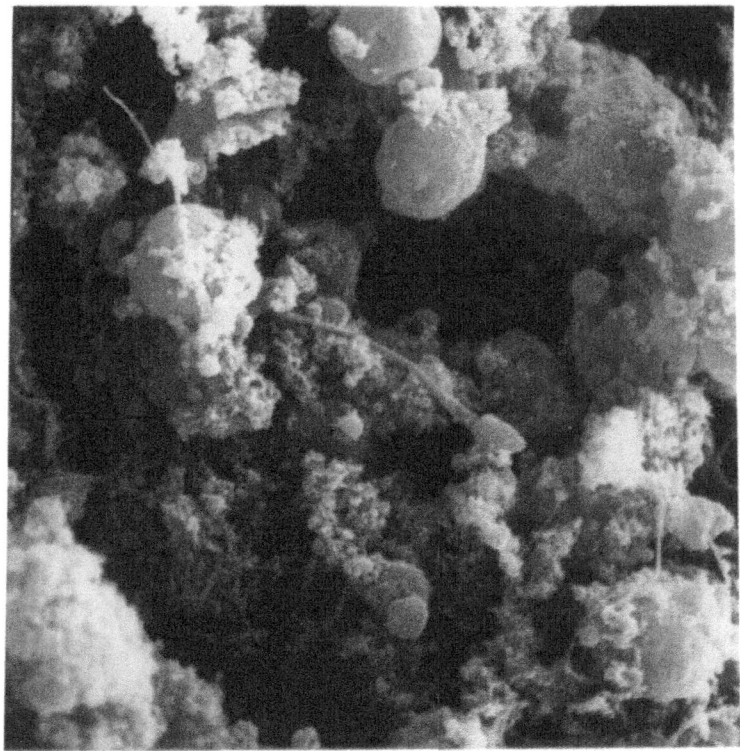

Figure 4. The same preparation as in Figure 3. View through one of the perforations in the lamina propria into the inside of a seminiferous tubule. A spermatozoon is clearly visible in the middle of the picture, which has become released from the germinal epithelium and is moving towards the perforation in the lamina propria.

In 23 patients (62%) the enzymatic test-TESE was positive, that is, a sufficient number of gametes suitable for ICSI could be isolated. Of the remaining 14 men (38%), no spermatozoa could be recovered from the testis material. Moreover, the corresponding histological analyses indicated a complete absence of germ cells. In these cases the cryoconserved tissue portions were successively thawed, and in the event that spermatozoa were after all detected then the tissue could be again cryoconserved. However, in none of

Table 1. Modified Johnsen score of testicular biopsies (De Kretser and Holstein 1976)

1. No seminiferous epithelial cells, tubular sclerosis
2. No germ cells, Sertoli cells only
3. Spermatogonia only
4. No spermatids, few spermatocytes, arrest of spermatogenesis at the primary spermatocyte stage
5. No spermatids, many spermatocytes
6. No late spermatids, few early spermatids, arrest of spermatogenesis at the spermatid stage, disturbance of spermatid differentiation
7. No late spermatids, many early spermatids
8. Few late spermatids
9. Many late spermatids, disorganized tubular epithelium
10. Full spermatogenesis

Table 2. Sperm retrieval by
enzymatic TESE in cases
with low scores

Score	Positive TESE-result (left testis / right testis)
1	0/0
2	7/4
3	1/0
4	0/3
5	0/0
6	6/5

the 56 tissue portions (four portions per patient), could germ cells be identified. This finding impressively demonstrates the prognostic reliability of the combined preliminary diagnosis, based on histology and the test-TESE, concerning the composition of the cryopreserved biopsy fragments.

Our studies also showed that there was a clear negative correlation between TESE result and serum FSH concentration of the patient. However, there appears to be no upper limit for the FSH levels, above which no spermatogenetic activity can be expected. Thus it was still possible to extract sufficient sperm for an ICSI treatment from the patient with the highest serum FSH level (62.3 IU/l). Similar observations have recently been reported by Kim et al. (1997).

The experience to date has shown that microinjection of testicular spermatozoa yields very similar fertilization and pregnancy rates to those obtained with spermatozoa from the ejaculate (Devroey et al., 1995; Al-Hasani et al., 1996; Fischer et al., 1996; Tournaye et al. 1996). Up to June 1996, we had performed cryo-TESE-ICSI procedures in 59 couples. In two cases the treatment had to be repeated, so that altogether there were 61 treatment cycles. The detailed results are indicated in Table 3, and show that neither the process of freezing and thawing nor the enzymatic preparation of the testicular biopsy have any negative effect on the pregnancy rates following ICSI treatment.

Table 3. Experiences with Cryo-TESE/ICSI 1995 - 6/96

n	%
61 cycles	
58 transfers	95.1% of cycles
22 pregnancies	37.9% per transfer
3 miscarriages	13.6 % per pregnancy
14 singletons	73.7% per intact pregnancy
4 twins	21.1% per intact pregnancy
1 triplet	5.3% per intact pregnancy
799 oocytes	13.1 per cycle
638 mature oocytes	79.8% of oocytes
612 injected	
569 intact	93.0% of injected
287 fertilized	50.4% of intact
21 3PN fertilization	7.3% of fertilized
156 embryos transferred	58.6% of normally fertilized

The cryo-TESE/ICSI concept offers optimal conditions for a successful clinical treatment of very severe male fertility disturbances. In addition, the preliminary histological examination offers an opportunity to gain more information about the pathology of the male gonad. In our laboratory we have now examined over 500 bilateral testicular biopsy samples, either from our own patients or delivered for evaluation from other centres of reproductive medicine which have adopted our cryo-TESE procedure. In the course of these analyses, we were able to identify six cases with early stage, treatable testicular tumour (carcinoma in situ; CIS). This is equivalent to a frequency of about 1% which is similar to that suggested in the literature (Skakkebaek, 1978; Schütte, 1984; Nieschlag et al., 1996). In our opinion, this preventive aspect offers another important argument in favour of the cryo-TESE procedure described here, as opposed to those methods which are directed solely towards sperm extraction. Our results also show that the "classical" diagnosis of "azoospermia" must be supplemented by a qualifying term indicating the level of diagnosis. In the past this was already necessary to differentiate obstruction from impaired spermatogenesis, but in the presence of an increased FSH level a biopsy was most often omitted. With the new tool of IVF/ICSI any azoospermia must be evaluated further by the method described above, if the couple is willing to include an IVF-treatment in their infertility work-up. If the term azoospermia is used to describe the male gamete production, the andrologist involved must clearly state whether this refers only to the semen analysis result or to further diagnosis by TESE. To clarify this we suggest to use the terms testicular vs. ejaculatory azoospermia for distinguishing these categories.

LITERATURE

Al-Hasani S, Ludwig W, Küpker W, Fornara P & Diedrich K 1996 Intrazytoplasmatische Spermieninjektion nach MESA oder PESA und TESE. Gynäkologe 1:58–65.

Devroey P, Liu J, Nagy Z, Tournaye H, Silber SJ & Van Steirteghem AC 1994 Normal fertilization of human oocytes after testicular sperm extraction and intracytoplasmic sperm injection. Fertility and Sterility 62: 639–641.

Devroey P, Liu J, Nagy Z, Goossens A , Tournaye H, Camus M, Van Steirteghem AC & Silber SJ 1995 Pregnancies after testicular sperm extraction (TESE) and intracytoplasmic sperm injection (ICSI) in non-obstructive azoospermia. Human Reproduction 10: 1457–1460.

De Kretser DM & Holstein AF 1976 Testicular biopsy and abnormal germ cells. In Human Semen and Fertility Regulation in Men, pp 332–343. Ed Hafez ESE. Mosby, St. Louis.

Fischer R, Baukloh V, Naether OGJ, Schulze W, Salzbrunn A & Benson DM 1996 Pregnancy after intracytoplasmic sperm injection of spermatozoa extracted from frozen-thawed testicular biopsy. Human Reprodicton 11: 2197–2199.

Hedinger CE & Dhom G 1991 Pathologie des männlichen Genitale. In Spezielle pathologische Anatomie, Band 21. Eds Doerr W & Seifert G. Springer, London, Berlin, Heidelberg, New York.

Holstein AF, Roosen-Runge EC & Schirren C 1988 Illustrated Pathology of Human Spermatogenesis. Grosse, Berlin.

Holstein AF, Schulze W & Breucker H 1994 Histopathology of human testicular and epididymal tissue. In Male Infertility, edn 2, pp105–148. Ed TB Hargreave. Springer, London, Berlin, Heidelberg, New York.

Holstein AF & Wulfhekel U 1971 Die Semidünnschitt-Technik als Grundlage für eine cytologische Beurteilung der Spermatogenese des Menschen. Andrologia 3: 65–69.

Johnsen SG 1970 Testicular biopsy score count - a method for registration of spermatogenesis in human testis: normal values and results in 335 hypogonadal males. Hormones 1: 2–25.

Kim ED, Gilbaugh JH, Patel VR, Turek PJ & Lipshultz LI 1997 Testis biopsies frequently demonstrate sperm in men with azoospermia and significantly elevated follicle-stimulating hormone levels. Journal of Urology 1: 144–146.

Macleod J GRMC 1955 Correlation of the male and female factors in human infertility. Fertility and Sterility 6: 112–143.

Nagy Z, Liu J, Janssenswillen C, Silber SJ; Devroey P & Van Steirteghem AC 1995 Comparison of fertilization, embryo development and pregnancy rates after intracytoplasmatic sperm injection using ejaculated, fresh and frozen-thawed epididymal and testicular spermatozoa. Fertility and Sterility 63: 808–815.

Nieschlag E 1996 (a) Gonadotropintherapie in der Andrologie. TW Gynäkologie 9: 199- 202.

Nieschlag E 1996 (b) Therapieversuche bei idiopathischer Infertilität. In Andrologie, pp 333–338. Eds Nieschlag E & Behre HM. Springer, Berlin, Heidelberg, New York.

Nieschlag E, Behre HM, Meschede D & Kamischke A 1996 Störungen im Bereich der Testes. In Andrologie, pp139–166. Eds Nieschlag E & Behre HM. Springer, Berlin, Heidelberg, New York.

Palermo G, Joris H, Devroey P & Van Steirteghem AC 1992 Pregnancies after intracytoplasmic injection of single spermatozoon into an oocyte. Lancet 340: 17–18.

Salzbrunn A, Benson DM , Holstein AF & Schulze W 1996 A new concept for the extraction of testicular spermatozoa as a tool for assisted fertilization (ICSI). Human Reproduction 11: 752–755.

Schütte B 1984 Hodenbiopsie bei Subfertilität. In Fortschritte der Andrologie 9, p 135. Eds Schirren C & Holstein AF. Grosse, Berlin.

Sigg C & Hedinger CE 1981 Quantitative and ultrastructural study of germinal epithelium in testicular biopsy with "mixed atrophy". Andrologia 13: 412–424.

Silber SJ, Nagy Z, Liu J, Tournaye H, Lissens W, Ferec C, Liebaers I, Devroey P & Van Steirteghem AC 1995 (a) The use of epididymal and testicular spermatozoa for intracytoplasmic sperm injection: the genetic implications for male infertility. Human Reproduction 10: 2031–2043.

Silber SJ, Van Steirteghem AC & Nagy Z 1995 (b) High fertilization and pregnancy rate after intracytoplasmic sperm injection with spermatozoa obtained from testicle biopsy. Human Reproduction 10: 148–152.

Skakkebaek NE 1978 Carcinoma in situ of the testis: frequency and relationship to invasive germ cell tumors in infertile men. Histopathology 2: 157–170.

Tournaye H, Liu J, Nagy PZ, Camus M, Goossens A, Silber S, Van Steirteghem AC & Devroey P 1996 Correlation between testicular histology and outcome after intracytoplasmic sperm injection using testicular spermatozoa. Human Reproduction 11: 127–132.

Tucker MJ, Morton PC & Wright G 1995 Intracytoplasmic sperm injection of testicular and epididymal spermatozoa for treatment of obstructive azoospermia. Human Reproduction 10: 486–489.

Vandekerckhove P, Odonovan PA, Lilford RJ & Harada TW 1993 Infertility treatment - from cookery to science - the epidemiology of randomised controlled trials. British Journal of Obstetrics and Gynaecology 100(11): 1005–1036.

Van Steirteghem AC, Nagy Z, Joris H, Liu J, Staessen C, Smitz J, Wisanto A & Devroey P 1993 High fertilization and implantation rates after intracytoplasmic sperm injection. Human Reproduction 8: 1061–1066.

Wong T-W & Horvath KA 1987 Pathological changes of the testis in infertility. In Pathology of Infertility, pp 265–289. Eds Gondos B & Riddick DH. Thieme Medical Publishers, New York.

18

MOLECULAR PATHOPHYSIOLOGY OF THE PITUITARY-GONADAL AXIS

M. Simoni,[1] J. Gromoll,[1] W. Höppner,[2] and E. Nieschlag[1]

[1]Institute of Reproductive Medicine of the University
Domagkstr. 11, 48129 Münster, Germany
[2]Institute for Hormone and Fertility Research
University of Hamburg
Grandweg 64, 22529 Hamburg
Germany

1. SUMMARY

Mutations of gonadotropin β subunits or gonadotropin receptors are involved in some reproductive diseases leading to alterations of pubertal maturation or infertility. Homozygous inactivation of LH results in absence of pubertal maturation and hypogonadism in the male, whereas inactivation of FSH causes primary amenorrhea in females. Mutations of the gonadotropin receptors are classified into activating (the receptor is also active in the absence of the hormone: gain-of-function mutations) and inactivating types (the receptor is not properly processed and/or the hormone cannot bind: loss-of-function mutations). Activating mutations of the LH receptor have been described in familiar and sporadic forms of male-limited pseudoprecocious puberty, whereas they do not express any phenotype in females. The only activating mutation of the FSH receptor described to date was found in a hypophysectomized man who was fertile despite undetectable serum gonadotropin levels; the effects of constitutive FSH receptor activity occurring with normal pituitary function are not known. Homozygous inactivating mutations of the LH and FSH receptor invariably lead to amenorrhea in genotypically female subjects. In males, inactivation of the LH receptor in its more severe form results in a clinical picture similar to the syndrome of complete androgen resistance, but milder forms of hypoandrogenization have been described as well. The clinical consequences of homozygous inactivation of the FSH receptor in males are associated with subfertility. Finally, polymorphic variants of both the gonadotropin LH and the FSH receptor are present in the normal population.

The Fate of the Male Germ Cell, edited by Ivell and Holstein
Plenum Press, New York, 1997

2. INTRODUCTION

Gonadal function depends totally on pituitary gonadotropins. Luteinizing hormone (LH) and follicle-stimulating hormone (FSH) are produced in the pituitary gland under the control of the neuropeptide gonadotropin-releasing hormone (GnRH), secreted in a pulsatile fashion in hypophyseal portal blood (for review: Weinbauer et al., 1997). GnRH acts through binding to a specific receptor found on the plasma membrane of the gonadotropic cells of the adenohypophysis. The GnRH receptor is a member of the G protein-coupled receptor family, characterized by a large polypeptide structure organized into seven segments integrated in the plasma membrane and connected to each other by extracellular and intracellular loops (Spiegel et al., 1992). In this class of receptors hormone binding occurs through interaction with the extracellular domain and with the regions of the membrane-spanning domain facing the extracellular space, whereas signal transduction ensues from a structural modification of that part of the receptor that couples to the G protein, with its intracellular loops and domain. In the case of the GnRH receptor, both extracellular and intracellular domains are very short.

LH and FSH are glycoproteins consisting of two subunits. The α subunit is common to LH, FSH, choriogonadotropin (CG) and thyroid-stimulating hormone (TSH), is a 92 amino acid long peptide chain and is N-glycosylated at positions 52 and 78. The β subunit is slightly different among the glycoprotein hormones and confers specificity of action. The LH β subunit has a length of 115 amino acids and is N-glycosylated at position 30. The FSH β subunit is made up of 111 amino acids and is glycosylated at positions 7 and 24. The sugar moieties represent a consistent part of the gonadotropin structure and are relatively similar in all glycoprotein hormones. The main difference between LH and FSH is found in the terminal branching of the carbohydrates, which form bi- or triantennary structures terminating predominantly with sialic acid (FSH) or sulphate (LH) residues (Baenziger & Green, 1988; Moyle & Campbell, 1995). The differences in glycosylation represent the molecular basis of the microheterogeneity seen after chromatographic separation of the glycoprotein hormones (Simoni et al., 1995). The tertiary structure of gonadotropins is known (Lapthorn et al., 1994).

Gonadotropins control gonadal function by interacting with their specific receptors, i.e., special members of the G protein-coupled receptor family characterized by a very large extracellular domain. The FSH receptor is found exclusively in the Sertoli cells and in the granulosa cells in males and females, respectively. The LH receptor is located on the surface of Leydig cells in the male and on the surface of of granulosa-lutein cells in the female. Recently, LH receptor transcripts have been described in several other tissues, but since there is no evidence that organs other than the gonad respond to LH stimulation, the meaning of these observations remains obscure (Lei & Rao, 1994). The tertiary structure of the gonadotropin receptors is not known, but can be inferred by analogy to the spatial arrangement of the ribonuclease inhibitor (Kobe & Deisenhofer, 1993) and of rhodopsin (Shertler et al., 1993), whose configuration is known and which possess a primary structure very similar to the extracellular domain and to the transmembrane domain of the gonadotropin receptors, respectively (Moyle et al., 1995). According to the current view, the interaction of the gonadotropin with its specific receptor would lead to a spatial rearrangement of the transmembrane domain resulting in coupling of the receptor to the G protein and activation of adenylate cyclase. G protein-coupled receptors, however, are rather unquiet structures and have the natural tendency to undergo this conformational modification even in the absence of the ligand. Signal transduction depends on the equilibrium between inactive and active receptor forms (Samana et al., 1993), whereby this balance can be influenced by interaction with the ligand, mutations resulting in amino acid substitution in key positions and, possibly, by other yet unknown factors.

The gonadotropin-receptor interaction results in the biological effects of gonadotropins, which can be summarized in stimulation of sex steroid synthesis and gamete maturation (Weinbauer et al., 1997). Although these effects have been known for many years, the relative roles of LH and FSH in gametogenesis and the molecular events specifically involved in these processes are still poorly understood. The artificial introduction of structural modifications in gonadotropins and their receptors and the natural occurrence of mutations in patients, resulting in loss- or gain-of-function provide an important contribution to the current knowledge of gonadal function.

3. MOLECULAR PATHOPHYSIOLOGY OF GONADOTROPINS

3.1. Mutations of β Gonadotropin Genes

An isolated deficit of LH and FSH has been sporadically described in the literature. In men, isolated LH deficiency was reported in the fifties in subjects with qualitatively normal spermatogenesis and hypogonadism. This syndrome, the „fertile eunuch" or Pasqualini syndrome, characterized by normal serum FSH levels and Leydig cell hypoplasia, has been tentatively ascribed to a hypothalamic defect, since the administration of GnRH normalized gonadotropin secretion (Hornstein et al, 1974), but poor performance of the gonadotropin assays in those years makes correct pathogenetic interpretation difficult. The isolated defect of LH bioactivity was described in a 17-year-old boy with pubertal delay, normal serum immunoreactive LH levels and prepubertal testosterone concentrations normally responsive to hCG administration (Axelrod et al., 1979). Only more than a decade thereafter the cause of this peculiar hypogonadism could be identified as due to a homozygous point mutation in the gene encoding the β LH subunit, exchanging Gln 54 to Arg. Testicular histology showed no evidence of Leydig cells but, when treated with hCG, the patient experienced normal virilization and spermatogenesis could be induced. Three of four heterozygous males from the same family had slightly increased serum LH levels without clinical evidence of hypogonadism and were infertile, suggesting a decrease of intratesticular testosterone concentrations. Female homozygotes were normal (Weiss et al., 1992). This remains the only documented case of a mutation of the LH β chain resulting in loss of gonadotropin function and shows that LH is not necessary for the development of the male phenotype and testicular descent and that, most probably, testosterone production can be stimulated by maternal and placental gonadotropins during fetal life. Although not evident at the time of testicular biopsy, Leydig cells must have been present during early development and/or could be induced to differentiate from existing precursor cells upon hCG administration. Homozygous, loss-of-function mutations of LH in women have not yet been reported but, by analogy with the phenotype observed in cases of inactivating mutations of the LH receptor, one might expect them to result in primary amenorrhea.

Selective FSH deficiency due to a frameshift mutation of the FSH β chain leading to premature termination of the protein has been described in a woman with primary amenorrhea (Matthews et al., 1993). The resulting FSH β protein was probably unable to couple to the α subunit and both immunoreactive and bioactive FSH levels were undetectable in the patient's serum. The elimination of FSH function did not lead to gonadal dysgenesis and when the patient was treated with FSH, fertility was achieved. The patient's mother was heterozyogous for the mutation and, interestingly, displayed serum FSH levels rather low for her postmenopausal age and reported a history of amenorrhea and subfertility. Apart from this, the only case existing in the literature with a documented loss of function due to a mutation of the FSH β subunit gene, only sporadic cases in infertile subjects were reported in the seventies (Maroulis et al., 1977; Rabinowitz et al., 1979).

Obviously, homozygous mutations of the gonadotropin β genes are very rare and the question arises whether heterozygous defects could be responsible for some milder forms of infertility/subfertility in both genders. If the mutation does not affect immunoreactivity, LH and FSH in vitro bioassays might be useful to recognize these cases that could, therefore, have been underestimated until now. Identification of these mutations is especially important because the patients can be cured by exogenous gonadotropin administration.

3.2. Polymorphic Gonadotropin Variants

The complete lack of immunoreactive LH in serum of normally fertile, non-hypogonadal subjects (Petterson et al., 1992) led to the identification of a genetic variant of LH β resulting from two point mutations (Trp 8 Arg and Ile 15 Thr) which abolished one epitope recognized by a monoclonal antibody routinely used in an immunofluorimetric assay. The in vitro bioactivity of the mutated gonadotropin was, however, preserved and could be exploited for the identification of new cases (Haavisto et al., 1995). In Finland about 3.6 % of the population is homozygous for this mutation and 24.1 % is heterozygous. The same variant has been also described in other countries, although at lower frequency (Huhtaniemi et al., 1996). Amino acid substitution at position 15 introduces a consensus sequence for glycosylation and renders the primary structure more similar to that of hCG, which is actually glycosylated at that site. The LH variant seems to be more sulphated than the wild type LH and thus has a shorter half-life in circulation (Haavisto et al., 1995) and is more bioactive in vitro (Suganuma et al., 1996). This prompted the investigation of whether subjects possessing at least one mutated allele show indices of lower LH bioactivity in vivo. In boys followed longitudinally for three years during puberty, no differences could be documented in serum gonadotropin and testosterone levels, but those subjects possessing the variant LH showed a lower growth rate and lower serum levels of IGFBP-3 compared to the boys with wild type, an indication that LH polymorphism can indeed have some impact on functions depending on sex steroids, yet remaining within normal ranges (Raivio et al., 1996). Polymorphic genetic variants of FSH are not known.

4. MOLECULAR PATHOPHYSIOLOGY OF THE GONADOTROPIN RECEPTORS

4.1. LH Receptor

The clinical picture of familial, male-limited pseudoprecocious puberty was first described in 1981. The syndrome is characterized by precocious sexual development and growth spurt due to autonomous activation of Leydig cell function. Serum gonadotropins remain at the prepubertal level and the tubular compartment of the testis is immature, whereas Leydig cells are hyperplastic. "True" puberty is sustained by pituitary activation at the physiological time (Rosenthal et al., 1996) and, as adults, affected patients have short stature and are normally fertile. The disease is sex-limited and dominant. In 1991 a putative testis-stimulating factor was described in serum from these patients but this find-

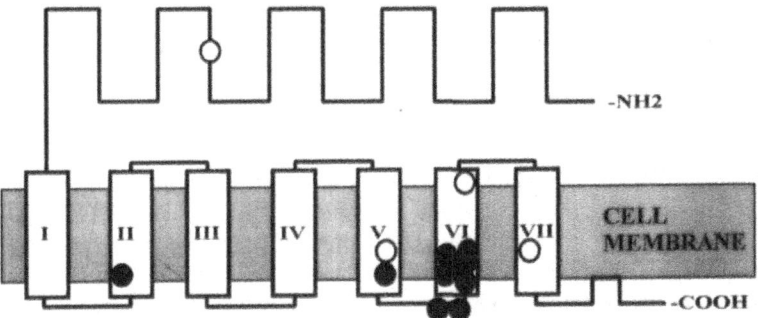

Figure 1. Location of the activating (filled circles) and inactivating (open circles) of the LH receptor, as summarized by Themmen & Brunner (1996) and Hutaniemi et al. (1996).

ing was not confirmed thereafter (Manasco et al., 1991). The first descriptions of constitutive activity of in vitro mutagenized G protein coupled-receptors suggested that the LH receptor might be involved in the pathogenesis of the disease and, in fact, activating mutations of the LH receptor were described in several affected families (Shenker et al., 1993; Kremer et al., 1993). The mutated receptor was shown to be constitutively active, i.e. to be responsible for a slightly but constantly increased cAMP production in the absence of the ligand compared to the wild type. Many other cases of activating mutations of the LH receptor have been described since then, both in familial and sporadic forms of male pseudoprecocious puberty (Fig. 1) (reviewed by Themmen & Brunner, 1996 and by Hutaniemi et al., 1996). Activating mutations are mainly but not exclusively located in the third intracellular loop and fifth transmembrane segment, obviously a hot spot for receptor activation. It also appears that the type of amino acid substitution is relatively unimportant, since activation sometimes ensues from an exchange of amino acids very similar in structure and charge characteristics to the wild type. This suggests that, similarly to the TSH receptor, the LH receptor is rather "unquiet" and has the natural tendency to turn into the activated state in the presence of relatively minor perturbations.

When mutations lead to receptor inactivation, the clinical picture resembles androgen insensitivity. The first homozygous inactivating mutation of the LH receptor was described in two genotypically male, phenotypically female siblings presenting for primary amenorrhea and lack of breast development (Kremer et al., 1995). Unlike patients with androgen resistance, patients with functional impairment of the LH receptor have very low serum testosterone levels and no Leydig cells (Leydig cell hypoplasia) but, similarly to that condition, the phenotype can vary from the complete lack of androgens to more moderate forms of hypoandrogenization. The functional consequences of the mutations described so far are not completely characterized but they could affect hormone binding as well as signal transduction and/or receptor synthesis and processing. It is interesting to note that, in the most severe form of LH receptor inactivation, there is complete sex reversal of the phenotype, whereas the loss of LH β function is compatible with a normal male phenotype. This indicates that maternal and placental gonadotropins can overcome the lack of LH and support Leydig cell development and function during fetal life, if a functional LH receptor is present.

In females, the LH receptor inactivation results in primary amenorrhea with normal breast development and pubertal feminization, suggesting that LH is not important for the development of secondary sexual characteristics at puberty (Latronico et al., 1996). Finally, heterozygous parents of affected children are normal, indicating that heterozygous loss-of-function mutations remain functionally silent.

4.2. FSH Receptor

Unlike the cognate TSH and LH receptors, to date very few mutations of the FSH receptor have been shown to have a pathophysiological correlate. Due to the absolutely essential role of FSH for female gametogenesis, the homozygous lack of receptor function is expected to result in primary amenorrhea, whereas the male phenotype might be subtly hidden in the large pot of "idiopathic" infertility with normal androgenization. The genetic characterization of several Finnish families with an autosomal recessive form of pure ovarian dysgenesis with normal karyotype (Aittomäki et al., 1994) led to the discovery of the first inactivating mutations of the FSH receptor (Aittomäki et al., 1995). The mutated receptor presents an Ala → Val substitution at position 189, in exon 7 in the extracellular domain. This amino acid belongs to a stretch of five highly conserved amino acids which also includes a glycosylation site probably necessary for receptor processing and/or transportation to the plasma membrane. Most probably, the mutation drastically reduces the number of receptors expressed at the plasmalemma without reducing hormone affinity (Aittomäki et al., 1995). Clinical characterization of the affected women showed that the individuals homozygous for the mutation have primary or early secondary amenorrhea, variable development of secondary sex characteristics and high serum gonadotropin levels. The ovaries are hypoplastic, but primordial follicles are present, suggesting some residual receptor activity (Aittomäki et al., 1996). The FSH receptor mutation described, however, is responsible for only a proportion of cases of pure gonadal dysgenesis, 29% in the survey of Finnish patients (Aittomäki et al., 1996), and it is conceivable that mutations in other receptor regions might be involved in other cases.

Heterozygous loss-of-function mutations are compatible with normal ovarian function, as suggested by the phenotype of the heterozygous parents of the Finnish patients and by a recent case observed by us. In a fertile woman we incidentally found a heterozygous Asn 191 Ile transition in exon 7 of the FSH receptor, i.e. in the same region affected by the inactivating mutation described in the Finnish patients with ovarian dysgenesis. When expressed in vitro, the mutated receptor was completely inactive, underlining the importance of the glycosylation site disrupted by the mutation (unpublished observation). In males, homozygous inactivating mutations of the FSH receptor are associated with variable degrees of spermatogenic failure and reduced fertility but, surprisingly, not with azoospermia or absolute infertility (Tapanainen et al., 1997).

The description of activating mutations of the TSH and LH receptor generated the hypothesis that similar mutations should affect the FSH receptor as well. However, since clinical pictures of autonomous hyperfunction of Sertoli and granulosa cells have not yet been identified, the phenotype related to permanent, constitutive activation of the FSH receptor could remain masked if reproductive functions are otherwise normal. This is suggested by the only example of an activating mutation of the FSH receptor described by us in a hypophysectomized patient who was normally fertile despite the complete lack of gonadotropins (Gromoll et al., 1996). The mutation introduces a Asp 567 Gly transition in the third intracytoplasmic loop of the FSH receptor, a position perfectly conserved among the glycoprotein hormone receptors that has been reported to be mutated in cases of constitutive activation of the LH (Laue et al., 1995) and TSH receptors (Parma et al., 1993). When transfected in COS-7 cells, the mutated receptor shows a low level of constitutive activity, only 1.5-fold higher compared to the wild type (Gromoll et al., 1996). This indicates that, unlike the other glycoprotein hormone receptors, the FSH receptor is rather stable in its conformation (Kudo et al., 1996). The constitutive activity caused by the Asp 567 Gly transition becomes more obvious when the receptor is transiently expressed in a cell line derived from mouse Sertoli cells (Walther et al., 1996), where the unliganded mutated receptor induces an accumulation of cAMP four-fold higher than the wild type (not shown). Obviously, a line derived from cells naturally lodging the receptor is better suited to demonstrate its functional properties, as it probably possesses a more physiological signal transduction machinery and can thereby mimic the in vivo situation more closely.

The search for mutations soon revealed that the FSH receptor is polymorphic in exon 10 in the general population (Aittomäki et al., 1996, our unpublished observation). Two amino acid positions are concerned: position 307 can be occupied either by Thr or by Ala and position 680 can be occupied either by Asn or by Ser. In two large groups of men of proven fertility and infertile patients, we observed that the two polymorphic sites occur in the form of two allelic variants, characterized by the combination Thr 307/Asn 680 and Ala 307/Ser 680, respectively. The frequency of the two alleles is similar in normal and infertile men (our unpublished observation). It will be interesting to investigate whether the two variants have the same functional properties in vitro and if, by analogy with the polymorphism observed in the gonadotropin LH, the occurrence of the two alleles is related to some phenotypical attribute, albeit within the physiological range.

5. CONCLUDING REMARKS

The study of the molecular pathophysiology of reproductive functions has shown that some forms of infertility can be genetically transmitted. Although it is expected that genetic alterations leading to infertility are auto-limiting in their expansion, recent results show that gonadotropin and gonadotropin receptor mutations may not be infrequent. Since the existing cases suggest that heterozygous, loss-of-function mutations of the gonadotropin β subunits might be related to infertility, this eventuality should be considered, especially in those cases where serum gonadotropin levels are incongruous with the symptoms. Moreover, so far gonadotropin receptor mutations have been described only in rather dramatic clinical pictures and it is possible that mutations leading to decrease rather than complete loss of function are involved in less drastic forms of reproductive dysfunction. Finally, the phenotype associated with activating mutations of the FSH receptor in persons with normal pituitary function remains to be identified.

6. ACKNOWLEDGMENTS

This work was supported by a grant of the Deutsche Forschungsgemeinschaft (Ni 130–15), Confocal Research Group "The male gamete: production, maturation, function".

7. REFERENCES

Aittomäki K 1994 The genetics of XX gonadal dysgenesis. American Journal of Human Genetics 54: 844–851.

Aittomäki K, Dieguez Lucena JL, Pakarinen P, Sistone P, Tapanainen J, Gromoll J, Kaskikari R, Sankila E-M, Lehvaslaiho H, Reyes Engel A, Nieschlag E, Huhtaniemi I & de la Chapelle A 1995 Mutation in the follicle-stimulating hormone receptor gene causes hereditary hypergonadotropic ovarian failure. Cell 82: 959–968.

Aittomäki K, Herva R, Stenman U-H, Juntunen K, Ylöstalo P, Hovatta O & de la Chapelle A 1996 Clinical features of primary ovarian failure caused by a point mutation in the follicle-stimulating hormone receptor gene. Journal of Clinical Endocrinology and Metabolism 81: 3722–3726.

Axelrod L, Neer RM & Kliman B 1979 Hypogonadism in a male with immunologically active, biologically inactive luteinizing hormone: an exception to a venerable rule. Journal of Clinical Endocrinology and Metabolism 48: 279–287.

Baenziger J & Green ED 1988 Pituitary glycoprotein hormone oligosaccharides: structure, synthesis and function of the asparagine-linked oligosaccharides on lutropin, follitropin and thyrotropin. Biochimica Biophysica Acta 947: 287–306.

Gromoll J, Simoni M, & Nieschlag E 1986 An activating mutation of the follicle-stimulating hormone receptor autonomously sustains spermatogenesis in a hypophysectomized man. Journal of Clinical Endocrinology and Metabolism 81: 1367–1370.

Haavisto A-M, Pettersson K, Bergendhal M, Virkamäki A & Huhtaniemi I 1995 Occurrence and biological propeties of a common genetic variant of luteinizing hormone. Journal of Clinical Endocrinology and Metabolism 80: 1257–1263.

Hornstein OP, Becker H, Hofmann N & Kleiß HP 1974 Pasqualini-Syndrom („fertiler Eunuchoidismus"). Klinische, histologische und hormonanalytische Befunde. Deutsche Medizinische Wochenschrift 99: 1907–1914.

Huhtaniemi I, Pakarinen P, Haavisto A-M, Nilsson C, Pettersson K, Tapanainen J & Aittomäki K 1996 The polymorphisms of gonadotropin action: molecular basis and clinical implications. In Signal Transduction in Testicular Cells, pp 319–341. Eds V Hansson, FO Levy & K Taskén. Springer, Berlin, Heidelberg, New York.

Kobe B & Deisenhofer J 1993 Crystal structure of porcine ribonuclease inhibitor, a protein with leucine-rich repeats. Nature 366: 751–755.

Kremer H, Kraaij R, Toledo SPA, Post M, Fridman JB, Hayashida CY, van Reen M, Milgrom E, Ropers H-H, Mariman E, Themmen APN & Brunner HG 1995 Male pseudohermaphroditism due to a homozygous missense mutation of the luteinizing hormone receptor gene. Nature Genetics 9: 160–164.

Kremer H, Mariman E, Otten BJ, Moll Jr GW, Stoelinga GBA, Wit J-M, Janse M, Drop SL, Faas B, Ropers H-H & Brunner HG 1993 Cosegregating of missense mutations of the luteinizing hormone receptor gene with familial male-limited precocious puberty. Human Molecular Genetics 2: 1779–1783.

Kudo M, Osuga Y, Kobilka BK, Hsueh AJW 1996 Transmembrane regions V and VI of the human luteinizing hormone receptor are required for constitutive activation by a mutation in the third intracellular loop. Journal of Biological Chemistry 271: 22470–22478.

Lapthorn AJ, Harris DC, Littlejohn A, Lustbader JW, Canfield RE, Machin KJ, Morgan FJ & Isaacs NW 1994 Crystal structure of human chorionic gonadotropin. Nature 369: 455–461.

Latronico AC, Anasti J, Arnhold IJP, Rapaport R, Mendonca BB, Bloise W, Castro M, Tsigos C & Chrousos GP 1996 Brief report: testicular and ovarian resistance to luteinizing hormone caused by inactivating mutations of the luteinizing hormone-receptor gene. New England Journal of Medicine 334: 507–512.

Laue L, Chan WY, Hsueh A, Kudo M, Hsu SY & Wu SM 1995 Genetic heterogeneity of constitutively activating mutations of the human luteinizing hormone receptor in familial male limited precocious puberty. Proceedings of the National Academy of Science of the USA 92 1906–1910.

Lei ZM & Rao ChV 1994 Novel presence of luteinizing hormone/human chorionic gonadotropin (hCG) receptors and the down-regulating action of hCG on gonadotropin releasing hormone gene expression in immortalitzed hypothalamic GT1–7 neurons. Molecular Endocrinology 8: 1111–1121.

Manasco PK, Girton ME, Diggs RL, Doppman JL, Feuillan PP, Barnes KM, Cutler Jr GB, Loriaux DL & Albertson BD 1991 A novel testis-stimulating factor in familial male precocious puberty. New England Journal of Medicine 324: 227–231.

Maroulis GB, Parlow AF & Marshall JR 1977 Isolated follicle-stimulating hormone deficiency in man. Fertility and Sterility 28: 818–822.

Matthews CH, Borgato S, Beck-Peccoz P, Adams M, Tone Y, Gambino G, Casagrande S, Tedeschini G, Benedetti A & Chatterjee VKK 1993 Primary amenorrhoea and infertility due to a mutation in the β-subunit of follicle-stimulating hormone. Nature Genetics: 5 83–86.

Moyle WR & Campbell RK 1995 Gonadotropins. In Endocrinology, pp 230–241. Eds. DeGroot JL, Besser M, Burger HG, Jameson LJ, Loriaux DL, Marshall JC, Odell WD, Potts Jr JT, Rubenstein AH, Saunders, Philadelphia, London, Toronto, Montreal, Sydney, Tokyo.

Moyle WR, Campbell RK, Venkateswara Rao SN, Ayad NG, Berbard MP, Han Y & Wang Y 1995 Model of human chorionic gonadotropin and lutropin receptor interaction that explains signal transduction of the glycoprotein hormones. Journal of Biological Chemistry 270: 20020–20031.

Parma J, Duprez L, Van Sande J, Cochaux P, Gervy C, Mockel J, Dumont J, Vassart G 1993 somatic mutations of the thyrotropin receptor gene cause hyperfunctioning thyroid adenomas. Nature 365: 649–651.

Petterson K, Ding Y-Q & Huhtaniemi I 1992 An immunologically anomalous luteinizing hormone variant in a healthy woman. Journal of Clinical Endocrinology and Metabolism 74: 164–171.

Rabinowitz D, Benveniste R, Lindner J, Lober D & Daniell J 1979 Isolated FSH deficiency rivisited. New England Journal of Medicine 300: 126–128.

Raivio T, Huhtaniemi I, Anttila R, Siimes MA, Hagenäs L, Nilsson C, Pettersson K & Dunkel L 1996 The role of luteinizing hormone-β gene polymorphism in the onset and progression of puberty in healthy boys. Journal of Clinical Endocrinology and Metabolism 81: 3278–3282.

Rosenthal IM, Refetoff S, Rich B, Barnes RB, Sunthornthepvarakul T, Parma J & Rosenfield RL 1996 Response to challenge with gonadotropin-releasing hormone agonist in a mother and her two sons with a constitutively activating mutation of the luteinizing hormone receptor-A clinical research center study 1996 Journal of Clinical Endocrinology and Metabolism 81: 3802–3806.

Samana P, Cotecchia S, Cosat T & Lefkowitz RJ 1993 A mutation-induced activated state of the β2-adrenergic receptor. Extending the ternary complex model. Journal of Biological Chemistry 268: 4625–4636.

Shenker A, Laue L, Kosugi S, Merendino Jr JJ, Minegishi T & Cutler Jr GB 1993 A constitutively activating mutation of the luteinizing hormone receptor in familial male precocious puberty. Nature 365: 652–654.

Shertler GFX, Villa C & Henderson R 1993 Projection structure of rhodopsin. Nature 362 770–772.

Simoni M & Nieschlag E 1995 FSH in therapy: physiological basis, new preparations and clinical use. Reproductive Medicine Reviews 4: 163–177.

Spiegel AM, Shenker A & Weinstein LS 1992 Receptor-effector coupling by G proteins: implications for normal and abnormal signal transduction. Endocrine Reviews 13: 536–565.

Suganuma N, Furui K, Kikkawa F, Tomoda Y & Furuhashi M 1996 Effects of the mutations ($Trp^x \rightarrow Arg$ and $Ile^{15} \rightarrow Thr$) in human luteinizing hormone (LH) β-subunit on LH bioactivity in vitro and in vivo. Endocrinology 137: 831–838.

Tapanainen JS, Aittomäki K, Min J, Vaskivuo T & Huhtaniemi I 1997 Men homozygous for an inactivating mutation of the follicle-stimulating hormone (FSH) receptor gene present variable suppression of spermatogenesis and fertility. Nature Genetics 15: 205–206.

Themmen APN & Brunner HG 1996 Luteinizing hormone receptor mutations and sex differentiation. European Journal of Endocrinology 134: 533–540.

Walther N, Jansen M, Ergün S, Kascheike B & Ivell R 1996 Sertoli cell lines established from H-H-2kb-TsA58 transgenic mice differentially regulate the expression of cell-specific genes. Experimental Cell Research 225: 411–421.

Weinbauer GF, Gromoll J, Simoni M & Nieschlag E 1997 Physiology of testicular function. In Andrology, Male Reproductive Health and Dysfunction, pp 25–57. Eds. Nieschlag E, Behre HM. Springer, Berlin, Heidelberg, New York.

Weiss J, Axelrod L, Whitcomb RW, Harris PE, Crowley WF & Jameson JL 1992 Hypogonadism caused by a single amino acid substitution in the β subunit of luteinizing hormone. New England Journal of Medicine 326: 179–183.

19

FETAL AND PERINATAL INFLUENCE OF XENOESTROGENS ON TESTIS GENE EXPRESSION

P. T. K. Saunders,[1] G. Majdic,[2] P. Parte,[3] M. R. Millar,[1] J. S. Fisher,[1] K. J. Turner,[1] and R. M. Sharpe[1]

[1]MRC Reproductive Biology Unit
37 Chalmers Street
Edinburgh, EH3 9EW, Scotland
[2]Faculty of Veterinary Medicine
Gerbiceva 60, 1000 Ljubljana
Slovenia
[3]Institute for Research in Reproduction
Bombay, India

1. SUMMARY

The incidence of reproductive abnormalities in the male has been reported to have increased during the past 50 years. It has been suggested that these changes may be attributable to the presence of chemicals with oestrogenic activity in our environment. The aim of the experiments described in this chapter was to investigate the effects of acute exposure to high levels of xenoestrogens either indirectly during fetal life, or directly during neonatal life, on gene expression in the testis and pituitary. Fetal treatment involved administration of diethylstilbestrol (DES), 4-octylphenol (OP) or vehicle (oil, control) to pregnant rats on days 11.5 and 15.5 post coitum; fetuses were recovered on day 17.5. There was no difference between fetuses from control and treated mothers in either the overall histology of the testes or numbers of Leydig cells as determined by immunohistochemistry with an antibody directed against 3ß-HSD. However there was a consistent and striking reduction in the amount of P450 17-a hydroxylase C17, 20 lyase (P450c17) and steroidogenic factor 1 (SF-1) detected by immunocytochemistry in testes from treatment groups given the higher doses of OP and DES. Oestrogen receptors (ERα) were present in the fetal Leydig cells of all animals. Neonatal treatment involved direct injection of oil (control), DES, OP or Bisphenol A (Bis A) on days 2, 4, 6, 8, 10 and 12; pituitaries and testes were recovered on day 18. Testis weights and seminiferous tubule diameters were significantly reduced in animals treated with DES. In these same animals immunocytochemical localisation revealed that the amounts of FSH ß subunit and inhibin α subunit

The Fate of the Male Germ Cell, edited by Ivell and Holstein
Plenum Press. New York. 1997

99

were reduced in their pituitaries and testes respectively. OP did not appear to have an acute, measurable effect on testis gene expression but a reduction in testis weight was noted in adult animals given the same treatment regime. The effects observed are consistent with negative feedback by oestrogens on pituitary production of FSH resulting in retarded maturation of seminiferous tubules and reduced Sertoli cell numbers.

These studies have demonstrated that administration of high levels of oestrogens can affect gene expression in the testis early in life. However, the relevance of these findings to observations in man await a) a greater understanding of the physiological role(s) of oestrogens in normal males, b) an evaluation of the sources, routes of exposure, concentrations in vivo and bioavailability of xenoestrogens.

2. INTRODUCTION

In the 1950s and 60s, several million women received treatment with diethylstilbestrol (DES), a potent synthetic oestrogen, during their pregnancy as a preventive to miscarriage (Stillman 1982). Follow-up studies have shown that DES was not effective (Dieckmann et al. 1953) and that the male offspring had an increased incidence of cryptorchidism, hypospadias and smaller testes with lower sperm output and lower sperm quality (Bibbo et al. 1977; Gill et al. 1979; Stillman 1982). Studies in mice by McLachlan and coworkers (McLachlan et al. 1975) have shown that DES can affect the fertility of male mice of treated mothers. In recent years, several studies have reported an increasing incidence of testicular cancer, cryptorchidism and hypospadias in the normal male population in the western world (Toppari et al. 1996). In addition several papers have reported falling sperm counts in some (Auger et al. 1995; Carlsen et al. 1992; Irvine 1994) but not all (Vierula et al. 1996) developed countries. Similarities between the problems recorded in offspring from DES-treated mothers and those occurring in the normal population have led to the suggestion of a possible connection to inappropriate oestrogen exposure (Sharpe 1993; Sharpe & Skakkebaek 1993). This argument has received support from the realisation that many chemicals in our environment have the ability to bind to the oestrogen receptor in vitro (White et al., 1994; Bittman & Cecil, 1970) but whether human exposure to such chemicals is sufficient to cause biological effects remains the subject of debate.

Classical oestrogen actions are mediated by specific receptors, members of a superfamily of transcription factors (Carson-Jurnica et al. 1990). Upon interaction with oestrogens the conformation of the receptor changes and the resulting hormone-receptor complex becomes able to bind specific regions of DNA (oestrogen response elements) usually resulting in alterations in transcription of target genes (reviewed by Parker 1991). As a first step in our understanding of how and where endogenous oestrogens and/or exogenous xenoestrogens might influence reproductive function in the male we can consider the pattern of expression of oestrogen receptors. The human oestrogen receptor (ER) was first cloned in 1985 (Walter et al. 1985) and until two publications in 1996 (Kuiper et al. 1996; Mosselman et al. 1996) was thought to represent the only class of ER. The protein sequences for the two ERs so far identified, now known as ERα and ERß, have regions of homology in the DNA and ligand binding domains but differ substantially in their N-terminal (A/B) and hinge regions (D) (see Mangelsdorf et al. 1995for summary of domains). Immunohistochemistry has identified ERα in fetal mouse testes on days 13.5 to 17.5 (Greco et al. 1992) and in fetal and adult Leydig cells of the rat (Majdic and Fisher unpublished). Oestrogens are known to exert a negative feedback effect on the synthesis and se-

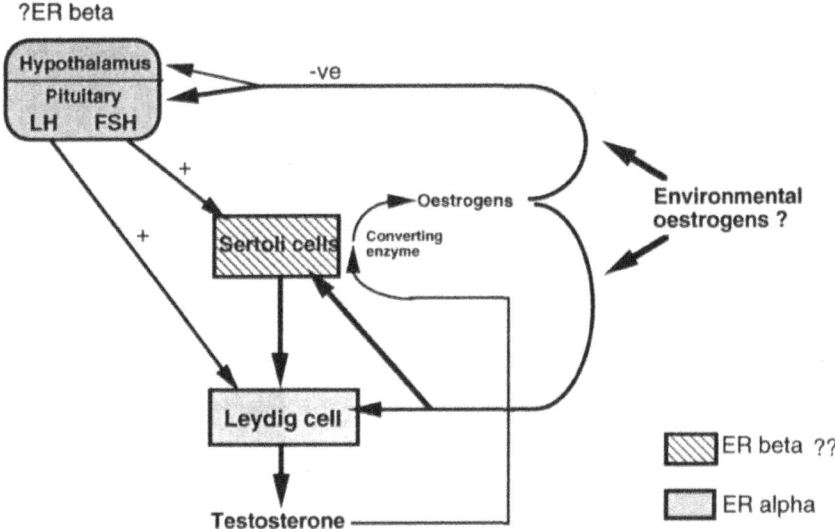

Figure 1. Possible sites of action for oestrogens, and chemicals with oestrogenic activity, on the reproductive axis. Note that only preliminary data is available to indicate that Sertoli cells express ER ß.

cretion of LH and FSH in adult males (Griffin & Wilson 1992; Haisenleder et al. 1994). Possible sites of oestrogenic action in the reproductive axis are presented in figure 1.

The aim of the studies described below was to determine what the consequences of exposure to xenoestrogens, indirectly via the mother, or directly via injection, have on testis size and gene expression in an animal model (the Wistar rat).

3. MATERIALS AND METHODS

3.1. Animals and Fixation of Tissues

All rats used were of the Wistar strain and were maintained under standard animal house conditions; pups remained in the same cage as their mothers until weaning (day 21) even when receiving treatment. Tissues obtained from immature animals (day 18) or fetuses (day 17.5 p.c.) were immersion-fixed in Bouins for 4–5h before transfer into 70% ethanol. Adult rats were perfusion-fixed with Bouins as described previously (Millar et al. 1993). All fixed tissues were processed on a Shandon 2LE tissue processor (Runcorn, Cheshire, UK) using a standard 17.5 h cycle and embedded in paraffin wax. To perform immunohistochemistry sections (5 microns) were cut and floated onto slides coated with TESPA (Sigma Chemical Co. Poole Dorset).

3.2. Treatments

Adult female rats were placed in individual cages with males and checked for the presence of copulatory plugs each morning. The day when the plug was found was taken as day 0.5 post coitum (p.c.). Pregnant females were injected subcutaneously on day 11.5 and 15.5 p.c. with diethylstilbestrol (DES; 100 or 500µg/Kg) or 4-octylphenol (OP,

Aldrich Chemical Co.; 100 or 600 mg/Kg) in corn oil (1ml/Kg) or with the vehicle alone (controls). Pregnant females were killed on day 17.5 p.c. by inhalation of carbon dioxide followed by cervical dislocation. Fetuses were examined under a dissecting microscope and the testes recovered. Fetal age was confirmed by morphological examination of the fetuses (see Kaufman 1992).

Neonatal male rats were given injections of either vehicle (oil), diethylstilbestrol (DES, Sigma 10µg/20µl), Bisphenol A (Bis A, Sigma, 0.5mg/20µl) or octylphenol (OP, Aldrich, Gillingham, Kent; 2mg/20µl). The doses of DES, Bis A and OP administered were chosen to reflect their relative oestrogenic potency in vitro (White et al. 1994) Treatments were administered on days 2, 4, 6, 8 10 and 12 of life. On day 18 animals were killed by inhalation of carbon dioxide followed by cervical dislocation. The bodyweight of each animal was recorded, testes and pituitaries were recovered and placed in Bouins. Measurement of the diameter of seminiferous tubules on day 18 utilised sections stained with haematoxylin and eosin evaluation using image analysis.

3.3. Immunohistochemistry

Rabbit IgGs directed against recombinant human P450c17 (Imai et al. 1993) were a gift from Professor Michael Waterman (Vanderbilt University, Nashville, Tennessee); rabbit anti-human 3ß-HSD (Lorence et al. 1990) was a gift from Professor Ian Mason (Department of Clinical Biochemistry, University of Edinburgh). Purified monoclonal antibodies directed against a inhibin subunit (Groome et al. 1990, code 173/9K) were used at a concentration of 2µg/ml and were a gift from Professor Nigel Groome (Brookes University, Oxford). An antiserum directed against rat FSH ß subunit (ßFSH IC-1) was obtained from NIDDK, NIH, Bethesda, USA and was used at a dilution of 1 in 5000.

Sections were prepared as described previously (Majdic et al. 1996) and except for those to be incubated with anti-3ß-HSD were subjected to antigen retrieval (Shi et al. 1993) by microwaving in 0.01M citrate buffer (pH 6.0) on full power for 20 min, and thereafter standing for 20 min without disturbance Sections were blocked using normal swine serum (Dako, High Wycombe, Bucks) diluted 1:5 in TBS; specific antibodies were diluted in normal swine serum in TBS and incubated on sections under plastic coverslips overnight at 4°C. For detection of bound antibodies, sections were first incubated with avidin-biotin complex conjugated with horseradish peroxidase for 30 min and washed 2 times in TBS (5 min each). Colour reaction product was developed by incubating sections in a mixture of 0.05% (w/v) 3,3'-diaminobenzidine tetra-hydrochloride (DAB, Sigma) in 0.05M Tris-HCl, pH 7.4 and 0.01% hydrogen peroxide. After 5–15 min, sections were washed in water, counterstained with haematoxylin, dehydrated in graded ethanols, cleared in xylene and coverslipped using Pertex mounting medium (Cell path, Hemel Hempstead, UK). Specificity of the antibodies was confirmed by using normal rabbit serum instead of the primary antibodies; no staining was observed in these controls.

4. RESULTS

4.1. Effect of Indirect Exposure to Xenoestrogens on Immunoexpression in the Fetal Testis at Day 17.5 P.C

The gross histology of the testes recovered on day 17.5 of gestation from treated and control animals appeared identical; testes of fetuses from controls contained numerous

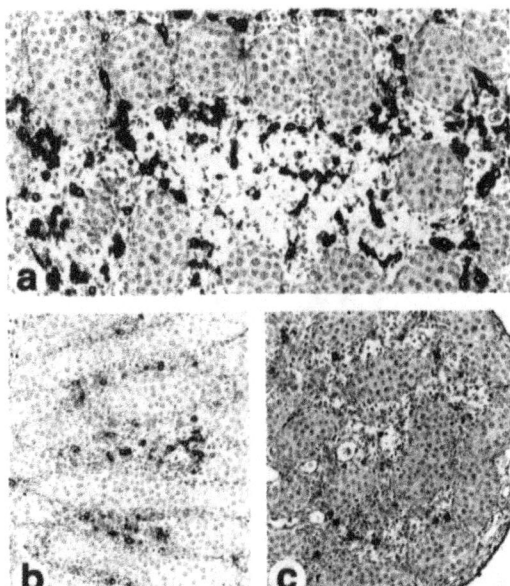

Figure 2. Immunocytochemical staining of fetal Leydig cells using antibodies against P450c17. a) testis from fetus of control b) testis from fetus of mother treated with 500µg/Kg DES, c) testis from fetus of mother treated with 600mg/Kg OP.

Leydig cells which were stained strongly following immunohistochemistry with primary antibodies directed against the steroidogenic enzymes P450c17 (Fig. 2a) and 3ß-HSD. No difference in the abundance of Leydig cells staining positively for 3ß-HSD was noted following any of the treatments (not shown). However immunostaining for P450c17 in the fetal testes was markedly influenced by the doses of oestrogenic chemicals administered to their mothers. A striking reduction in immunostaining was noted in those given the higher doses of DES (Fig.2b) or OP (Fig.2c) but testes from the treatment group administered 100 µg/Kg of DES or 100 mg/Kg OP appeared broadly similar to controls (not shown).

Maternal treatment with high doses of DES or OP also resulted in reduced SF-1 immunostaining in fetal rat testes (Fig. 3); the amount of SF-1 protein was reduced in all of the cell types in which it was normally expressed (Sertoli cells, Leydig cells and other interstitial cells). The observation that Sertoli cells and some non-Leydig interstitial cells appeared immunonegative after maternal treatment with DES appeared to reflect the inherently lower level of expression of SF-1 in these cells, compared with the amount of SF-1 in the Leydig cells.

4.2. Effect of Neonatal Exposure to Oestrogens on Testis Size and Seminiferous Tubule Diameter

At day 18 testis weights were significantly reduced compared with controls in animals administered DES (Table 1); this difference remained significant even when results were corrected for the reduced body weight observed in these animals. Although testis weights also tended to be lower in OP-treated males this did not reach statistical significance. No effect of treatment was observed in Bis A-treated rats at day 18.

The results obtained at day 18 were mirrored by those obtained for animals treated during the neonatal period but thereafter left untreated until adulthood; testis weights of

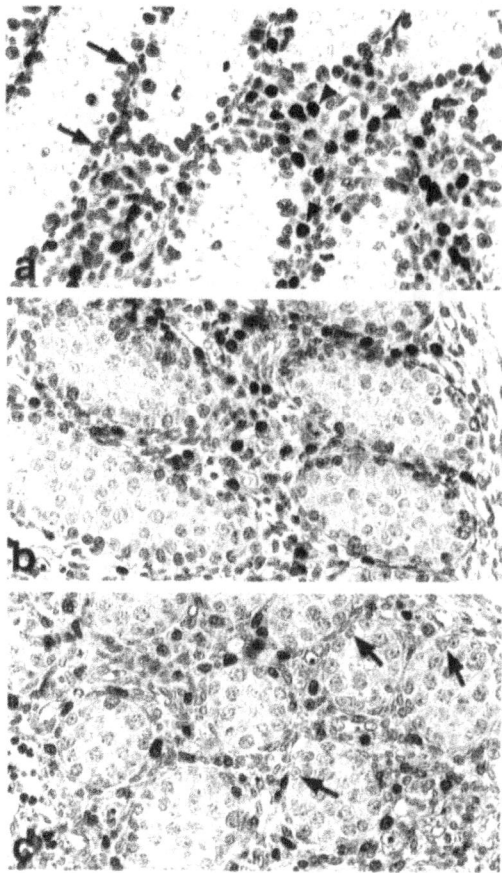

Figure 3. Immunolocalisation of SF-1 in gonads removed from the fetuses of control (a) and DES-treated (b) mothers on day 17.5 of gestation. In testes from control mothers immunostaining was present in Sertoli cell nuclei (arrows) within the seminiferous cords and in the majority of interstitial cells, some of which were very heavily stained (arrowheads). When the amount of SF-1 was determined in sections from the testes of fetuses from DES-treated mothers processed at the same time on the same slides, a marked decrease in the intensity of immunostaining was noted in the treated animals.

OP-treated and DES-treated animals were significantly reduced compared with those animals in the same litters that were left untreated (Fig 4).

On day 18 analysis of seminiferous tubule diameter revealed that those in testes of animals treated with Bis A were indistinguishable from controls, animals treated with OP tended to have tubules with a slightly reduced diameter, which was not statistically different from controls, and those from animals treated with DES were significantly reduced in size (Table 2).

Table 1. Effect of treatment on testis and body weight on day 18 (mean ± SD)

Treatment (n=)	Right testis weight (mg)	Body weight (g)	Corrected testis weight (mg/10g BW)
Cobtrol (16)	71.9±20.3	39.5±6.6	17.8±2.8
DES (12)	21.8±6.5*	35.3±4.4	6.8±1.7*
OP (12)	59.9±25.8	35.5±7.4	16.5±5.2
Bis A (8)	68.0±24.7	39.4±8.0	16.9±3.4

*p<0.001 by t test using pooled variance

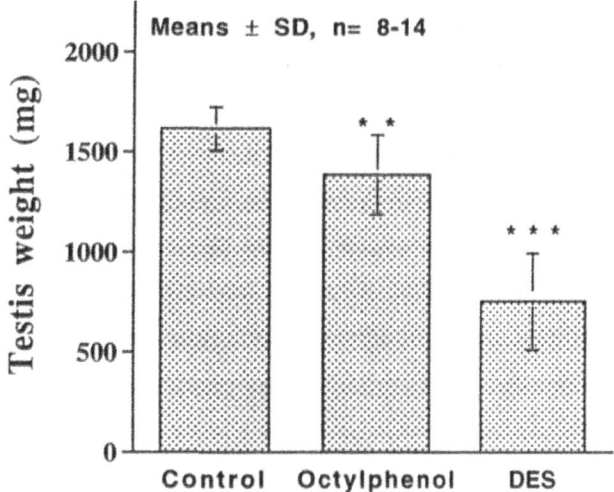

Figure 4. Testis weights of adult rats treated with vehicle (control), DES or OP only during neonatal life.

4.3. Effect of Treatment on Immunoexpression of FSH ß Subunit and ER α in Pituitaries of Day 18 Rats

The intensity of immunopositive staining for FSHß was reduced in pituitaries of males treated with DES compared with controls (Fig 5) whereas those from animals treated with OP or Bis A were indistinguishable from controls (not shown).

4.4. Effect of Treatment on Immunoexpression of Inhibin in Testes of Day 18 Rats

In testes from control animals intense immunopositive staining for the α subunit of inhibin was detected in the cytoplasm of Sertoli cells and the few Leydig cells present at this age (Fig 6a). The intensity of immunostaining for the α subunit did not appear changed in animals that had received treatments with OP or Bis A (not shown) but the intensity of immunostaining in testes from DES-treated animals appeared reduced (Fig 6b).

5. DISCUSSION

The similarities between the observations made on the male offspring of DES-treated mothers (Gill et al. 1979) and the increase in the incidence of reproductive abnor-

Table 2. Effect of treatment on diameter of the seminiferous tubules on day 18

Treatment (n=)	ST diameter (μm)
Control (16)	99.34±8.9
DES (12)	74.6±6.2*
OP (12)	94.3±9.1
Bis A (8)	97.0±17.3

*p<0.05 by 2 table t-test

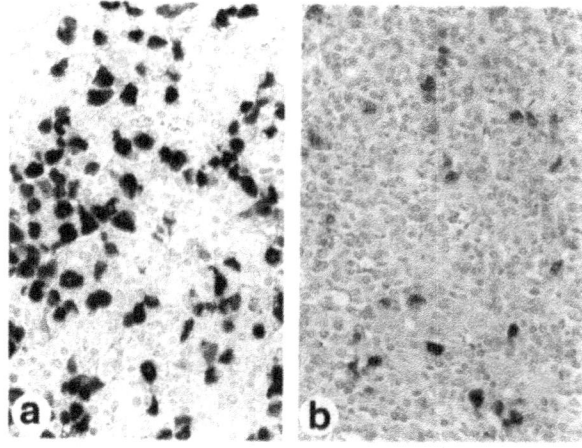

Figure 5. Effect of neonatal treatment on immunoexpression of FSH ß subunit. FSH ß is expressed in gonadotrophs and in pituitaries from controls (a) intense immunostaining was observed. In contrast treatment with DES (b) was associated with reduced numbers of immunopositive cells.

malities observed in the general population (Toppari et al. 1996) has led to the hypothesis that one potential cause of the rise in male reproductive problems might be inappropriate exposure to oestrogens during fetal and/or neonatal life (Sharpe & Skakkebaek 1993). This proposal has become the topic of considerable debate in both the scientific (Stone 1994) and general press and has fuelled concern that not only are the sources and routes of exposure to substances with oestrogenic activity poorly understood but that the mechanisms by which they might affect gonadal function are not fully elucidated.

We have used a variety of treatment regimes to explore the consequences of oestrogen exposure during fetal and neonatal life on development of the testis in the rat. In parallel studies we have administered very low levels (1mg/L) of xenoestrogens to rats prior to conception, throughout pregnancy and for the duration of lactation and examined the consequences for the male offspring. In these animals testis weights were found to be reduced by 8–15% compared with offspring of control (untreated) mothers (Sharpe et al. 1995) and the conclusion was that Sertoli cell number was reduced. FSH is known to be an important regulator of Sertoli cell number and suppression of FSH during the prepubertal period has been shown to result in reduced Sertoli cell number (Huhtaniemi et al. 1986;

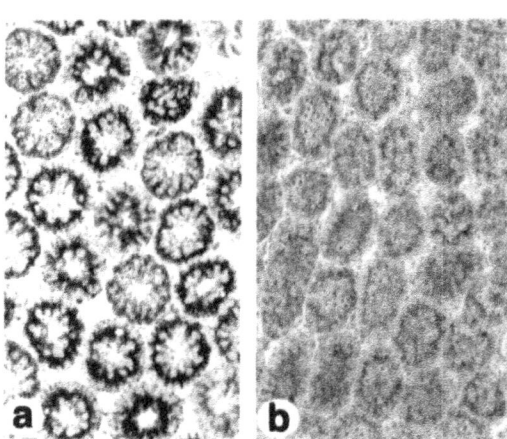

Figure 6. Immunodetection of inhibin a subunit in testes on day 18. In a testis from a control rat (a) intense immunostaining is present within the tubules in a pattern consistent with the location of the Sertoli cells; some Leydig cells are also immunopositive. In contrast following neonatal treatment with DES (b) the amount of inhibin α subunit detected is reduced and most tubules did not contain a clearly defined lumen.

Orth 1984) whereas administration of FSH stimulates proliferation (Griswold et al. 1977). In the present study acute administration of high levels of DES was associated with reduced levels of FSH in the pituitary and of the FSH modulated protein, inhibin, in the testis on day 18. Notably a reduction in testis size of adult rats was observed following this method of exposure to xenoestrogens.

The reduction in P450c17 immunostaining observed in the fetuses of treated mothers was associated with reduced expression of its mRNA consistent with the effect of the oestrogenic chemicals being mediated at the level of transcription of the 17α-hydroxylase gene (Majdic et al. 1996). Steroidogenic factor 1 (SF-1/Ad4BP) was initially isolated and characterised as a transacting transcription factor that regulated expression of several steroidogenic enzymes (Lala et al. 1992; Morohashi et al. 1992). Consistent with this finding, SF-1 was found to be expressed abundantly in the Leydig cells in the testis and in the cells of the theca and corpora lutea of the ovary (Morohashi et al. 1994). The reported role of SF-1 in the regulation of P450 enzymes (Lala et al. 1992; Morohashi et al. 1992) led us to examine its expression in testes of fetuses from treated mothers. The parallel reductions in expression of SF-1 and P450c17 which we observed, appear to fit with the data showing that the transcription factor SF-1 is involved in modulation of P450 gene expression. However, although we have observed a reduction in SF-1 mRNA (unpublished) to date information on control of expression of the SF-1 gene is restricted to the identification of a putative binding site with homology to the consensus E box element which binds proteins with a helix-loop-helix motif. Mobility shift analysis has suggested that a protein(s) capable of binding to this element is present in the fetal testis (Nomura et al. 1995). We have no idea if the SF-1 gene has an oestrogen response element but have established that ER α protein is present in the nuclei of fetal Leydig cells at this age.

Whilst exposure to DES has always been via deliberate administration (see Gill et al. 1979; McLachlan & Newbold 1975) OP is a hydrophobic alkylphenolic compound formed as a metabolite of some non-ionic surfactants which tend to accumulate in sewage sludge and river sediment and to persist in the environment (Naylor et al. 1992). OP has been shown to bind to the ER and to stimulate oestrogen-dependent growth of breast cancer cells and gene expression in vitro to a similar extent to 17ß-estradiol albeit at 1000 fold greater concentration (White et al. 1994). Male trout exposed to low levels of OP (38.5 μg/l) in their water synthesize vitellogenin, an oestrogen inducible protein normally only expressed in the female, and had a reduced testis size (Jobling et al. 1996). Recently it has been reported that the major and persistent metabolite of DDT, p,p'-DDE, which had previously been implicated in induction of abnormalities in male sex development in wildlife, bound more efficiently to the androgen receptor (AR) than it did to the oestrogen receptor (Kelce et al. 1995) and that DES could also bind to the AR. In the present study we do not believe that the effects on Leydig cell P450c17 expression observed following treatment with DES or OP reflect interaction at the level of the AR as our own studies have shown that the few AR detected within the testis at this time are not present in Leydig cells (Majdic et al. 1995).

These studies have involved administration of high levels of xenoestrogens to rats but the effects observed within the testis are presumed to reflect over-activation of the physiological effects of oestrogens produced locally within the testis by Sertoli cells (Dorrington & Khan 1993). Information emerging on the effects of ERα knockout on reproductive function in male mice (Donaldson et al. 1996; Eddy et al. 1996) and the description of the phenotype of patients with naturally occurring mutations of ERα (Smith et al. 1994) have all led to a greater appreciation of the role(s) of oestrogens in the male. The discovery of a "new" ß-form of the receptor (Kuiper et al. 1996) which is expressed in the human testis (Mosselman et al. 1996) means that the potential targets of the action of endogenous and exogenous oestrogens now requires further evaluation.

6. ACKNOWLEDGMENTS

We thank Jim McDonald for expert animal husbandry and Julie Wilson and Joseph Gaughan for skilled technical assistance. GM was supported by a grant from the Ernst Schering Foundation, an Overseas Research Student Award and financial assistance from the Ministry of Sciences and Technology, Slovenia. PP was the recipient of a fellowship under the Rockefeller Foundation Contraceptive Initiative Programme. KJT is supported by EU contract BMH4-CT96–0314. JSF is a recepient of an MRC Research Studentship.

7. REFERENCES

Auger J, Kunstmann JM, Czyglik F & Jouannet P 1995 Decline in semen quality among fertile men in Paris during the past 20 years. The New England Journal of Medicine 332: 281–285.

Bibbo M, Gill WB, Azizi F, Blough R, Fang V, Rosenfield RL, Schumacher GFB, Sleeper K, Sonek MG & Weid GL 1977 Follow-up study of male and female offspring of DES-exposed mothers. Journal of Obstetrics and Gynecology 49: 1–8.

Bittman J & Cecil H 1970 Estrogenic activity of DDT analogs and polychlorinated biphenyles. Journal of Agriculture and Food Chemistry 18:1108–1112.

Carlsen E, Giwercman A, Keiding N & Skakkebaek NE 1992 Evidence for decreasing quality of semen during past 50 years. British Medical Journal 305: 609–613.

Carson-Jurnica MA, Schrader WT & O'Malley BW 1990 Steroid receptor superfamily: structure and functions. Endocrine Reviews 11: 209–220.

Dieckmann WJ, Davis ME & Rynkiewicz LM 1953 Does administration of diethylstilbestrol during pregnancy have therapeutic value? American Journal of Obstetrics and Gynecology 66: 1062–1081.

Donaldson KM, Tong SYC, Washburn T, Lubahn DB, Eddy EM, Hutson JM & Korach KS 1996 Morphometric study of the gubernaculum in male estrogen receptor mutant mice. Journal of Andrology 17: 91–95.

Dorrington JH & Khan SA 1993 Steroid production, metabolism, and release by Sertoli cells. In The Sertoli Cell pp 538–549. Eds L. Russell & M. Griswold. Clearwater: Cache River Press.

Eddy EM, Washburn TF, Bunch DO, Goulding EH, Gladen BC, Lubahn DB & Korach KS 1996 Targeted disruption of the estrogen receptor gene in male mice causes alteration of spermatogenesis and infertility. Endocrinology 137: 4796–4805.

Gill WB, Schumacher GFB, Bibbo M, Straus FH & Schoenberg HW 1979 Association of diethystilbestrol exposure in utero with cryptorchidism, testicular hypoplasia and semen abnormalities. Journal of Urology 122: 36–39.

Greco TL, Furlow JD, Duello TM & Gorski J 1992 Immunodetection of estrogen receptors in fetal and neonatal male mouse reproductive tracts. Endocrinology 130: 421–429.

Griffin JE & Wilson JD 1992 Disorders of the testes and the male reproductive tract. In Williams Textbook of Endocrinology pp 799–852. Eds J. D. Wilson & D. W. Foster. Philadelphia, London, Toronto, Montreal, Sydney, Tokyo: Harcourt Brace Jovanovich, Inc.

Griswold M, Solari A, Tung P & Fritz I 1977 Stimulation by follicle-stimulating hormone of DNA synthesis and of mitosis in cultured Sertoli cells prepared from testes of immature rats. Molecular and Cellular Endocrinology 7: 151–165.

Groome NP, Hancock J, Betteridge A, Lawrence M & Craven R 1990 Monoclonal and polyclonal antibodies reactive with the 1–32 amino terminal sequence fo the alpha subunit of human 32K inhibin. Hybridoma 9: 31–35.

Haisenleder DJ, Dalkin AC & Marshall JC 1994 Regulation of gonadotropin gene expression. In The Physiology of Reproduction pp 1793–1813. Eds E. Knobil & J. D. Neill. New York: Raven Press.

Huhtaniemi IT, Nevo N, Amsterdam A & Naor Z 1986 Effect of postnatal treatment with a gonadotropin-releasing hormone antagonist on sezual maturation of male rats. Biology of Reproduction 35: 501–509.

Imai T, Globerman H, Gertner J, Kagawa N & Waterman M 1993 Expression and purification of functional human 17α-hydroxylase/17,20-lyase (P450c17) in Escherichia Coli. Journal of Biological Chemistry 268: 19681–19689.

Irvine DS 1994 Falling sperm quality. British Medical Journal 309: 1

Jobling S, Sheahan D, Osborne JA, Matthiessen P & Sumpter JP 1996 Inhibition of testicular growth in trout exposed to environmental estrogens. Environmental Toxicology and Chemistry 15: 194–202.

Kaufman M 1992 Atlas of mouse development London: Academic Press.

Kelce W, Stone C, Laws S, Gray L, Kemppainen J & Wilson E 1995 Persistent DDT metabolite p,p'-DDE is a potent androgen receptor antagonist. Nature 375: 581–585.

Kuiper GGJM, Enmark E, Pelto-Hukko M, Nilsson S & Gustafsson J-A 1996 Cloning of a novel estrogen receptor expressed in rat prostate. Proceedings of the National Academy of Sciences, USA 93: 5925–5930.

Lala DS, Rice DA & Parker KL 1992 Steroidogenic factor I, a key regulator of steroidogenic enzyme expression, is the mouse homolog of fushi tarazu-factor I. Molecular Endocrinology 6: 1249–1258.

Lorence M, Murry B, Trant J & Mason J 1990 Human 3beta-hydroxysteroid dehydorgenase/delta (5)>(4) isomerase from placenta: expression in non steroidogenic cells of a protein that catalyses the dehydrogenation/isomerization of C21 and C19 steroids. Endocrinology 126: 2493–2498.

Majdic G, Millar MR & Saunders PTK 1995 Immunolocalisation of androgen receptor to interstitial cells in fetal rat testes and to mesenchymal and epithelial cells of associated ducts. Journal of Endocrinology 147: 285–293.

Majdic G, Sharpe RM, O'Shaughnessy PJ & Saunders PTK 1996 Expression of cytochrome P450 17α-hydroxylase/C17–20 lyase (P450c17) in the fetal rat testis is reduced by maternal exposure to exogenous estrogens. Endocrinology 137: 1063–1070.

Mangelsdorf D, Thummel C, Beato M, Herrlich P, Schutz G, Umesono K, Blumberg B, kastner P, Mark M, Chambon P & Evans R 1995 The nuclear receptor superfamily: the second decade. Cell 83: 835–839.

McLachlan JA, Newbold R & Bullock B 1975 Reproductive tract lesions in male mice exposed prenatally to diethylstilbestrol. Science 190: 991–992.

McLachlan JA & Newbold RR 1975 Reproductive tract lesions in male mice exposed prenatally to diethylstilbestrol. Science 190: 991–992.

Millar MR, Sharpe RM, Maguire SM & Saunders PTK 1993 Cellular localisation of messenger RNAs in rat testis: application of digoxigenin labelled probes to embedded tissue. Cell and Tissue Research 273: 269–277.

Morohashi K, Honda S, Inomata Y, Handa H & Omura T 1992 A common trans-acting factor, Ad4-binding protein, to the promoters of steroidogenic P-450s. Journal of Biological Chemistry 267: 17913–17919.

Morohashi K, Iida H, Nomura M, Hatano O, Honda S, Tsukiyama T, Niwa O, Hara T, Takakusu A, Shibata Y & Omura T 1994 Functional difference between Ad4BP and ELP, and their distributions in steroidogenic tissues. Molecular Endocrinology 8: 643–653.

Mosselman S, Polman J & Dijkema R 1996 ERbeta: identification and characterization of a novel human estrogen receptor. FEBS letters 392: 49–53.

Naylor G, Mierure J, Weeks J, Castaldi F & Romano R 1992 Alkylphenol ethoxylates in the environment. Journal of American Oil Chemists Society 69: 695–703.

Nomura M, Bartsch S, Nawata H, Omura T & Morohashi K 1995 An E box element is required for the expression of the ad4bp gene, a mammalian homologue of ftz-f1 gene, which is essential for adrenal and gonadal development. Journal of Biological Chemistry 270: 7453–7461.

Orth JM 1984 The role of FSH in controlling Sertoli cell proliferation in testes of fetal rats. Endocrinology 115: 1248–1255.

Parker MG 1991 Nuclear Hormone Receptors. London: Academic Press.

Sharpe RM 1993 Declining sperm counts in men – is there an endocrine cause? Journal of Endocrinology 136: 357–360.

Sharpe RM, Fisher J, Millar MR, Jobling S & Sumpter JS 1995 Gestational and/or neonatal exposure of rats to environmental estrogenic chemicals results in reduced testis size and daily sperm production in adulthood. Environmental Health Perspectives 103: 1136–1143.

Sharpe RM & Skakkebaek NE 1993 Are oestrogens involved in falling sperm counts and disorders of the reproductive tract? Lancet 341: 125–126.

Shi S-R, Chaiwun B, Young L, Cote RJ & Taylor CR 1993 Antigen retrieval technique utilizing citrate buffer or urea solution for immunohistochemical demonstration of androgen receptor in formalin-fixed paraffin sections. Journal of Histochemistry and Cytochemistry 41: 1599–1604.

Smith EP, Boyd J, Frank GR, Takahashi H, Cohen RM, Specker B, Williams TC, Lubahn DB & Korach KS 1994 Estrogen resistance caused by a mutation in the estrogen-receptor gene in a man. New England Journal of Medicine 331: 1056–1061.

Stillman RJ 1982 In utero exposure to diethylstilbestrol: adverse effects on the reproductive tract and reproductive performance in male and female offspring. American Journal of Obstetrics and Gynecology 142: 905–921.

Stone R 1994 Environmental estrogens stir debate. Science 265: 308–310.

Toppari J, Larsen JC, Christiansen P, Giwercman A, Grandjean P, Guillette LJ, Jegou B, Jensen TK, Jouannet P, Keiding N, Leffers H, McLachlan JA, Meyer O, Muller J, Rajpert-De Meyts E, Scheike T, Sharpe RM, Sumpter J & Skakkebaek NE 1996 Male reproductive health and environmental xenoestrogens. Environmental Health Perspectives 104: 741–803.

Vierula M, Niemi M, Keiski A, Saatanen M, Saarikoski S & Suominen J 1996 High and unchanged sperm counts of Finnish men. International Journal of Andrology 19: 11–17.

Walter P, Green S, Greene G, Krust A, Bornert J-M, Jeltsch J-M, Straub A, Jensen E, Scrace G, Waterfield M & Chambon P 1985 Cloning of the human estrogen receptor cDNA. Proceedings of the National Academy of Sciences, USA 82: 7889–7893.

White R, Jobling S, Hoare S, Sumpter J & Parker M 1994 Environmentally persistent alkylphenolic compounds are estrogenic. Endocrinology 135: 175–182.

20

PROTEASE-PROTEASE INHIBITOR INTERACTIONS IN SERTOLI CELL-GERM CELL CROSSTALK

T. K. Monsees, W. B. Schill, and W. Miska

Center of Dermatology and Andrology
Justus Liebig University
Gaffkystr. 14, 35385 Giessen
Germany

1. SUMMARY

Peritubular cells, Sertoli cells, and germ cells of the seminiferous tubule synthesize and secrete several proteases and protease inhibitors. Experimental evidence suggests that the complex network of proteolytic enzyme activity and their regulation by protease inhibitors play an important role in male reproduction. Interaction between protease and protease inhibitors seems to play an important role in remodeling and restructuring of the seminiferous tubule during spermatogenesis. Controlled proteolytic activity is also involved in the migration of germ cells from the basal compartment to the lumen of the seminiferous epithelium, and in the release of spermatids during spermiation. The recently reported occurrence of Sertoli cell membrane-associated proteases indicate the possible involvement of regulatory peptide systems within the testis. This view is supported by the detection of all components of one of these paracrine systems, the kallikrein-kinin system, in cells of the seminiferous tubule.

2. INTRODUCTION

Proteases and protease inhibitors are ubiquitously distributed in all biological tissues and fluids and play important roles in various biological systems (Fig 1). Based on a comparison of active sites, mechanism of action, and three-dimensional structure, four mechanistic classes of proteases are recognized to date: (i) serine proteases, (ii) cysteine proteases, (iii) aspartic proteases, and (iv) metallo proteases.

In the seminiferous tubules of the testis, interactions between Sertoli cells, peritubular cells, and developing germ cells play a central role in spermatogenesis. A growing body of experimental evidence suggests that proteases and proteases inhibitors are an im-

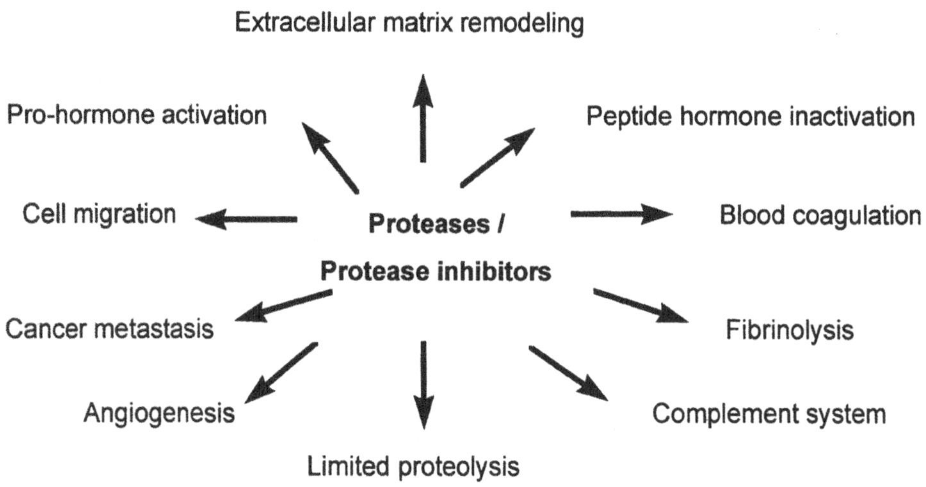

Extracellular matrix remodeling

Pro-hormone activation

Peptide hormone inactivation

Cell migration ← **Proteases /** → Blood coagulation

Protease inhibitors

Cancer metastasis

Fibrinolysis

Angiogenesis

Complement system

Limited proteolysis

Figure 1. Multiple physiological functions of proteases and protease inhibitors.

portant part of these interactions. This article will briefly review the proteolytic enzymes and their inhibitors that have been characterized in the seminiferous tubule. Their physiological functions in the interaction between Sertoli cells and developing germ cells will be discussed. This paper will focus on the detection of membrane-associated metallo proteases by means of the high performance liquid chromatography technique. These metallo proteases may be a part of paracrine peptide hormone systems such as the kallikrein-kinin system or the renin angiotensin system.

3. PROTEASES AND PROTEASE INHIBITORS IN THE SEMINIFEROUS TUBULE

To date only a limited number of proteases and protease inhibitors have been characterized in the seminiferous tubule. Sertoli cells are known to secrete matrix metalloproteases, plasminogen activators, and cyclic protein 2. On the other hand, they secrete protease inhibitors such as TIMP-2, α_2-macroglobulin and cystatin C. Matrix metalloproteinases and plasminogen activators are also synthesized by peritubular cells. In addition, these cells secrete the protease inhibitors TIMP-2 and PAI-1. The inhibitors α_2-macroglobulin and cystatin C have been detected in germ cells. Several germ cell proteases have been investigated, including metalloendopeptidases, cathepsins and acrosin, which are mostly located within the acrosome. These proteases would appear to play a role during acrosome reaction and fertilization but probably not during spermatogenesis.

3.1. Plasminogen Activators

Plasminogen activators (PA) are serine proteases that liberate the protease plasmin from the inactive precursor plasminogen, which is present in plasma. Both types of plasminogen activators are synthesized and secreted by rat Sertoli cells, which are the primary cellular sources of PAs in the testis. The 45 kD urokinase type (UK) is predominantly secreted under basal conditions, whereas the 70 kD tissue-type (tPA) is mainly produced after stimulation with FSH or cAMP derivatives (Hettle et al., 1986). Highest rates in secretion of PAs occur at stages VII and VIII of the cycle, in which the Sertoli cells are most responsive to FSH stimulation (Lacroix et al., 1981). This enhanced PA activity is dependent upon the presence of preleptotene spermatocytes (Vihko et al., 1984). These findings suggest that PAs are involved in tissue restructuring and cell migration at these stages of spermatogenesis.

The involvement of PAs in germ cells is unclear. Faint staining of tPA was observed by immunocytochemical techniques in rat spermatocytes (Vihko et al., 1988). However, no PA activity could be detected in elongated spermatids.

3.2. Cyclic Protein 2/Cathepsin L

Sertoli cells in culture secrete cyclic protein 2 (CP-2). Based upon analysis of cDNA sequences for CP-2 mRNA and cathepsin L mRNA, cyclic protein 2 is regarded as the pro-enzyme form of the lysosomal cysteine protease cathepsin L (Erickson-Lawrence et al., 1991). Cathepsin L requires an acidic pH for optimal enzyme activity and is involved in intracellular proteolytic processes. Cathepsin L can, however, also be secreted, and may then degrade proteins in the extracellular environment (Thomas and Davies, 1989). Sertoli cells are the only cells in the rat seminiferous tubule that contains detectable amounts of CP-2 mRNA or CP-2 protein. Synthesis and secretion rates of CP-2 were shown to be stage- and development-dependent. Highest concentrations of CP-2 mRNA and CP-2 protein were detected at stages VI-VII of the seminiferous cycle; none were found at stages XII to II (Zabludoff et al., 1990; Erickson-Lawrence et al., 1991). Synthesis rates of CP-2 increase during gonadal maturation, with highest rates in Sertoli cells obtained from 38–45 days-old rats. During this period, the number of spermatids rapidly increases. The expression of CP-2 is controlled by the presence of elongated spermatids (Maguire et al., 1993).

3.3. Matrix Metalloproteinases

Matrix metalloproteinases (MMPs) are implicated in processes of tissue development and restructuring. All known MMPs have some degree of homology, contain a zinc ion at their active site, are calcium-dependent, and are inhibited by specific tissue inhibitors of metalloproteinases (TIMPs). However, they differ in their substrate-specificity. MMPs are secreted in a latent form that needs a regulated activation before protease activity can be expressed. The most predominant matrix metalloproteinases secreted by cultured rat Sertoli cells are the 72 kD and 92 kD type IV pro-collagenases. They have been identified with MMP-2 and MMP-9 (gelatinase B), respectively (Sang et al., 1990; Hoeben et al., 1996). MMP-2 (type IV collagenase, gelatinase A) degrades preferentially type IV collagen, the predominant collagen in basement membrane. The expression of MMP-2 mRNA by Sertoli cells is regulated by peritubular cells and by specific components of the extracellular matrix. MMP-2 and MMP-9 are also secreted by peritubular cells, but the pattern of the secreted matrix metalloproteinases is different from that secreted by Sertoli cells (Ailenberg et al., 1991). Also, the factors controlling the synthesis and secretion of type IV pro-collagenase differ in both cell types.

3.4. Metalloproteinases

Metalloproteinases are endopeptidases that contain a metal ion (generally zinc) in their active center. They can cleave small bioactive peptides and are usually bound to cell membranes, although the active enzyme can also be released. *Neutral metalloendopeptidase* (NEP, EC 3.4.24.11) is detected in seminal fluid, and in homogenates of testis, epididymis and prostate gland (Erdös et al., 1985). Rat testis homogenate also contains high amounts of NEP isoforms: namely NEP 24.15 (70 kD MW) and NEP 24.16 (72 kD MW), which were first identified in the brain (Orlowski et al., 1989; Rodd and Hersh, 1995). The amino acid sequences of the NEPs have been deduced from the respective cDNAs (Malfroy et al., 1987; Pierotti et al., 1990; Dauch et al., 1996). They all contain a HEXXH motif at their active site, in which the two histidine residues coordinate the zinc ion and the glutamate is involved in the bond-breaking process.

Two isoenzymes of the *angiotensin-converting enzyme* (ACE, kininase II) are known: a 140 kD form synthesized in several somatic tissues including testis (Erdös et al., 1985). The testicular 80 kD isoform is step-specific expressed in rodent germinal cells during spermatogenesis. Testicular ACE mRNA and protein are observed only after completion of meiosis and are first detected in round spermatids of 23 days-old rats. Maximal expression is occurred during the acrosome phase (steps 8–12), whereas ACE mRNA is no longer expressed in spermatids beyond step 14. In contrast, the gene product ACE is detectable during the end of spermatogenesis and also in ejaculated spermatozoa (Sibony et al., 1994). Testicular ACE production in rodents is dependent on sexual maturation and appears be controlled by the pituitary (Strittmatter et al., 1985). Somatic and testicular ACE are encoded by the same gene, which is transcribed into two different mRNAs (Hubert et al., 1991). In testis, the isoenzymes are expressed in a 1:4 ratio (Lanzillo et al., 1985). ACE is important in male reproduction because homozygous male mice lacking both ACE isozymes have reduced fertility (Esther et al., 1996).

N-arginine dibasic convertase (NRD convertase) is a zinc containing metalloendopeptidase that cleaves at the N-terminus of arginine residues. NRD convertase was detected in round and elongating spermatids of the rat seminiferous epithelium. In situ hybridization demonstrated NRD convertase mRNA in earlier stages of spermatogenesis (Chesneau et al., 1994).

3.5. Tissue Inhibitors of Metalloproteinases

Tissue inhibitors of metalloproteinases (TIMPs) are proteins that inhibit members of the group of matrix metalloproteinases (MMPs). The 21 kD protein TIMP-2 is secreted by rat Sertoli cells in culture (Ailenberg et al., 1991). TIMP-2 mRNA levels increased steadily from 3 to 60 days of age during testicular development. However, this increase is not the result of an up-regulation by germ cells as demonstrated in Sertoli cell-Germ cell cocultures (Grima et al., 1996). TIMP-2 inhibited both the latent and the active form of type IV collagenase. In contrast, no other metalloproteinases secreted by Sertoli cells or by peritubular cells were inhibited by TIMP-2, demonstrating the specificity of this inhibitor. TIMP-2 is also secreted by rat peritubular cells in culture.

3.6. α_2-Macroglobulin

α_2-Macroglobulin is a nonspecific protease inhibitor and binding protein for peptide hormones. Sertoli cells synthesize and secrete an α_2-macroglobulin which has the same properties as the liver-derived α_2-macroglobulin present in serum (Cheng et al., 1990). Sertoli cell-derived α_2-macroglobulin was also found in seminiferous tubular fluid and in rete testis fluid at concentrations close to those in rat serum. However, the regulatory mechanism controlling the synthesis and secretion of α_2-macroglobulin by Sertoli cells are different from those controlling the formation of α_2-macroglobulin by liver (Stahler et al., 1991). α_2-Macroglobulin was further detected around the heads of elongated spermatids. Staining was most intense in stages I-VI and only faint in all other stages except stages VIII-X, where no immunoreactivity was detectable. α_2-Macroglobulin was first present in Sertoli cells of 21 days and its concentration increased with age of the rat, suggesting a developmental controlled expression. No immunostaining was observed adjacent to round spermatids and spermatocytes (Zhu et al., 1994).

3.7. Cystatin C

Sertoli cells synthesized and secreted cystatin C, a 12 kD protein that inhibits cysteine proteases. Sections of rat testis showed intense immunohistochemical staining of cystatin C in Sertoli cells but not in spermatogonia and spermatocytes. Sertoli cells contained a 700-nucleotide cystatin C mRNA, whereas a mixed population of spermatids and spermatocytes contained a 550-nucleotide transcript. Northern blot analysis from stage-synchronized testis demonstrated that steady-state RNA levels were lowest in stages VI-VIII of the cycle (Tsuruta et al., 1993). It is suggested that the role of cystatin C in the testis may be to inhibit the proteolytic activity of the cysteine protease cathepsin L in all stages of the testis except stages VI-VII.

3.8. Plasminogen Activator Inhibitors

Three immunologically distinct plasminogen activator inhibitors (PAI-1, PAI-2 and PAI-3) have been identified so far. The 55 kD protein, PAI-1, inhibits urokinase and tissue plasminogen activator, but does not inhibit the activities of other proteases such as trypsin or plasmin. The cDNA sequences coding for human PAI-1 and rat PAI-1 have been identified (Zeheb and Gelehrter, 1988). PAIs belong to the superfamily of SERPINS (SERine Proteinase INhibitors) that interact with their target proteinases at a reactive site located within a loop structure that is exposed on the surface of the enzyme. This loop structure is only accessible after a conformational change (activation) of the latent form of the PA (Potempa et al., 1994).

Cultured testicular peritubular cells synthesized and secreted PAI-1, which inhibits the activity of the active plasminogen activators released into the medium by Sertoli cells (Hettle et al., 1988). Levels of mRNA for PAI-1 in peritubular cells are highly increased after treatment with TGF-β, which was similar to the effect of TGF-β reported in lung fibroblasts (Nargolwalla et al., 1990). In contrast, PAI-1 mRNA levels in peritubular cells declined in the presence of cyclic AMP derivatives.

4. HPLC TECHNIQUE: ONE WAY TO EXPLORE CELLULAR PROTEASES

A different approach to identify proteases involved in Sertoli cell metabolism was recently introduced by Monsees et al. (1996a, 1996b) by using the HPLC technique. High performance liquid chromatography (HPLC) is an established method for metabolic studies of peptides. The HPLC cleaving pattern provides information about type and relative amounts of the formed metabolites. To gain information concerning the involved proteases, specific protease inhibitors are employed in peptide degradation experiments. Regarding their specificity, protease inhibitors can be divided into three groups: (i) those, that react with more than one class of proteases, (ii) those that are specific for one of the classes, and (iii) those that are highly selective for a single protease.

In brief, the experimental setup was as follows: Sertoli cells were prepared from 18-day-old Sprague-Dawley rats according to published procedures. After 3 days in culture, the Sertoli cells were treated with a hypotonic Tris-HCl solution (20 mM, pH 7.5, 5 min) to remove germ cell contaminants. On day six of culture, cells were washed and incubated at 34°C in 10 mM phosphate buffered saline, pH 7.3, supplemented with 1.42 mM $CaCl_2$ and 1 mM $MgCl_2$. Sertoli cells (500,000/cm^2) were preincubated with protease inhibitors for 20 min, then 50 μM of peptide was added. At several time intervals, 200-μl aliquots were removed, supplemented with 50 μl of 10% aqueous trifluoroacetic acid and analyzed by means of HPLC. Degradation products were applied to a reversed phase column (Merck LiChrosorb RP-8, 250 x 4 mm, 5 μm particle size) and separated isocratically. The degradation products were identified at 214 nm by co-elution with peptide standards.

Several physiologically active peptides were degraded by proteases of cultured Sertoli cells. Cleaved peptides include bradykinin, the angiotensins I, II, and III, substance P, met- and leu-enkephaline. In contrast, no peptidolysis of oxytocin was observed during the incubation period. Fig. 2 shows a representative HPLC chromatogram demonstrating bradykinin degradation and appearance of metabolites. BK(1–4), BK(1–5), BK(8–9), BK(1–7), and BK(6–9) were identified as the main cleavage products. The metabolite BK(1–8) was detected in small amounts under different HPLC conditions (Fig. 2b). The results indicated that the main primary cleavages of bradykinin occurred at the Pro^7-Phe^8, Phe^5-Ser^6, and Gly^4-Phe^5 bonds. Using purified enzymes it was previously shown that neutral metalloendopeptidase (NEP 24.11) can hydrolyze bradykinin at the Pro^7-Phe^8 and Gly^4-Phe^5 bonds (Stephenson and Kenny, 1987). A candidate for successive cleavage of the terminal dipeptides of Pro^7-Phe^8 and Phe^5-Ser^6 bonds is the kininase type II (Boettger et al., 1993), which is also known as angiotensin converting enzyme (ACE). To demonstrate the occurrence of the proteases NEP and ACE, Sertoli cells were incubated with bradykinin and specific protease inhibitors. Phosphoramidon and thiorphan, which are specific inhibitors of NEP, strongly suppressed BK degradation. The ACE specific inhibitors, captopril and enalapril, showed a small but significant reduction of BK peptidolysis (Monsees et al., 1996b). The release of BK(6–9) by cleavage of the Phe^5-Ser^6 bond might result from the action of metalloendopeptidases 24.15 and/or 24.16 which are phosphoramidon-insensitive. Both enzymes were previously detected in testis homogenate (Orlowski et al., 1989; Rodd and Hersh, 1995). The metabolite BK(1–8) occurred by cleavage of the C-terminal arginine by the action of a carboxypeptidase such as CPN or CPM. CPN (kininase type I) is usually found in plasma, whereas CPM is a membrane-bound carboxypeptidase found in many tissues and cultured cells (Tan et al., 1989).

The effect of a variety of group- or enzyme-specific protease inhibitors on the degradation of angiotensins by cultured Sertoli cells in shown in Fig. 3. Angiotensin I and II were mainly cleaved by metallo proteases which are inhibited by oPhe. These metallo proteases include ACE (inhibited by captopril) and NEP 24.11 (inhibited by phosphoramidon). A bestatin-sensitive aminopeptidase is further involved in the degradation of AII. AII was also cleaved by aspartic proteases whose activities could be blocked by pepstatin. In addition, PMSF-sensitive cysteine or serine proteases were also involved in the peptidolysis of all angiotensins. This protease activity is probably not caused by cathepsin L, which should be sensitive towards the cysteine class-specific inhibitor E64. However, this was not observed. Moreover, the degradation experiments were performed at pH 7.3 but cathepsin L requires an acid pH for optimal activity. It has been shown that cell-free, Sertoli cell-conditioned media contain no or only very limited peptidolytic activity (Monsees

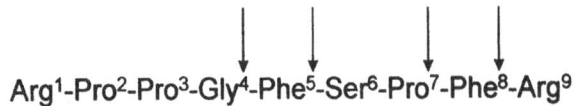

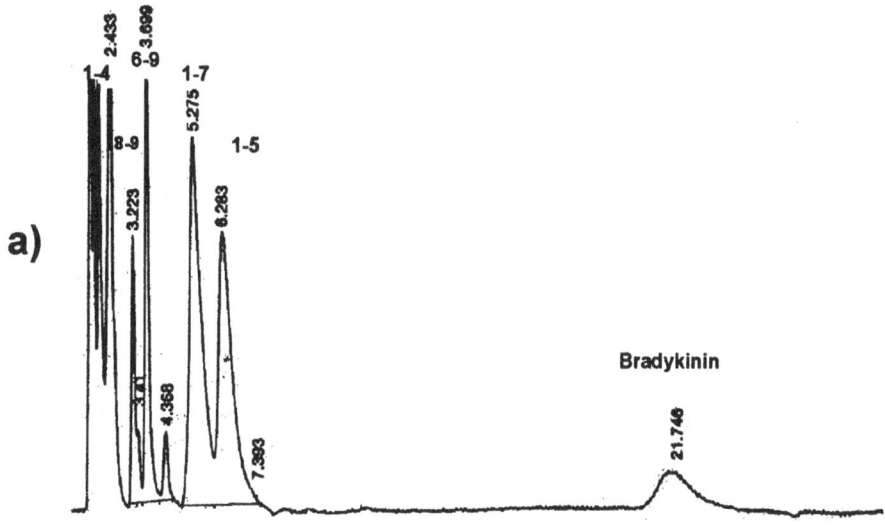

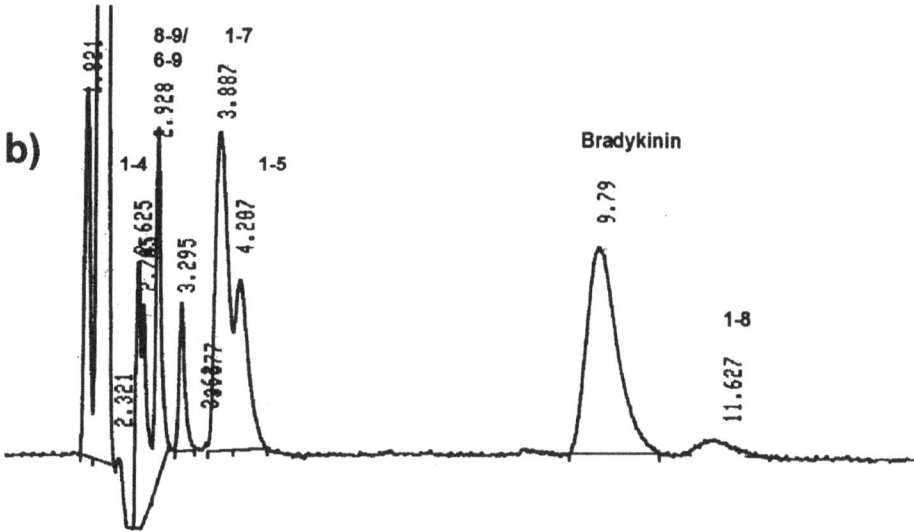

Figure 2. Molecular structure of bradykinin (BK) with identified cleaving sites (arrows) and HPLC chromatography of rat Sertoli cell supernatants obtained after incubation with BK. Reversed phase HPLC was performed isocratically with 24% acetonitrile and 74% 75 mM NaClO$_4$, NaH$_2$PO$_4$ as indicated, pH 2.2, at a flow rate of 1.3 mL/min. The degradation products were monitored at 214 nm. A) 2 h incubation, 50 µM BK, 5 mM NaH$_2$PO$_4$, b) 3 h incubation, 100 µM BK, 50 mM NaH$_2$PO$_4$.

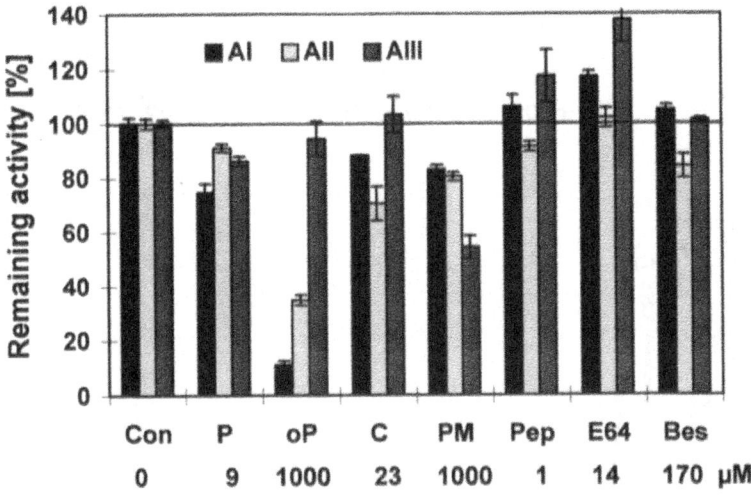

Figure 3. Influence of protease inhibitors on degradation of angiotensins (50 μM) by cultured Sertoli cells (500,000/cm²) after 6 h incubation at 34°C. Abbreviation, used concentration and specificity are: control (Con), phosphoramidon (P; 9 μM; NEP 24.11), o-phenanthroline (oPhe; 1000 μM; metallo proteases), captopril (C; 23 μM; ACE), PMSF (PM; 1000 μM; serine and cysteine proteases), pepstatin (Pep; 1 μM; aspartic proteases), E64 (14 μM; cysteine proteases), bestatin (Bes; 170 μM; aminopeptidases). Control = 100%; data are mean ± SD (bars) values; (n = 2–3).

et al., 1996b). This indicates that the proteases responsible for peptide degradation were not to any great degree released by Sertoli cells, which is consistent with the fact that the above mentioned metallo proteases are membrane-associated.

These results indicate that degradation of biological active peptides by Sertoli cells is controlled by the combined action of several membrane-bound metalloproteases such as neutral metalloendopeptidases, kininase type II, aminopeptidases, and carboxypeptidases. In addition, activities of serine and aspartic proteases were also detected.

5. POSSIBLE PHYSIOLOGICAL FUNCTIONS OF PROTEASES AND PROTEASE INHIBITORS IN SPERMATOGENESIS

In the testis, proteolytic enzymes seem to be involved in (i) remodeling of the seminiferous tubule during testicular development and maturation, (ii) maintenance of the integrity of the Sertoli cell barrier, (iii) migration of germ cells during spermatogenesis, (iv) release of spermatids in the process of spermiation, and (v) activation and inactivation of paracrine peptide hormone systems. The maintenance of these physiological functions requires controlled activation and inactivation of the proteases involved. Proteases in solution are usually effectively inhibited by corresponding protease inhibitors. In contrast, receptor-bound proteases retain activities as long as concentrations of substrates in the immediate environment exceed concentrations of inhibitors.

Activities of many proteases are regulated via a cascade. In the plasminogen activator system (Fig. 4), the transduction of a physiological signal led to the conversion of the precursor pro-plasminogen activator to the protease plasminogen activator, which in turn converts plasminogen to plasmin. Plasmin then catalyses the conversion of pro-col-

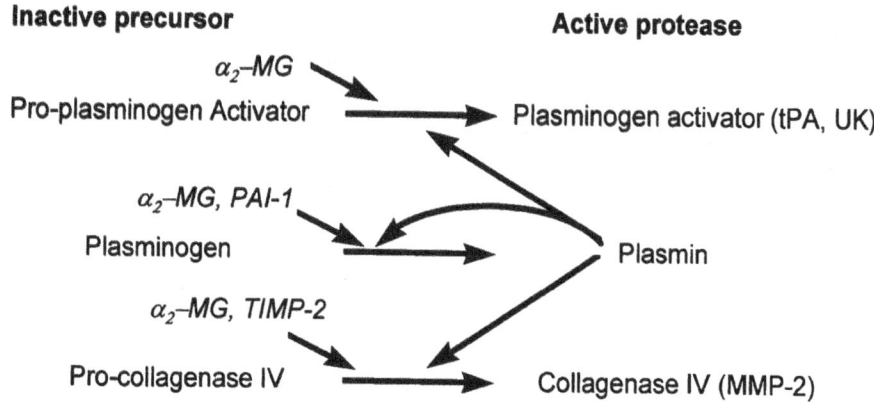

Figure 4. Scheme of the plasminogen activator cascade.

lagenase IV to active collagenase IV (Fritz et al., 1991). However, plasmin can also cata-lyze the activation of PA, and even the release of more plasmin from plasminogen (auto-catalysis). The duration and breadth of the cascade are fine-tuned at each step by inhibitors. The degree of amplification of the physiological signal is determined by the number of steps of the cascade. In the seminiferous epithelium, PA activity is regulated by PAI-1, secreted by peritubular cells, and by Sertoli cells derived α_2-macroglobulin, which also inhibits other proteases such as plasmin and collogenase IV. Net PA activity is further modulated by tissue inhibitors of metalloproteinases (TIMPs) and factors such as FSH, cAMP or TGF-β.

Tight junctions between neighboring Sertoli cells divide the seminiferous epithelium into a basal and an adluminal compartment. Besides this so-called "blood-testis barrier" or "Sertoli cell barrier" there are desmosome-like junctions and gap junctions between Ser-toli cells and germ cells. Ectoplasmic specializations and tubulobulbar complexes are formed between Sertoli cells and elongated spermatids (Russell et al., 1990). During sper-matogenesis the developing germ cells are moving from the basal compartment to the lu-men of the seminiferous tubule. During this translocation process, the various junctions between Sertoli cells and germ cells must be cleaved and reformed periodically. Sertoli cell derived proteases and protease inhibitors such as TIMP-2 and α_2-macroglobulin are believed to play an important role in the initiation and regulation of this process (Russell 1993; Grima et al., 1996). The involvement of controlled proteolysis has been demon-strated in other tissue remodeling events such as embryogenesis, angiogenesis and metas-tasis (Rifkin et al., 1984).

In the seminiferous epithelium, the activities of several proteases and protease in-hibitors display pronounced cyclic changes. Highest activities of plasminogen activator (PA, stages VII-VIII) and cyclic protein-2 (CP-2, stages VI-VII) are observed at stages where germinal cells are initially detached from the basement membrane and transported through the Sertoli cell barrier into the adluminal compartment. On the other hand, lowest activities of the protease inhibitors cystatin (stages VI-VIII) and α_2-macroglobulin (VII-X) are detected at these timepoints of the cycle. Thus, net protease activity is highest at stages VI-VIII. Therefore, the plasminogen activator cascade and CP-2/cathepsin L are suggested to play a role in the degradation of adhesion molecules that (i) anchor prelep-totene spermatocytes to the basement membrane of the seminiferous tubule (Fritz and Ailenberg, 1991), and (ii) bind spermatids to Sertoli cells in preparation for the release of spermatids into the lumen (Vihko et al., 1984; Erickson-Lawrence et al., 1991). Inhibition of proteolytic activity by α_2-macroglobulin could favor the maintenance of the integrity of the Sertoli cell barrier (Ailenberg and Fritz, 1989). α_2-Macroglobulin also binds many proteases; thus it may protect the seminiferous epithelium from damage by acrosomal proteases released from elongated spermatids during the maturation process (Zhu et al., 1994).

Membrane-associated metalloproteases such as neutral metalloendopeptidases, carboxy-peptidases, aminopeptidases and angiotensin converting enzyme (kininase type II) are a central part of paracrine peptide hormone systems. The kallikrein kinin system (KKS) and the renin angiotensin system (RAS) control many physiological and pathological processes such as regulation of blood pressure, fluid and tissue homeostasis, and inflammation. In the case of the KKS, the highly specific serine protease tissue kallikrein liberates the peptide kinin from the inactive precursor kininogen. The physiological effects of kinins are mediated by specific, membrane-bound receptors. Kinins are rapidly inactivated by membrane-associated proteases (kininases). Besides peptide-degrading enzymes located on Sertoli cells, all other components of the KKS seem to be present within the rat testis. Kininogen is ubiquitously distributed in all body fluids and tissues including the testis (Hossain et al., 1995). The protease kallikrein has been detected in spermatocytes and early spermatids (Monsees et al., 1996c). mRNA levels for the kallikrein-like protease P1 were first detectable in testis homogenate of 30-day old rats and reach maximal levels in adult animals (Clements et al., 1990). The mRNA encoding the bradykinin B_2 receptor was detected in testis homogenate, which indicates the expression of the bradykinin receptor in testicular cells (McEachern et al., 1991). Recently, we demonstrated the presence of the B_2 mRNA in cultured Sertoli cells (Fig. 5). Therefore, the KKS may play a role in the regulation of Sertoli cell function and probably in the crosstalk between Sertoli and germ cells. However, mice that are homozygous for the disruption of the B_2 receptor are fertile and have normal litter size (Borkowski et al., 1995). Hence, further investigation must be performed to clarify the possible physiological role of the bradykinin B_2 receptor and the KKS in spermatogenesis.

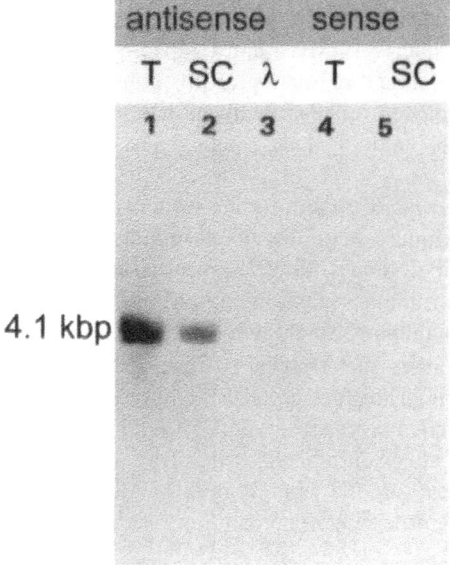

Figure 5. Chemiluminogram of Northern blot hybridized with DIG-labeled bradykinin B_2 receptor probe (395 bp) .15 μg of total RNA was loaded in every lane. Lanes: 1 and 4 adult rat testis homogenate; 2 and 5 cultured Sertoli cells; 3 marker λ HindIII. Lanes 1 and 2 were hybridized with the antisense probe. Lanes 4 and 5 were hybridized with the sense probe.

All components of the RAS including renin (a proteinase that liberates angiotensin I from angiotensinogen), ACE, and the angiotensin peptides I, II and III have been detected in testicular Leydig cells (Pandey et al., 1984). It is suggested that angiotensin I, secreted by Leydig cells may also act on Sertoli cells. Targeted disruption of both ACE genes led to homozygous mice that produced significantly smaller litters than wild-type animals. However, these mice are not infertile and testis morphology, sperm counts and motility are normal. Also, the male fertility defect is possibly not related to the markedly reduced low blood pressure observed in these mice (Esther et al., 1996).

6. ACKNOWLEDGMENTS

The authors thank Mrs. G. Thiele for her excellent technical assistance, Drs. A. Winkler and WE Siems for their help and advise with the B_2 receptor probe, and the Deutsche Forschungsgemeinschaft for financial support (DFG Mo 693/3-1).

7. REFERENCES

Ailenberg M & Fritz IB 1989 Influences of follicle-stimulating hormone, proteases and antiproteases on permeability of the barrier generated by Sertoli cells in a two-chambered assembly. Endocrinology 124:1399–1407.

Ailenberg M, Stetler-Stevenson WG & Fritz IB 1991 Secretion of latent type IX procollagenase and active type IV collagenase by testicular cells in culture. Biochemical Journal 279:75–80.

Boettger A, Kertscher U, Steinmann C, Baeger U, Siems WE & Heder G 1993 Degradation of bradykinin in semen of ram and boar. Biochemical Pharmacology 45:1983–1988.

Borkowski JA, Ransom RW, Seabrook GR, Trumbauer M, Chen H, Hill RG, Strader CD & Hess JF 1995 Targeted disruption of a B_2 bradykinin receptor gene in mice eliminates bradykinin action in smooth muscle and neurons. The Journal of Biological Chemistry 270:13706–13710.

Cheng CY, Grima J, Strahler MS, Guglielmotti A, Silvestrini B & Bardin CW 1990 Sertoli cell synthesize and secretes a protease inhibitor, α_2-macroglobulin. Biochemistry 29:1063–1068.

Chesneau V, Pierotti AR, Prat A, Gaudoux F, Foulon T & Cohen G 1994 N-Arginine dibasic convertase (NRD convertase): A newcomer to the family of processing endopeptidases. An overview. Biochimie 76:234–240.

Clements JA, Matheson BA & Funder JW 1990 Tissue-specific developmental expression of the kallikrein gene family in the rat. The Journal of Biological Chemistry 265:1077–1081.

Dauch P, Vincent JP & Checler F 1996 Molecular cloning and expression of rat brain endopeptidase 3.4.24.16 The Journal of Biological Chemistry 270:27266–27271.

Erdös EG, Schulz WW, Gafford JT & Defendini R 1985 Neutral metalloendopeptidase in human male genital tract. Comparison to angiotensin I-converting enzyme. Laboratory Investigation 52:437–447.

Erickson-Lawrence M, Zabludoff SD & Wright WW 1991 Cyclic protein-2, a secretory product of rat Sertoli cells, is the proenzyme form of cathepsin L. Molecular Endocrinology 5:1789–1798.

Esther CR, Howard TE, Marino EM, Goddard JM, Capecchi MR & Bernstein KE 1996 Mice lacking angiotensin-converting enzyme have reduced low blood pressure, renal pathology, and reduced male fertility. Laboratory Investigation 74:953–965.

Fritz IB & Ailenberg M 1991 Plasminogen activators and metalloproteinases in the male reproductive system. In Plasminogen Activators: From Cloning to Therapy, pp 67–80. Eds R Abbate, T Barni & A Tsafriri. Raven Press, New York.

Grima J, Calagno K & Cheng CY 1996 Purification, cDNA cloning, and developmental changes in the steady-state mRNA level of rat testicular tissue inhibitor of metalloproteases-2 (TIMP-2). Journal of Andrology 17: 263–275.

Hettle JA, Waller EK & Fritz IB 1986 Hormonal stimulation alters the type of plasminogen activator produced by Sertoli cells. Biology of Reproduction 34:895–904.

Hettle JA, Balekjian E, Tung PS & Fritz IB 1988 Rat testicular peritubular cells in culture secrete an inhibitor of plasminogen activator activity. Biology of Reproduction 38:359–371.

Hoeben E, Aelst IV, Swinnen JV, Opdenakker G & Verhoeven G 1996 Gelantinase A secretion and its control in peritubular and Sertoli cells cultures: effects of hormones, second messengers and inducers of cytokine production. Molecular and Cellular Endocrinology 118:37–46.

Hossain AM, Whitman GF & Khan I 1995 Kininogen present in rat reproductive tissues is apparently synthesized by the liver, not by the reproductive system. American Journal of Obstetrics and Gynecology 173:830–834.

Hubert C, Houot AM, Corvol P & Soubrier F 1991 Structure of the angiotensin I-converting enzyme gene. Two alternative promoters correspond to evolutionary steps of a duplicated gene Journal of Biological Chemistry 266:15377–15383.

Lacroix M, Parvinen M & Fritz IB 1981 Localization of testicular plasminogen activator in discrete portions (stage VII and VIII) of the seminiferous tubule. Biology of Reproduction 25:143–146.

Lanzillo JJ, Stevens J, Dasarathy Y, Yotsumoto H & Fanburg BL 1985 Angiotensin-converting enzyme from human tissues. Journal of Biological Chemistry 260:14938–14944.

Maguire SM, Millar MR, Shape RM & Saunders PTK 1993 Stage-dependent expression of mRNA for cyclic protein 2 during spermatogenesis is modulated by elongate spermatids. Molecular and Cellular Endocrinology 94:79–88.

Malfroy B, Schofield PR, Kuang WJ, Seeburg PH, Mason AJ & Henzel WJ 1987 Molecular cloning and amino acid sequence of rat enkephalinase. Biochemical and Biophysical Research Communications 144:59–66.

McEachern AE, Shelton ER, Bhakta S, Obernolte R, Bach C, Zuppan P, Fujisaki J, Aldrich RW & Jarnagin K 1991 Expression cloning of the rat B$_2$ bradykinin receptor. Proceedings of the National Academy of Sciences USA 88:7724–7728.

Monsees TK, Miska W & Schill WB 1996a Characterization of kininases in testicular cells. Immunopharmacology 32:169–171.

Monsees TK, Miska W & Schill WB 1996b Enzymatic digestion of bradykinin by rat Sertoli cell cultures. Journal of Andrology 17:375–381.

Monsees TK, Pena P, Schill WB & Miska W (1996c) Localization of kallikrein in rat testis and epididymis. Reprod Dom Anim 30:387

Nargolwalla C, McCabe D & Fritz IB 1990 Modulation of levels of messenger RNA for tissue-type plasminogen activator in rat Sertoli cells, and levels of messenger RNA for plasminogen activator inhibitor in testis peritubular cells. Molecular and Cellular Endocrinology 70:73–80.

Orlowski M, Reznik S, Ayala J & Pierotti AR 1989 Endopeptidase 24.15 from rat testes. Biochemical Journal 261:951–958.

Pandey KN, Misono KS & Inagami T 1984 Evidence for intracellular formation of angiotensins: coexistence of renin and angiotensin-converting enzyme in Leydig cells of rat testis. Biochemical and Biophysical Research Communications 122:1337–1343.

Pierotti A, Dong KW, Glucksman MJ, Orlowski M & Roberts JL 1990 Molecular cloning and primary structure of rat testis metalloendopeptidase EC 3.3.24.15. Biochemistry 29:10323–10329.

Potempa J, Korzus E & Travis J 1994 The serpin superfamily of proteinases inhibitors: structure, function, and regulation. The Journal of Biological Chemistry 269:15957–15960.

Rifkin DB, Moscatelli D, Gross J & Jaffe E 1984 Proteases, angiogenesis, and invasion. In Cancer invasion and metastasis: Biologic and therapeutic aspects, pp 187–194. Eds LG Nicholson & L Milas. Raven Press, New York.

Rodd D & Hersh LB 1995 Endopeptidase 24.16B. A new variant of endopeptidase 24.16. The Journal of Biological Chemistry 270:100056–100061.

Russell LD, Ettlin RA, Hikim AS & Clegg ED (eds) 1990 Histological and histopathological evaluation of the testis, pp 1–38. Cache River Press, Clearwater.

Russell LD, 1993 Morphological and functional evidence for Sertoli-germ cell relationships. In The Sertoli Cell, pp 365–390. Eds LD Russell & MD Griswold. Cache River Press, Clearwater.

Sang QX, Stetler-Stevenson WG, Liotta LA & Byers SW 1990 Identification of type IV collagenase in rat Testicular cell culture: influence of peritubular-Sertoli cell interactions. Biology of Reproduction 43:956–964.

Sibony M, Segretain D & Gasc JM 1994 Angiotensin-converting enzyme in murine testis: Step-specific expression of the germinal isoform during spermatogenesis. Biology of Reproduction 50:1015–1026.

Stahler MS, Cheng CY, Morris PL, Cailleau J, Verhoeven G & Bardin CW 1991 α_2-macroglobulin, a multifunctional protein of the seminiferous tubule. Annals of the New York Academy of Sciences 626:73–80.

Stephenson SL & Kenny AJ 1987 Hydrolysis of angiotensins, bradykinin, substance P and oxytocin by pig kidney microvillar membranes. Biochemical Journal 241:237–247.

Strittmatter SM, Thiele EA, De-Souza EB & Snyder SH 1985 Angiotensin-converting enzyme in the testis and epididymis: differential development and pituitary regulation of isoenzymes. Endocrinology 117:1374–1379.

Tan F, Chan SJ, Steiner DF, Schilling JW & Skidgel RA 1989 Molecular cloning and sequencing of the cDNA for human membrane-bound carboxypeptidase M. The Journal of Biological Chemistry 22:13165–13170.

Thomas GJ & Davies M 1989 Potential role of human kidney cortex cysteine proteinases in glomerular basement membrane degradation. Biochemical et Biophysical Acta 990:246–253.

Tsuruta JK, O'Brien DA & Griswold MD 1993 Sertoli cell and germ cell cystatin C: stage-dependent expression of two distinct messenger ribonucleic acid transcripts in rat testis. Biology of Reproduction 49:1045–1054.

Vihko KK, Suominen JJO & Parvinen M 1984 Cellular regulation of plasminogen activator secretion during spermatogenesis. Biology of Reproduction 31:383–389.

Vihko KK, Kristensen P, Dano K & Parvinen M 1988 Immunohistochemical localization of urokinase-type plasminogen activator in Sertoli cells and tissue-type plasminogen activator in spermatogenic cells in the rat seminiferous epithelium. Developmental Biology 126:150–155.

Zabludoff SD, Erickson-Lawrence M & Wright WW 1990 Sertoli cells, proximal convoluted tubules in the kidney, and neurons in the brain contain cyclic protein-2. Biology of Reproduction 43:15–24.

Zeheb R & Gelehrter TD 1988 Cloning and sequencing of cDNA for the rat plasminogen activator inhibitor-1. Gene 73:459–468.

Zhu LJ, Cheng CY, Phillips DM & Bardin CW 1994 The immunohistochemical localization of α_2-macroglobulin in rat testis is consistent with its role in germ cell movement and spermiation. Journal of Andrology 15:575–582.

21

NEW ASPECTS OF LEYDIG CELL FUNCTION

R. Middendorff,[1] D. Müller,[2] H. J. Paust,[2] A. F. Holstein,[1] and M. S. Davidoff[1]

[1]Institute of Anatomy
University of Hamburg (UKE)
Martinistraße 52, 20246 Hamburg, Germany
[2]Institute for Hormone and Fertility Research
University of Hamburg
Grandweg 64, 22529 Hamburg, Germany

1. SUMMARY

Previous studies indicated that the Leydig cells of the human testes show similarities to neuroendocrine cells. In this context, the local synthesis of two neuroactive signaling molecules, namely nitric oxide (NO) and C-type natriuretic peptide (CNP), both acting via the second messenger, cyclic guanosine monophosphate (cGMP), might be of physiological relevance. By immunoblotting, immunohistochemical analyses and affinity crosslinking experiments, respectively, the presence of soluble guanylate cyclase (sGC), the NO receptor, and of guanylate cyclase B (GC-B), representing the CNP receptor, was demonstrated in Leydig cells, seminiferous tubules and blood vessels of the human testis. Moreover, cGMP and its binding protein cGMP-dependent protein kinase type I (GK I) were found in these structures. The functional activity of the two receptors was proved by generation of cGMP in response to treatments with the NO donor, sodium nitroprusside (SNP), and with CNP, respectively. As indicated by immunohistochemical analyses and by treatments of cells with either SNP or CNP, human Leydig tumour cells and MA10 cells, representing a mouse Leydig tumour cell line, were found to be distinguished by a reduced expression of the receptors for NO and CNP. Furthermore, expression levels of the components of the two cGMP-generating systems were found to be widely unchanged in Leydig cells during different ontogenetic stages. Though cGMP has been shown to influence testosterone release, the constant developmental expression patterns of NO and CNP apparently independent of differences in androgen production, the down-regulation of their receptors in tumorous cells, and the presence of GK I, may point to additional autocrine functions of these factors and of cGMP in Leydig cells. Moreover, possible paracrine actions of NO and CNP may include relaxation of seminiferous tubules and blood vessels in order to modulate sperm transport and testicular blood flow, respectively. These findings suggest that Leydig cell-derived factors may exert activities different from or in addition to those involved in the regulation of testosterone production.

2. INTRODUCTION

Androgen production is the primary function of testicular Leydig cells. In this context, the presence of Leydig cell-specific organelles such as smooth endoplasmic reticulum is well established (Schulze, 1984). Recently, Leydig cells of the human testis have been shown to possess cytoplasmic vesicles and storage granules similar to those found in neuroendocrine and nerve cells (Davidoff et al., 1993). Moreover, a series of nerve cell-specific substances has been detected in Leydig cells (for review see: Saez, 1994; Davidoff et al., 1997a). Two of the agents recently demonstrated in these cells (Davidoff et al., 1995; Middendorff et al., 1996), namely nitric oxide (NO) and C-type natriuretic peptide (CNP), bind to and activate guanylate cyclases, resulting in elevated intracellular levels of the second messenger cyclic guanosine monophosphate (cGMP). This factor regulates a variety of complex and hitherto not completely understood cellular functions acting through binding to different molecular targets (Lincoln & Cornwell, 1993).

CNP belongs to the family of natriuretic peptides, which also includes atrial natriuretic peptide (ANP) and brain natriuretic peptide (BNP). Whereas ANP and BNP are mainly secreted by cardiac cells to act as hormones in the regulation of blood pressure and fluid volume homeostasis (for review see: Drewett & Garbers, 1994), CNP is produced and of particular physiological relevance in the brain (Komatsu et al., 1991; Minamino et al., 1993; Langub et al., 1995). NO, on the other hand, serves as a neurotransmitter in the nervous system, as a mediator of endothelium-dependent relaxation of blood vessels and mediates the tumoricidal and bactericidal actions of macrophages (Moncada et al., 1991; Bredt & Snyder, 1994; Schmidt & Walter, 1994). NO can be produced by either a neuronal (nNOS), an endothelial (eNOS) or an inducible (iNOS) isoform (Förstermann et. al., 1994) of NO synthase (NOS).

Whereas NO, which diffuses freely across membranes (Moncada et al., 1991), binds to and activates a soluble guanylate cyclase (sGC), CNP is the specific ligand of a plasma membrane receptor, designated as GC-B (Drewett & Garbers, 1994).

The question, which processes are influenced by cGMP-dependent mechanisms in the human testis, has not yet been elucidated.

The data presented here, show the presence and activity of receptors for NO and CNP in the human testis. We also provide evidence that cGMP may act by interaction with cGMP protein kinase I (GK I). The occurrence of the receptors in testicular vasculature and in the peritubular lamina propria suggests relaxation of vessels and tubules in order to modulate testicular blood flow and sperm transport, respectively. Autocrine activities of NO and CNP in Leydig cells may affect cellular functions other than the production of testosterone.

3. MATERIALS AND METHODS

3.1. Isolation of Leydig Cells, Seminiferous Tubules and Testicular Blood Vessels

Testes were obtained from 11 patients aged 30 - 86 years who were undergoing orchiectomy as the primary treatment of prostatic carcinoma. One to 2 h after surgery chilled human testes were cut into 4 pieces and transferred to dishes containing Ham's F12/DMEM culture medium (Gibco, Eggenstein, Germany) supplemented with 15 mM

NaHCO$_3$, 20 mM HEPES pH 7.4, 100 IU/ml penicillin, 100 µg/ml streptomycin, 2.5 µg/ml amphotericin B, 10µg/ml transferrin, 5 µg/ml hydrocortisone and 2% fetal bovine serum. Seminiferous tubules and Leydig cells were then exposed as described (Seidl & Holstein, 1990; Davidoff et al., 1997b).

Blood vessels of testicular pieces were identified under a stereo microscope. After pulling the connective tissue gently away, vessels of different size that appeared to be intact were isolated with fine forceps and scissors. Isolated Leydig cells, tubules and vessels were frozen in liquid nitrogen or transferred to 24-well microtiter plates containing the above mentioned culture medium.

3.2. RT-PCR Assays

RT-PCR assays were performed as described (Middendorff et al., 1996).

3.3. Protein Preparation

Membrane and cytosolic protein fractions for use in Western blot and affinity crosslinking experiments were prepared as described previously (Müller et al., 1991).

3.4. Western Blot Analyses

Immunoblotting was carried out essentially as described elsewhere (Weiner et al., 1997) using the following antibodies: the polyclonal rabbit antisera against the α_1-subunit of sGC (kindly provided by D. Koesling, Berlin, Germany, diluted 1:200) and against cGMP-dependent protein kinase Iß (GK I, kindly provided by D. Pöhler, Würzburg, Germany, 1: 400) as well as the monoclonal mouse anti-nNOS (Transduction, Lexington, KY, 1:500).

3.5. Immunohistochemistry

Blocks of normal human testicular tissue (n=11) and of human Leydig cell tumours (n=8) were fixed by immersion in Bouin's fluid for 24 h at 20°C and embedded subsequently in paraffin. Afterwards, 6 µm thick sections were cut. All sections were mounted onto chrome-gelatin precoated slides. Testes of 5-, 10-, 15-, 20-, 24-, 27-, 60- and 90-day-old Wistar rats were fixed by immersion in Bouin's fluid immediately after removal and embedded in paraffin as described above.

Using an immunohistochemical peroxidase anti-peroxidase (PAP)-avidin-biotin-peroxidase complex (ABC) combination procedure with additional nickel-glucose oxidase amplification (Davidoff et al., 1995) the following rabbit polyclonal antisera were tested: anti-cGMP (Biogenesis, Sandown, NH, 1:300), anti-CNP (Peninsula, Belmont, CA, 1:500), anti-sGC (1:200, see above), anti-GK I (1:100, see above) and anti nNOS (Biomol, Hanburg, Germany, 1:500).

3.6. Affinity-Labeling of GC-B

Crosslinking experiments with ^{125}I-[Tyr^0]-CNP (purchased from Peptide Radioiodination Service Center, Pullman, WA) were performed as described previously (Middendorff et al., 1996).

3.7. Measurement of cGMP Production by MA10 and TM3 Cells

Mouse TM3 Leydig cells and mouse MA10 Leydig tumour cells were cultured as described (Mather, 1980; Ascoli, 1981). Plated cells (10^6 cells/well, 24 well plates) were preincubated at 34°C in 250 µl of Locke's salt solution (154 mM NaCl, 5.6 mM KCl, 2.2 mM $CaCl_2$, 1 mM $MgCl_2$, 6 mM $NaHCO_3$, 10 mM glucose, 2 mM HEPES, pH 7.4) containing 0.25 mM isobutyl-methylxanthine (IBMX, purchased from Sigma). The NO donor sodium nitroprusside (SNP, purchased from Sigma) as well as CNP (Bissendorf, Hannover, Germany) were then applied in 250 µl of 0.25 mM IBMX containing Locke's solution for 1h at 34°C. Incubations were terminated by the addition of -20°C ethanol. After one hour at -20°C the plates were centrifuged at 1200 x g for 10 min. Aliquots of the supernatants were transferred to glass tubes and evaporated to dryness. The samples were redissolved in Locke's solution before use in the cGMP radioimmunoassay (see below).

3.8. Measurement of cGMP Production by Isolated Seminiferous Tubules and Blood Vessels

After one day of culture at 34°C (5% CO_2 / 95% O_2), the medium was removed and tubules as well as vessels were washed twice in Locke's salt solution. To measure nitroprusside- or CNP-dependent cGMP production, the vessels and tubules, respectively, were incubated for 1 h at 34°C in 250 µl Locke's solution containing additionally 0.25 mM IBMX in the absence and then in the presence of 1 mM SNP or of 1 µM CNP. Solutions were removed after each incubation, immediately frozen in liquid nitrogen and stored at -70°C.

cGMP was measured by radioimmunoassay as described (Mukhopadhyay et al., 1986) with reagents kindly provided by IBL (Hamburg, Germany). The minimum detection limit was approximately 10 fmol/tube and cross-reactivity with cAMP was less than 0.001%.

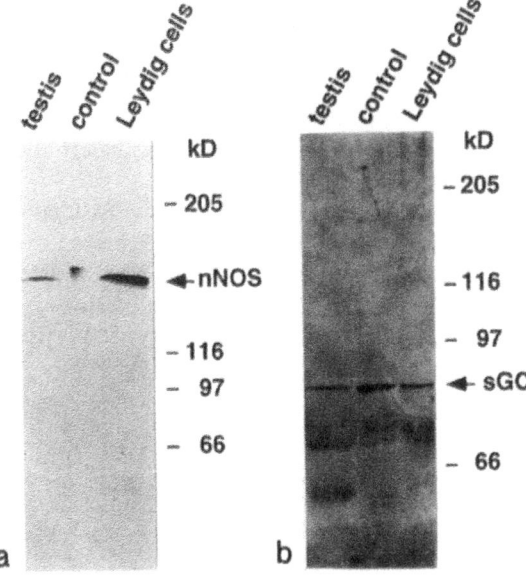

Figure 1. Western blot analyses for nNOS (a) and sGC (b) in testicular structures. A protein of 160 kD was demonstrated by anti-nNOS antibodies in extracts of isolated Leydig cells, whole testis and rat pituitary tumour cells, the latter used as a positive control (a). Immunoreactivity for the α_1-subunit of sGC, migrating at approximately 82 kD, was detected in extracts of Leydig cells and of whole testis using rat brain extracts for reference (b). Signals representing the antigens are marked by arrows. The migration of molecular weight markers (Sigma SDS-6H) is indicated.

The results shown in figures 4 and 8 are mean ± SE (SEM) of triplicate determinations. Each experiment was repeated at least four times. Treatment effects were statistically assessed using t test as installed in the GraphPad InStat Software (GraphPad Inc., CA) with $P \leq 0.05$ as the criterion of significance.

4. RESULTS

The presence of the NO-synthesizing enzyme nNOS was revealed by Western blot analyses in protein fractions of whole testis and isolated Leydig cells (Fig. 1a). Further immunoblot analyses demonstrated the receptor for NO, sGC, in testicular structures. The α_1-subunit of sGC, 82 kD in size (Koesling et al., 1991), was found in extracts of whole testis and isolated Leydig cells (Fig. 1b) as well as in testicular vessels and tubules (not shown).

Immunohistochemical investigations were carried out to localize more exactly sGC in blood vessels and seminiferous tubules. Immunoreactivity (IR) for sGC was observed in endothelial and smooth muscle cells of a subpopulation of arteries (Fig. 2a). Staining of testicular veins was not observed. A subpopulation of peritubular myofibroblasts (Fig. 2b) showed also sGC-IR.

As indicated by RT-PCR assays and immunohistochemical analyses CNP is present in human Leydig cells (not shown). The expression of the CNP receptor (GC-B) mRNA

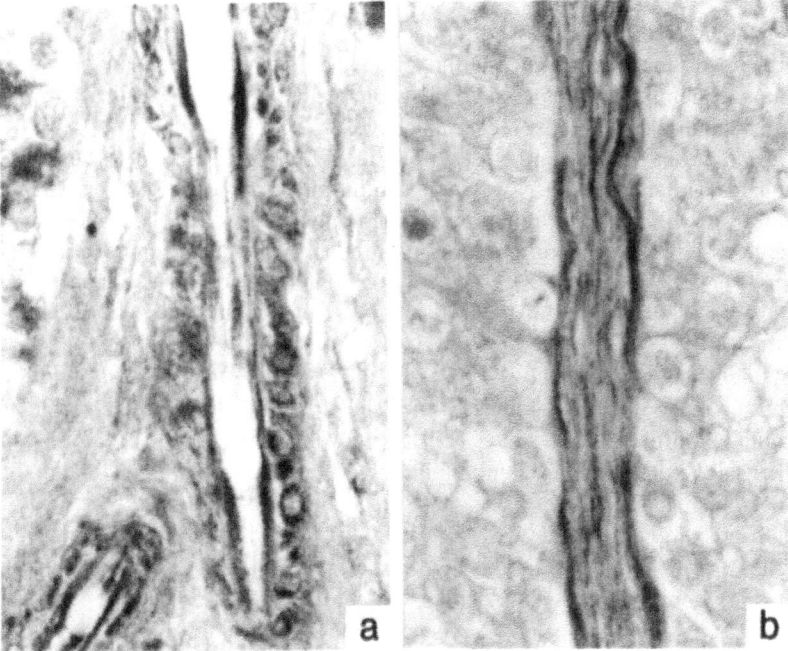

Figure 2. sGC-immunoreactivity in blood vessels and the peritubular lamina propria of the human testis (x820). A small artery (longitudinal section of the vessel) showed strong sGC-IR in endothelial cells as well as in smooth muscle cells (a). In peritubular lamina propria cells of two neighbouring tubules a positive staining for sGC was seen (b).

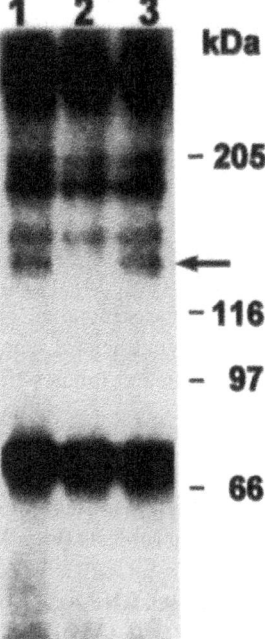

Figure 3. Detection of GC-B expression in human testis membranes by affinity labeling. Membranes were incubated with ^{125}I-labeled [Tyr[0]]-CNP in the absence (lane 1) or presence of unlabeled CNP (2) and ANP (3), respectively. After irradiation with UV light, samples were analyzed by SDS-PAGE and autoradiography. The migration of molecular weight markers (Sigma SDS-6H) is indicated. An arrow marks the position of a 135 kD protein whose labeling is specifically prevented in the presence of unlabeled CNP. All other bands represent unspecifically labeled proteins, the most prominent of which (at 66 kDa) is bovine serum albumin, present in the ^{125}I-peptide solution.

could be demonstrated by RT-PCR in human testis and isolated Leydig cells (not shown). Crosslinking experiments performed with ^{125}I-labeled [Tyr[0]]-CNP revealed the presence of the receptor in membranes prepared from whole testis (Fig. 3) as well as from isolated seminiferous tubules and blood vessels (not shown).

To examine the functional activity of sGC and GC-B, isolated vessels and tubules were incubated either with the NO donor sodium nitroprusside (SNP) or CNP, and the production of cGMP was determined by radioimmunoassays (Fig. 4). SNP- and CNP-induced increases of cGMP were measured in all experiments. SNP induced a 14-fold (tubules) and 47-fold (vessels), respectively, stimulation of sGC activity, whereas incubations with CNP resulted in 14-fold (tubules) and 28-fold (vessels), respectively, increases of cGMP levels.

Immunohistochemical analyses revealed the presence of cGMP in endothelial and smooth muscle cells of testicular arteries (not shown). In addition, Leydig cells and peritubular lamina propria cells showed cGMP-IR (not shown).

To reveal the presence of potential binding proteins of cGMP in testicular structures, immunohistochemical analyses of the cGMP-dependent protein kinase type I (GK I) were carried out. This target molecule was detected in Leydig cells and vascular smooth muscle cells (Fig. 5) as well as in peritubular myofibroblasts (not shown). The results were confirmed by immunoblot experiments, demonstrating a 76 kD protein (Pöhler et al., 1995) in testicular fractions of Leydig cells, vessels and tubules (Fig. 6).

To examine the development-dependent expression of components of cGMP pathways in Leydig cells, immunohistochemistry was performed with human (not shown) and rat testis sections at different pre- and postnatal developmental stages. All these factors, nNOS, sGC, CNP (Fig. 7), and cGMP were found to be constantly expressed during development.

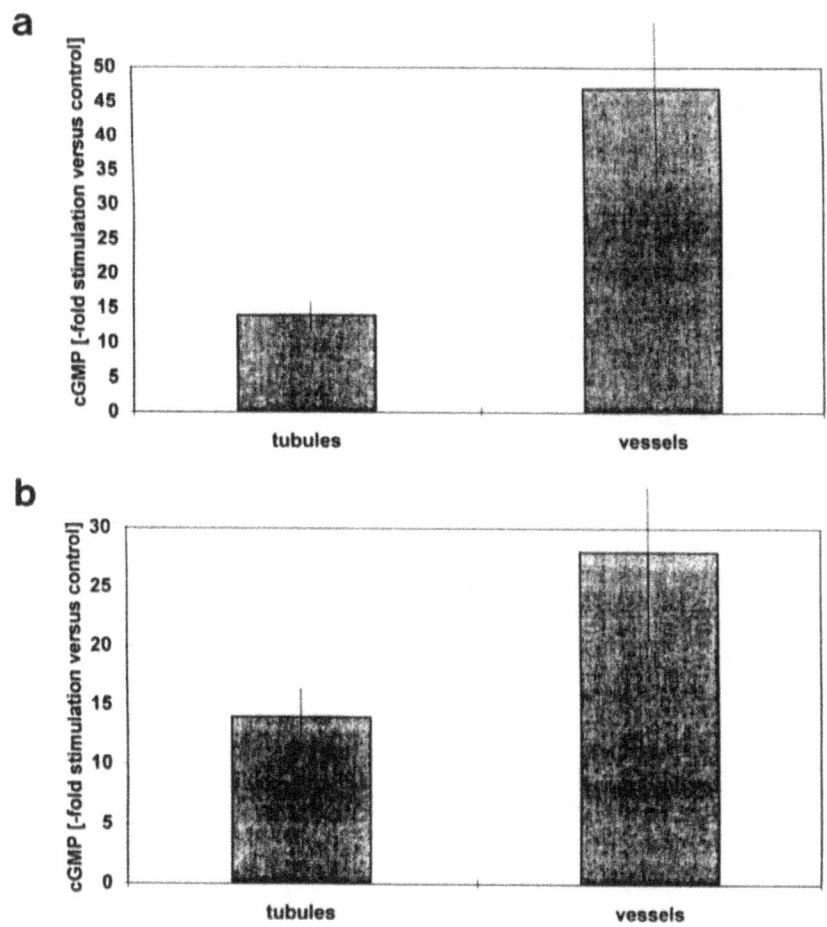

Figure 4. Effects of the NO donor SNP (a) and of CNP (b) on cGMP accumulation by isolated seminiferous tubules and blood vessels. Preparations of vessels and tubules were incubated in the presence of 0.25 mM IBMX with either 1 mM SNP (a) or 1μM CNP (b) for 1 h. Incubations in the absence of SNP and CNP were used to assess basal guanylate cyclase activities (control). SNP- and CNP-induced cGMP production is indicated as "-fold stimulation versus control". Vertical bars represent SEM.

Furthermore, components of the NO and CNP pathways were analysed in human and mouse Leydig tumour cells. Immunohistochemical analyses revealed the presence of nNOS and CNP in the mouse Leydig tumour cell line MA10 as well as in a control Leydig cell line (TM3), deriving from prepubertal mice (not shown). Most remarkably, incubations of these cell lines with either CNP (Fig. 8) or SNP (not shown) resulted in a dose-dependent accumulation of cGMP in TM3 cells, but not in the MA10 Leydig tumorous cells, suggesting a tumour-associated reduction of sGC and GC-B. In fact, as indicated by immunohistochemistry, sGC-IR was barely detectable in human Leydig tumour cells, whereas non-tumourous Leydig cells showed a positive staining for sGC (Fig. 9). In contrast, nNOS- as well as CNP-IR were present both in normal and tumourous human Leydig cells (not shown).

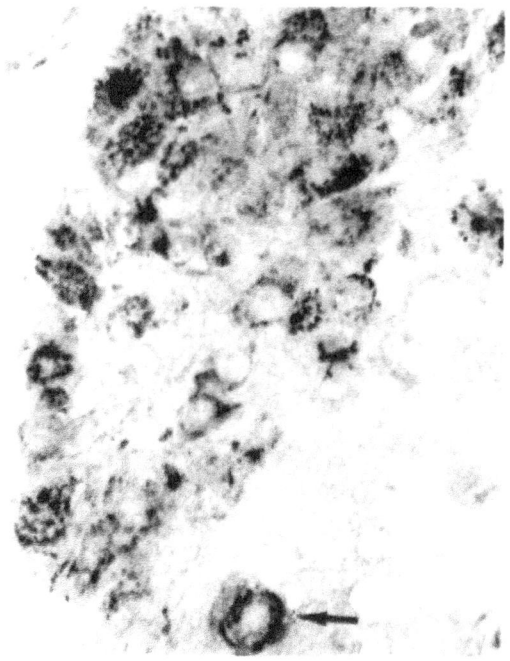

Figure 5. Immunohistochemical demonstration of the cGMP-dependent protein kinase type I (GK I) in the human testis (x800). GK I-IR is visible in the cytoplasm of most of the Leydig cells and in a small artery (arrow).

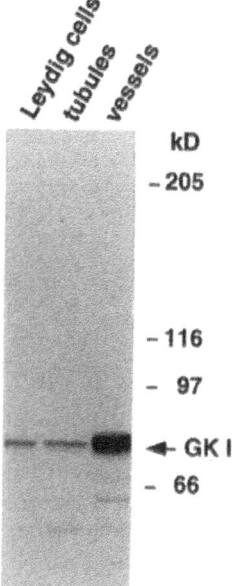

Figure 6. Western blot analyses for GK I in Leydig cells, seminiferous tubules and blood vessels of the human testis. GK I-IR, equivalent to 76 kD, was observed in extracts of Leydig cells, tubules and vessels. The migration of molecular weight markers (Sigma SDS-6H) is indicated.

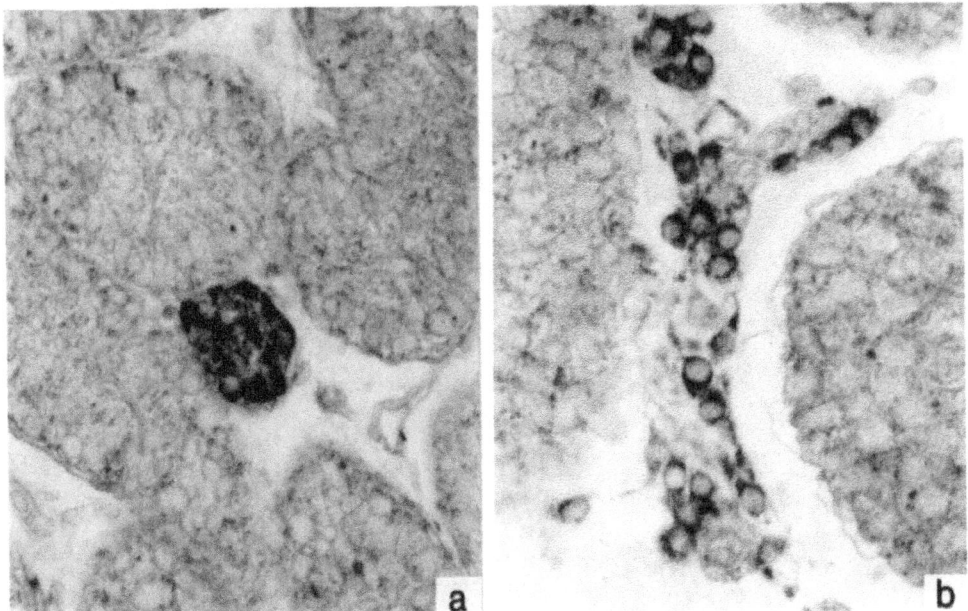

Figure 7. Immunohistochemical demonstration of CNP in rat Leydig cells during postnatal development (x820). At postnatal days 10 (a) and 24 (b) nearly all Leydig cells showed CNP-IR.

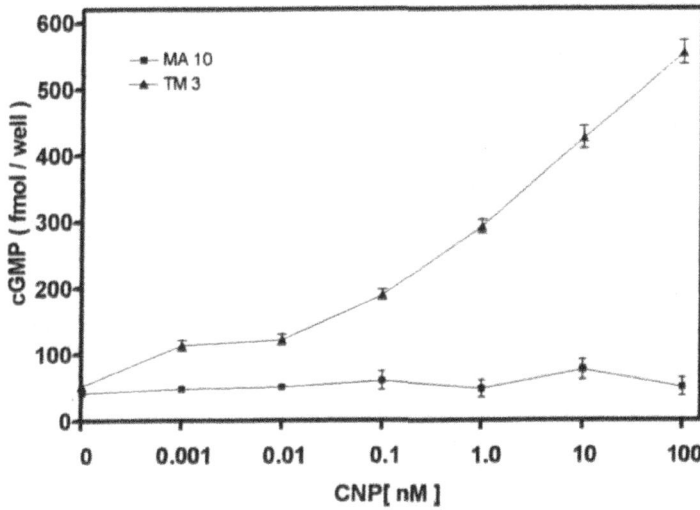

Figure 8. The effects of CNP on cGMP production by TM3 Leydig cells and MA10 Leydig tumour cells. CNP resulted in a dose-dependent increase of cGMP production in TM3 cells, but not in MA10 cells. Vertical bars represent SEM.

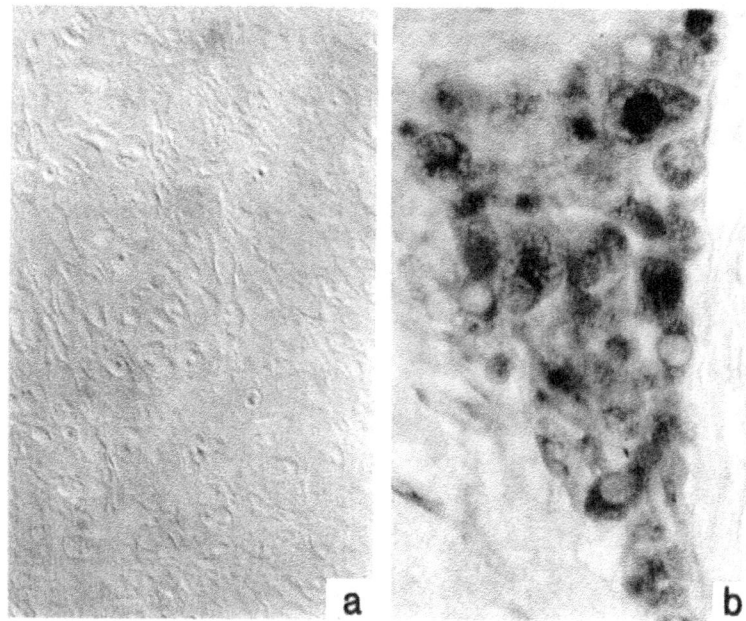

Figure 9. Immunohistochemical analyses of sGC in tumourous (a) and normal (b) human Leydig cells (x750). Whereas sGC-IR is barely detectable in tumourous cells (a), in non-tumourous tissue most of the Leydig cells showed a positive staining in the cytoplasm (b).

5. DISCUSSION

This study demonstrates the presence and activity of receptors for NO and CNP in Leydig cells, peritubular myofibroblasts and blood vessels of the human testis. Together with the recently shown production of the messenger molecules NO and CNP by human Leydig cells (Davidoff et al., 1995; Middendorff et al., 1996), these findings suggest a local biological role mediated by auto- and/or paracrine effects.

Because NO and CNP are associated with key neuroendocrine structures (Komatsu et al., 1991; Ceccatelli et al., 1993; Herman et al., 1993; Torres et al, 1993; Langub et al., 1995) and seem to exert autocrine effects especially in those neuroendocrine cells, which centrally regulate Leydig cell function, namely gonadotropes (Ceccatelli et al., 1993; McArdle et al., 1994a; McArdle et al., 1994b) and LHRH neurons (Belsham et al., 1996; Middendorff et al., 1997), the presence and activity of these substances in human Leydig cells seems to be consistent with the previously described neuroendocrine properties of Leydig cells (Davidoff et al., 1993).

Since testosterone production constitutes the primary function of Leydig cells, a possible physiological involvement of NO and/or CNP in androgen regulation has to be considered. In fact, NO and CNP have been shown to influence testosterone release. While both agents were found to accumulate cGMP (Khurana & Pandey, 1993; Welch et al., 1995; Davidoff et al., 1997b), the resulting effects were contradictory, showing an increase of testosterone by CNP in mice (Khurana & Pandey, 1993), but a decrease of testosterone by NO in rats (Welch et al., 1995). The underlying mechanisms for this apparent discrepancy have not yet been elucidated.

Established actions of cGMP involve binding to either cyclic nucleotide-gated ion channels (CNG channels), cGMP-dependent protein kinases or cGMP-dependent phosphodiesterases (reviewed in: Lincoln & Cornwell, 1993). Whereas direct effects of phosphodiesterases on testosterone production have been excluded (Hipkin & Moger, 1991), and CNG channels (Weiner et al., 1997) as well as cGMP-dependent protein kinases of type II (Orstavik et al., 1996) are not expressed in Leydig cells, the cytosolic cGMP-dependent protein kinase I (GK I) was detected both by immunoblotting and immunohistochemistry. In Leydig cells, however, stimulation of testosterone production by natriuretic peptides has been shown to result from a promiscuous activation of cAMP-dependent protein kinase by cGMP (Hipkin & Moger, 1991; Schumacher et al., 1992). The presence of GK I in this cell type might strongly suggest that cGMP, accumulated in response to NO and/or CNP, may affect—at least additionally—cellular functions unrelated to regulation of testosterone release.

In this context it is of particular interest that the expression levels of the components of the two cGMP-generating systems remain widely unchanged during each period of Leydig cell development, whereas, in contrast, the ability of these cells to produce testosterone dramatically changes (Chemes, 1996).

Our findings suggesting a reduced expression of functionally active NO and CNP receptors both in mouse (MA10) and human Leydig tumour cells may further support the existence of testosterone-independent physiological functions. Since ANP receptors (McArdle et al., 1994a) as well as CNP and nNOS are still present in tumourous Leydig cells, this change cannot be explained in the context of a general dedifferentiation. Therefore, down-regulation of receptors for CNP and NO might prevent autocrine actions possibly necessary for the maintenance of normal Leydig cell phenotype. Previous studies have shown that NO influences tumour growth (Jenkins et al., 1995) and triggers a switch to growth arrest during differentiation of neuronal cells (Peunova & Enikolopov, 1995). Natriuretic peptides, on the other hand, have been described as inducing hypertrophy of Leydig cells (Mazzocchi et al., 1990) and of astrocytes (Miyajima et al., 1995).

In addition to autocrine mechanisms, paracrine effects of NO and CNP are also conceivable. For example, Leydig cells are close to the peritubular lamina propria on the one hand (Davidoff et al., 1990) and to testicular blood vessels on the other hand (Ergün et al., 1994). Thus, Leydig cell-produced NO and CNP may have direct influence on these structures. Since both NO and CNP have been found to act as vasodilators (Moncada et al., 1991; Inagami et al., 1995), mediating their effects via cGMP and apparently cGMP-dependent protein kinases (Lincoln & Cornwell, 1993), the presence of sGC, GC-B, cGMP and GK I in testicular blood vessels as well as the observed NO- and CNP-mediated cGMP accumulations in these structures strongly suggest a local vasorelaxant function (Inagami et al., 1995).

In seminiferous tubules, NO- and CNP-induced cGMP production may mediate relaxation of myofibroblasts. Whereas endothelin, for example, has been shown to be involved in peritubular cell contraction in the rat (Tripiciano et al., 1996), the agents responsible for relaxation have not yet been defined. Peritubular myofibroblasts express filaments characteristic of fibroblasts and smooth muscle cells (Davidoff et al., 1990; Holstein et al., 1996). An influence of NO and CNP on myofibroblasts may be postulated in context of the well known effects of these factors on smooth muscle cells in other organs (Lincoln & Cornwell, 1993; Drewett and Garbers, 1994). In the peritubular lamina propria, NO and CNP may participate in the regulation of the peristaltic activity of the tubules, which in turn is necessary for sperm transport (reviewed in: Setchell et al., 1996). Furthermore, both agents may influence the permeability of the lamina propria and, by this, the transport of nutrients into the tubular lumen (Holstein et al., 1996).

In summary, our data indicate possible functions of two of the neuroactive factors produced by Leydig cells, pointing to possible auto- and/or paracrine effects of NO and CNP.

It is an attractive idea that NO and CNP may act locally (Fig. 10), to regulate (i) the distribution of oxygen, nutrients, and hormones by testicular vessels, (ii) the peristaltic activity of tubules in context of sperm transport, (iii) as well as the maintenance of normal Leydig cell phenotype.

6. ACKNOWLEDGMENTS

We are greatful to Dr. A.K. Mukhopadhyay for his very helpful discussion. We thank Mrs. S. Giehler, Mrs. M. Schwarz and Mrs. S. Schwarz for their excellent technical assistance as well as to Professor Becker, Marienkrankenhaus Hamburg, Professor Busch, Allgemeines Krankenhaus Hamburg-Eilbek, Dr. Hartmann, Bundeswehrkrankenhaus Hamburg and to Professor Tauber, Allgemeines Krankenhaus Hamburg-Barmbek for providing the surgical specimens. Furthermore, we are indebted to Dr. D. Koesling, Berlin,

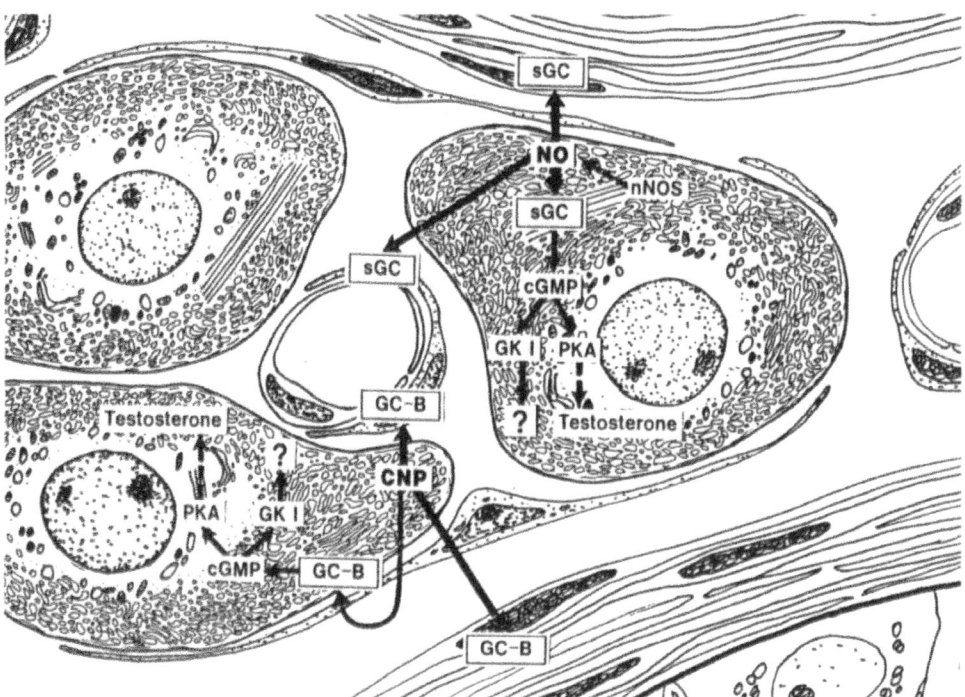

Figure 10. Schematic presentation of presumed testicular actions of NO and CNP. NO and CNP are produced in human Leydig cells. Based on the presence of specific receptors for NO and CNP, Leydig cells, peritubular myofibroblasts, and vascular smooth muscle cells represent potential sites of NO and CNP activity. This may result in relaxation of contractile cells of testicular blood vessels and of the lamina propria, presumably mediated by cGMP and GK I. In Leydig cells, autocrine actions of CNP and NO may influence testosterone production via a promiscuous activation of cAMP-dependent protein kinase (PKA) by cGMP (Hipkin & Moger, 1991; Schumacher et al. 1992). Additional hypothetical actions of NO and CNP, possibly involved in the maintenance of normal Leydig cell phenotype by interaction of cGMP with GK I, are indicated.

Germany, and to Dr. D. Pöhler, Würzburg, Germany, who donated the anti-sGC and the anti-GK I antiserum, respectively. The present study was supported by a grants from the Deutsche Forschungsgemeinschaft (Ho 388/6-3) and the Bundesministerium für Bildung, Wissenschaft, Forschung und Technologie (BMBF) as part of a larger concerted project "Fertilitätsstörungen" (01KY 9502).

7. REFERENCES

Ascoli M 1981 Regulation of gonadotropin receptors and gonadotropin responses in a clonal strain of Leydig tumor cells by epidermal growth factor. Journal of Biological Chemistry 256: 179–183.

Belsham DD, Wetsel WC & Mellon PL 1996 NMDA and nitric oxide act through the cGMP signal transduction pathway to repress hypothalamic gonadotropin-releasing hormone gene expression. EMBO Journal 15: 538–547.

Bredt DS & Snyder SH 1994 Nitric oxide: a physiological messenger molecule. Annual Reviews in Biochemistry 63: 175–195.

Ceccatelli S, Hulting AL, Zhang X, Gustaffsson L, Villar M & Hökfelt T 1993 Nitric oxide synthase in the rat anterior pituitary gland and the role of nitric oxide in regulation of luteinizing hormone secretion. Proceedings of the National Academy of Sciences USA 90: 11292–11296.

Chemes H 1996 Leydig cell development in humans. In The Leydig cell, edn 1, pp 176–201. Eds AH Payne, MP Hardy & LD Russel. Cache River Press, Vienna, IL.

Davidoff MS, Breucker H, Holstein AF & Seidl K 1990 Cellular architecture of the lamina propria of human seminiferous tubules. Cell and Tissue Research 262: 253–261.

Davidoff MS, Schulze W, Middendorff R & Holstein AF 1993 The Leydig cells of the human testis - a new member of the diffuse neuroendocrine system. Cell and Tissue Research 271: 429–439.

Davidoff MS, Middendorff R, Mayer B & Holstein AF 1995 Nitric oxide synthase (NOS-I) in Leydig cells of the human testis. Archives in Histology and Cytology 58: 17–30.

Davidoff MS, Middendorff R, Holstein AF 1997a Dual nature of Leydig cells of the human testis. BioMedical Reviews 9: in press.

Davidoff MS, Middendorff R, Mayer B. de Vente J, Koesling D & Holstein AF 1997b Nitric oxide/cGMP-pathway components in Leydig cells of the human testis. Cell and Tissue Research 287: 161–170.

Drewett JG & Garbers DL 1994 The family of guanylyl cyclase receptors and their ligands. Endocrine Reviews 15: 135–162.

Ergün S, Stingl J & Holstein AF 1994 Microvasculature of the human testis in correlation to Leydig cells and seminiferous tubules. Andrologia. 26: 255–262.

Förstermann U, Closs EI, Pollock JS, Nakane M, Schwarz P, Gath I & Kleinert H 1994 Nitric oxide synthase isozymes. Characterization, purification, molecular cloning, and functions. Hypertension 23: 1121–1131.

Herman JP, Langub MC, Watson J, & Watson Jr RE 1993 Localization of C-type natriuretic peptide mRNA in rat hypothalamus. Endocrinology 133: 1903–1906.

Hipkin RW & Moger WH 1991 Interaction between cyclic nucleotide second messenger systems in murine Leydig cells. Molecular and Cellular Endocrinology 82: 251–257.

Holstein AF, Maekawa M, Nagano T & Davidoff MS 1996 Myofibroblasts in the lamina propria of human seminiferous tubules are dynamic structures of heterogenous phenotype. Archives in Histology and Cytology 59: 109–125.

Inagami T, Naruse M & Hoover R 1995 Endothelium as an endocrine organ. Annual Reviews in Physiology 57: 171–189.

Jenkins DC, Charles IG, Thomsen LL, Moss DW, Holmes LS, Baylis SA, Rhodes P, Westmore K, Emson PC & Moncada S 1995 Roles of nitric oxide in tumor growth. Proceedings of the National Academy of Sciences USA 92: 4392–4396.

Khurana ML & Pandey KL 1993 Receptor-mediated stimulatory effect of atrial natriuretic factor, brain natriuretic peptide, and C-type natriuretic peptide on testosterone production in purified mouse Leydig cells: activation of cholesterol side-chain cleavage enzyme. Endocrinology 133: 2141–2149.

Koesling D, Böhme E & Schulz G 1991 Guanylyl cyclases, a growing family of signal-transducing enzymes. FASEB Journal 5: 2785–2791.

Komatsu Y, Nakao K, Suga S, Ogawa Y, Mukoyama M, Arai H, Shirakami G, Hosoda K, Nakagawa O, Hama N, Kishimoto I & Imura H 1991 C-type natriuretic peptide (CNP) in rats and humans. Endocrinology 129: 1104–1106.

Langub Jr MC, Dolgas CM, Watson Jr RE & Herman JP 1995 The C-type natriuretic peptide receptor is the pre-dominant natriuretic peptide receptor mRNA expressed in rat hypothalamus. Journal of Neuroendocrinology 7: 305–309.

Lincoln TM & Cornwell TL 1993 Intracellular cyclic GMP receptor proteins. FASEB Journal 7: 328–338.

Mather JP 1990 Establishment and characterization of two distinct mouse testicular epithelial cell lines. Biology of Reproduction 23: 243–251.

Mazzocchi G, Malendowicz LK, Rebuffat P, Kasprzak A & Nussdorfer GG 1990 Effects of acute and chronic treatments with atrial natriuretic factor (ANF) on the Leydig cells of the rat testis. Endocrine Research 16: 323–331.

McArdle C.A, Ivell R, Käppler K, Müller D, Schmidt C, Poch A & Kratzmeier M 1994a Production and action of C-type natriuretic peptide in the gonadotrope-derived αT3–1 cells. Endocrine 2: 849–856.

McArdle CA, Olcese J, Schmidt C, Poch A, Kratzmeier M & Middendorff R 1994b C-type natriuretic peptide (CNP) in the pituitary: Is CNP an autocrine regulator of gonadotropes? Endorinology 135: 2794–2801.

Middendorff R, Müller D, Paust HJ, Davidoff MS & Mukhopadhyay AK 1996 Natriuretic peptides in the human testis: evidence for a potential role of C-type natriuretic peptide (CNP) in Leydig cells. Journal of Clinical Endocrinology and Metabolism 81: 4324–4328.

Middendorff R, Paust HJ, Davidoff MS & Olcese J 1997 Synthesis of C-type natriuretic peptide (CNP) by immor-talized LHRH cells. Journal of Neuroendocrinology 9: in press.

Minamino N, Aburaya M, Kojima M, Miyamoto K, Kangawa K & Matsuo H 1993 Distribution of C-type natri-uretic peptide and its messenger RNA in rat central nervous system and peripheral tissue. Biochemical and Biophysical Research Communications 197: 326–335.

Miyajima M, Ogura T, Nornes HO, Neuman T, Kiyoshi A & Sato Y 1995 C-type natriuretic peptide increases [Ca^{2+} level and changes cell volume in cultured astrocytes. Proceedings of the 25th Meeting of the Society for Neuroscience, San Diego, CA, 1995, part 3, p 89.

Moncada S, Palmer RMJ & Higgs EA 1991 Nitric oxide: physiology, pathophysiology and pharmacology. Phar-macological Reviews 43: 109–142.

Mukhopadhyay AK, Schumacher M & Leidenberger FA 1986 Steroidogenic effect of atrial natriuretic factor in isolated mouse Leydig cells is mediated by cyclic GMP. Biochemal Journal 239: 463–467.

Müller D, Baumeister H, Buck F & Richter D 1991 Atrial natriuretic peptide (ANP) is a high-affinity substrate for rat insulin-degrading enzyme. European Journal of Biochemistry 202: 285–292.

Orstavik S, Solberg R, Tasken K, Nordahl M, Altherr MR, Hansson V, Jahnsen T & Sandberg M 1996 Molecular cloning, cDNA structure, and chromosomal localization of the human type II cGMP-dependent protein ki-nase. Biochemical and Biophysical Research Communications 220: 759 765.

Peunova N & Enikolopov G 1995 Nitric oxide triggers a switch to growth arrest during differentiation of neuronal cells. Nature 375: 68–73.

Pöhler D, Butt E, Meißner J, Müller S, Lohse M, Walter U, Lohmann S & Jarchau T 1995 Expression, purifica-tion, and characterization of the cGMP-dependent protein kinases Iβ and II using the baculovirus system. FEBS Letters 374: 419–425.

Saez JM 1994 Leydig cells: endocrine, paracrine, and autocrine regulation. Endocrine Reviews 15: 574–626.

Schmidt HHW & Walter U 1994 NO at work. Cell 78: 919–925.

Schulze C 1984 Sertoli cells and Leydig cells in man. Advances in Anatomy, Embryology, and Cell Biology 88: 1–104.

Schumacher H, Müller D & Mukhopadhyay AK 1992 Stimulation of testosterone production by atrial natriuretic peptide in isolated mouse Leydig cells results from a promiscuous activation of cAMP-dependent protein kinase by cyclic GMP. Molecular and Cellular Endocrinology 90: 47–52.

Seidl K & Holstein AF 1990 Organ culture of human seminiferous tubules: A useful tool to study the role of nerve growth factor in the testis. Cell and Tissue Research 261: 539–547.

Setchell BP, Maddocks S & Brooks DE 1994 Anatomy, vasculature, innervation, and fluids of the male reproduc-tive tract. In The physiology of reproduction, edn 2, pp 1063–1175. Eds E Knobil & JD Neill. Raven Press, New York.

Torres G, Lee S & Rivier C 1993 Ontogeny of the rat hypothalamic nitric oxide synthase and colocalization with neuropeptides. Molecular and Cellular Neurosciences 4: 155–163.

Tripiciano A, Filippini A, Giustiniani Q & Palombi F 1996 Direct visualization of rat peritubular myoid cell con-traction in response to endothelin. Biology of Reproduction 55: 25–31.

Weiner J, Middendorff R, Kaupp UB & Weyand I 1997 Spatial expression of a CNG channel on sperm flagellum - evidence for homo- and heterooligomeric channels. The Journal of Cell Biology: submitted

Welch C, Watson ME, Poth M, Hong T & Francis GL 1995 Evidence to suggest nitric oxide is an interstitial regu-lator of Leydig cell steroidogenesis. Metabolism 44: 234–238.

22

SERTOLI CELL-SPECIFIC GENE EXPRESSION IN CONDITIONALLY IMMORTALIZED CELL LINES

N. Walther, M. Jansen, S. Ergün, B. Kascheike, G. Tillmann, and R. Ivell

Institute for Hormone and Fertility Research
University of Hamburg
Grandweg 64, 22529 Hamburg
Germany

A set of conditionally immortalized Sertoli cell lines has recently been established in this laboratory from $H\text{-}2K^h$-tsA58 transgenic mice (Walther et al., 1996). These cell lines have conserved the expression of a number of important Sertoli cell marker genes, and therefore provide ideal tools for detailed studies on the regulation of Sertoli cell-specific gene expression. The $H\text{-}2K^h$-tsA58 transgenic mice (Jat et al., 1991) carry a temperature-sensitive SV40 T antigen in their germline. Primary cell cultures derived from these mice can easily be grown at the permissive temperature of 33°C in the presence of γ-interferon in order to induce SV40 T antigen expression, and immortalized cell lines can be isolated from these cultures. Upon transfer to the non-permissive temperature of 39.5°C the temperature-sensitive SV40 T antigen is inactivated. The temperature-sensitive growth of the Sertoli cell lines was characterized by shifting one of two parallel cultures to 39.5°C and observing the effects on growth and expression of Sertoli cell-specific genes. After two days of culture at 39.5°C the cells begin to show a flat morphology and cell growth stops as shown for the immortalized cell line SK11 (Fig. 1). At this temperature the cells cannot be passaged and die after prolonged culture. SK11 cells were infected with a recombinant retrovirus expressing a mutant (175 His) of the cell cycle control protein p53 (Morgenstern and Land, 1990; Hollstein et al., 1991). The expression of the mutant p53 inactivates the cellular p53 and should result in unconditional immortalization of the cells. However, colonies of retrovirus-infected SK11 cells could not continue growth after shift from 33°C to 39.5°C. This experiment suggests that the immortalizing effect of SV40 T antigen on the Sertoli cells is not only due to inactivation of the cellular p53, but that other, possibly transcriptional, actions of SV40 T antigen (Damania and Alwine, 1996) contribute to Sertoli cell immortalization.

For the investigation of the effects of growth or hormonal induction on the expression of Sertoli cell-specific genes in the immortalized cell lines, parallel cultures were grown at 33°C (non-induced) or induced at 39.5°C in the presence of testosterone and the adenylate cy-

33°C + γ-IFN   39.5°C - γ-IFN

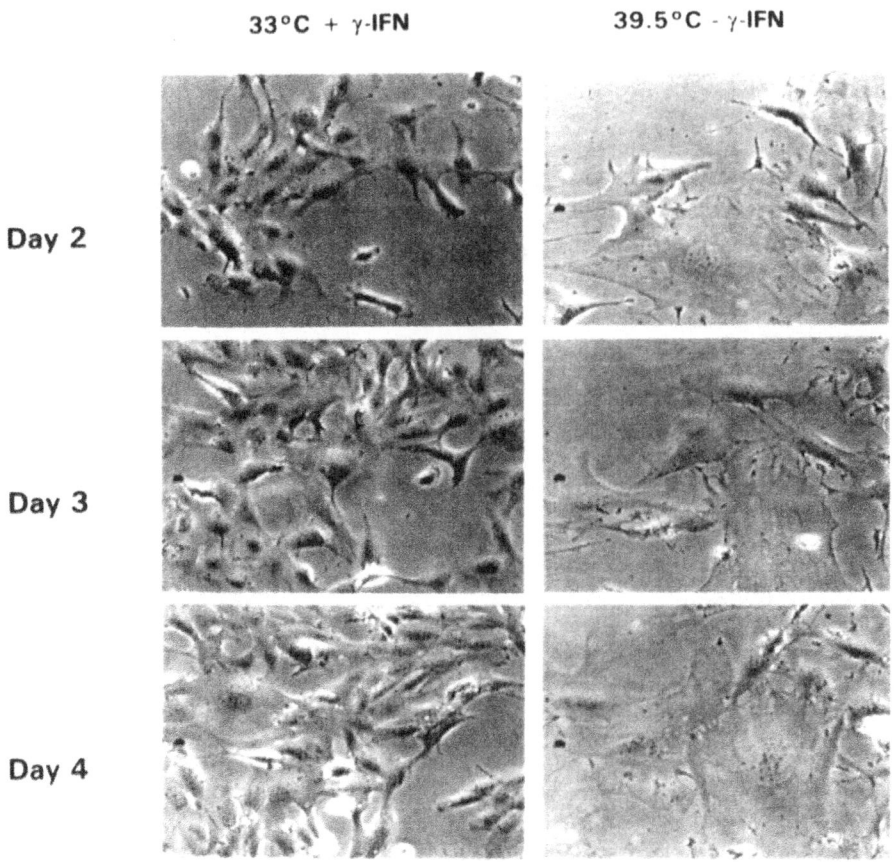

Figure 1. Morphology of the conditionally immortalized SK11 Sertoli cell line growing at 33°C in the presence of γ-interferon and after shift to the non-permissive temperature of 39.5°C.

clase activator forskolin in order to mimic the *in vivo* hormonal environment. For most of the Sertoli cell-specific genes investigated RT-PCR analysis could not detect significant differences in expression between induced and non-induced conditions. The expression of transferrin, however, appeared to be consistently upregulated in the induced cultures for all cell lines examined. Large variation between the different immortalized Sertoli cell lines were found by RT-PCR analysis in the expression of basic fibroblast growth factor (bFGF), a growth factor shown to be involved in the support of germ cell survival by Sertoli cells (van Dissel-Emiliani et al., 1996). The expression of bFGF was found to be the highest in SK49 cells (not shown). This result correlates well with the finding that, from all Sertoli cell lines tested, only SK49 cells are able to support survival of early rat germ cells in coculture studies (van den Ham et al., in press). RNase protection studies were performed in order to quantify the transcript levels of Sertoli cell-specific genes in the immortalized cell lines. Detailed investigation of expression of sulphated glycoprotein-2 (SGP-2) and Steel factor (Walther et al., 1996) revealed that the main stimulus acting on expression of these genes is not hormonal induction, but the temperature shift leading to cessation of growth due to inactivation of the temperature-sensitive SV40 T antigen.

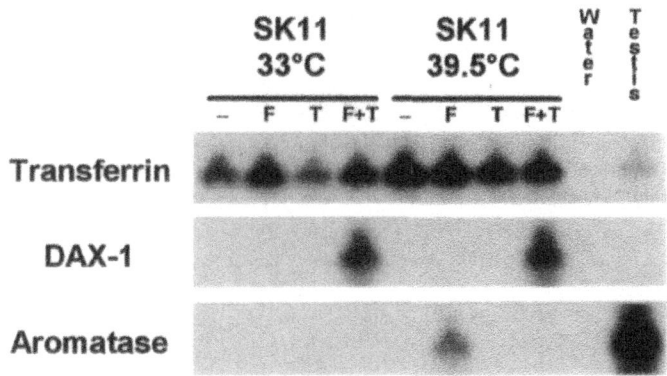

Figure 2. RT-PCR analysis of gene expression in the conditionally immortalized SK11 Sertoli cell line after induction by the adenylate cyclase activator forskolin (F) and/or testosterone (T) at the permissive (33°C) and the non-permissive temperature (39.5°C).

In order to test for hormonal induction of specific gene expression, SK11 cells were treated with the adenylate cyclase activator forskolin and/or with testosterone at 33°C as well as at 39.5°C. RT-PCR analysis of gene expression was performed as described (Walther et al., 1996) and revealed differential effects of increased cAMP levels or testosterone (Fig. 2). Transferrin transcript levels appear to be increased after treatment with forskolin at 33°C, but the shift to the higher temperature also has a major effect. Expression of the orphan nuclear receptor DAX-1 was investigated as this factor apparently is involved in modulation of SF-1-mediated transcriptional activation (Ito et al., 1997). DAX-1 transcripts were expressed in cultures treated both with forskolin and testosterone at either temperature. Unexpectedly however, DAX-1 transcripts could not be detected in adult mouse testis and brain (not shown), although in another study expression has been shown in these tissues (Bae et al., 1996). Low levels of aromatase transcripts could be detected in SK11 cells treated with forskolin at 39.5°C. As aromatase expression previously had not been detected in any of the immortalized Sertoli cell lines, this finding shows that under specific conditions of induction, Sertoli cell-specific genes can be expressed that are massively downregulated during growth of the cell lines. The results presented show that gene expression can be regulated by hormonal treatment in the conditionally immortalized Sertoli cell lines, but also that the growth conditions of the cells can have important effects on gene expression.

REFERENCES

Bae DS, Schaefer ML, Partan BW & Muglia L 1996 Characterization of the mouse DAX-1 gene reveals evolutionary conservation of a unique amino-terminal motif and widespread expression in mouse tissue. Endocrinology 137: 3821–3927.

Damani B & Alwine JC 1996 TAF-like function of SV40 large T antigen. Genes and Development 10: 1369–1381.

Hollstein M, Sidransky D, Vogelstein B & Harris CC 1991 p53 mutations in human cancers. Science 253: 49–53.

Ito M, Yu R & Jameson L 1997 DAX-1 inhibits SF-1-mediated transactivation via a carboxy-terminal domain that is deleted in adrenal hypoplasia congenita. Molecular and Cellular Biology 17: 1476–1483.

Jat PS, Noble MD, Ataliotis P, Tanaka Y, Yannoutsos N, Larsen L & Kioussis D 1991 Direct derivation of conditionally immortal cell lines from an H-2K^b-tsA58 transgenic mouse. Proceedings of the National Academy of Sciences USA 88: 5096–5100.

Morgenstern JP & Land H 1990 Advanced mammalian gene transfer: high titre retroviral vectors with multiple drug selection markers and a complementary helper-free packaging cell line. Nucleic Acids Research 18: 3587–3596.

van den Ham R, van Pelt AMM, de Miguel MP, van Kooten PJS, Walther N & van Dissel-Emiliani FMF 1997 Immunomagnetic isolation of fetal rat gonocytes. American Journal of Reproductive Immunology (in press).

van Dissel-Emiliani FM, de Boer-Brouwer M & de Rooij DG 1996 Effect of fibroblast growth factor-2 on Sertoli cells and gonocytes in coculture during the perinatal period. Endocrinology 137: 647–654.

Walther N, Jansen M, Ergün S, Kascheike B & Ivell R 1996 Sertoli cell lines established from H-$2K^h$-tsA58 transgenic mice differentially regulate the expression of cell-specific genes. Experimental Cell Research 225: 411–421.

23

FUNCTIONAL MARKERS FOR FETAL AND POSTNATAL DIFFERENTIATION OF RAT LEYDIG CELLS

S. G. Haider,[1] G. Servos,[1] S. Tajtaraghi,[1] G. Berthold,[1] A. K. Mukhopadhyay,[2] N. Kilic,[3] and S. Ergün[3]

[1]Institute of Anatomy II
Heinrich Heine University, POB 101007
40001 Düsseldorf, Germany
[2]Institute for Hormone and Fertility Research at the University Hamburg
Grandweg 64, 22529 Hamburg, Germany
[3]Institute of Anatomy
University Hospital Hamburg Eppendorf (UKE)
Martinistr. 52, 20246 Hamburg
Germany

The aim of the present work was to establish morphological markers for the functional differentiation of Leydig cells in rat in order to study the origin, development and maturation of these cells. Two different populations of Leydig cells can be observed in the rat testis from embryonic period till adulthood (Lording and de Kretser, 1972, Koupio et al. 1989; Teerds, 1989; Haider et al. 1986, 1995, Majdic et al. 1995)): 1. Fetal-Type Leydig cells (FTLC). 2. Adult-Type Leydig Cells (ATLC). Here is a summary of our results.

FETAL-TYPE LEYDIG CELLS

These cells are arranged in the form of round or oval compact clusters in the interstitial space between the seminiferous tubules and show 3-ß hydroxysteroid dehydrogenase (3ß-HSDH) reaction from fetal day (fd) 16 onwards with a maximum peak on fd 19. Light microscopically the FTLC can be clearly identified till postnatal day (pnd) 23–25. The androgen receptors are present in the interstitial cells, peritubular and elongated mesenchymal cells at fd 17. The LH receptors can be detected immunohistochemically in the FTLC from fd 16 onwards. The FTLC are completely surrounded by a basal lamina, show short microvilli, very often flat membrane foldings, abundant smooth endoplasmic reticulum, mitochondria, large lipid droplets and very small Golgi field. The development of FTLC remains largely unaffected by the treatment of animals with an antiandrogen or antiestrogen. The results strengthen the notion that the mesenchymal fibroblast like cells from mesonephros and /or gonadal ridge are the most probable precursors for FTLC.

ADULT-TYPE LEYDIG CELLS

These cells appear on about pnd 13 as indicated by histochemical reactions for LH receptors, androgen receptors and 3ß-HSDH. They are located around the seminiferous tubules without building clusters and give rise ultimately to the Leydig cells of the adult phase, as demonstrated by the histochemical reaction for 11ß-HSDH from pnd 35 onwards.. The ATLC show only a few flat membrane foldings, abundant smooth endoplasmic reticulum, mitochondria, only some small lipid droplets, a large well developed Golgi field and no basal lamina. The development of ATLC is delayed by the treatment of animals with an antiandrogen and largely inhibited by the treatment with an antiestrogen. Till pnd 12 the peritubular fibroblasts in the outer layer of Lamina propria contained ultrastructural features typical for fibroblasts; however, from pnd 13 onwards these cells showed additionally cell organelles typical for a steroid synthesizing cell. Employing H-3-thymidine, these fibroblasts showed the highest labeling index among the interstitial cells and a duration of DNA synthesis of 10 h. The data suggest that the peritubular fibroblasts lying in the outer layer of boundary tissue differentiate probably—after a phase of transition—ultimately ATLC. This theory is in accordance also with the results of Chemes et al. (1994). The androgens from FTLC seem to play a vital role in functional differentiation of ATLC. The role of neural crest in differentiation of Leydig cells, as suggested for man by Schulze et al. 1991, should be explored also for rat Leydig cells.

REFERENCES

Chemes H, Cigorraga S. Dym M., Musse M, Pellizari E. Schteingart H & Venara M 1994 Isolation and characterization of mesenchymal cell fraction of Leydig cell precursors from adult EDS-treated rats. 8th European Workshop on Molecular and Cellular Endocrinology of Testis. De Panne (Belgium), p. 9.

Haider SG, Passia D & Overmeyer G 1986 Studies on the fetal and postnatal development of rat Leydig cells employing 3ß-hydroxy steroid dehydrogenase activity. Acta Histochemica. 32: 187–202.

Haider SG, Laue D. Schwochau G & Hilscher B 1995 Morphological studies on the origin of adult-type Leydig cells in rat testis. Italian Journal of Anatomy and Embryology 100: 535–541.

Koupio T, Tapanainen J, Pelliniemi LJ & Huhtaniemi I. 1989 Development stages of fetal-type Leydig cells in prepuberal rats. Development 107: 213–220.

Lording DW & de Kretser DM 1972 Comparative ultrastructural and histochemical studies of the interstitial cells of rat testis during fetal and postnatal development. Journal of Reproduction and Fertility 29: 261–269.

Majdic G. Millar MR & Saunders PTK 1995 immunolocalization of androgen receptor to interstitial cells in fetal rat testes and to mesenchymal and epithelial cells of associated ducts. Journal of Endocrinology 147: 285 - 293.

Schulze W, Davidoff MS, Ivell R & Holstein AF 1991 Neuron-specific enolase-like immunoreactivity in human Leydig cells. Andrologia 23: 279–283.

Teerds K 1989 Aspects of Leydig cell development in the rat testis. Ph.D. thesis. Faculty of Biology, University of Utrecht (Netherlands): 1–135.

24

ENZYME HISTOCHEMICAL ADDITION TO MORPHOLOGICAL FEATURES OF LEYDIG CELLS IN SENIUM

D. Passia and B. Hilscher

Institute of Anatomy II
Heinrich-Heine-University Düsseldorf, POB 101007
40001 Düsseldorf, Germany

The present paper is based on the knowledge of the chemo-cyto-architecture of the human testis as determined by enzyme histochemical methods. Several papers (Goslar et al., 1982; Passia et al., 1983; Passia et al., 1985) have been published on the issue of normally functioning testicular tissue as well as that of patients with disturbed fertility. All these examinations were carried out in material from young men. The question was, whether there do exist differences in the enzyme histochemical reactions of the testicular tissue in older patients as compared to young men.

Cryostat sections were prepared of twelve specimens from patients suffering from prostatic cancer, aged 56 to 76 years. Only patients without any pretreatment were included in the study. Enzyme histochemical reactions for various hydrolases (acid phosphatases, alkaline phosphatases, thiamine pyrophosphatase, Ca++-activated adenosine triphosphatase, nonspecific esterases) and oxidoreductases (lactate dehydrogenase, 3ß-hydroxysteroid dehydrogenase) were carried out. Biopsy material collected from four young patients with fertility problems served as control.

The enzyme histochemical reactions of the oxidoreductases and most of the hydrolases in the biopsy material from both groups were very similar, depending on the functional state of the seminiferous epithelium, the boundary tissue, and the interstitial tissue.

However, there was one decisive difference in the reactions of the alkaline phosphatases (ALPases). In normal testicular tissue of young men, ALPases can be demonstrated in the seminiferous tubules in the contact zones of germ cells to Sertoli cells as well as in the basal lamina. Strong activity can be observed in the tunica intima of the small arteries and in the capillaries. The interstitial cells do not display any reactive activity. In young patients suffering from fertility problems, the activities of ALPases in the seminiferous epithelium may be reduced and with increasing hyalinosis of the adventitia of the arterial blood vessels additional enzyme activity may appear there. But there is no activity of ALPases in the interstitial cells.

In the cryostat sections of the material from the aged patients, ALPases were present in the seminiferous tubules depending on the functional state of spermatogenesis and in the arterial blood vessels. Furthermore, groups of Leydig cells with intermediate or strong reactions for ALPases could be observed. Two types of reaction could be distinguished: in some cases the whole cytoplasm was filled with the reaction product, in other cases the individual Leydig cells were surrounded by reacting zones. In the parallel sections incubated for 3ß-hydroysteroid dehydrogenase such Leydig cells reacted intermediately to weakly.

These findings fit quite well with the pictures presented by Holstein et al., 1988, on Leydig cells during senescence. The authors pointed out that Leydig cells of aged patients can present very different morphological features. Besides inconspicuous Leydig cells, cells with very dense cytoplasm or loose groups linked by connective tissue can be detected. As the Leydig cells which stain positively for ALPases also stain heavily for hematoxylin/eosin we suppose them to be identical with the dense type, while the large Leydig cells surrounded by active zones belong to the large scattered type depicted by Holstein et al. (1988). The meaning of these observations is still unclear. ALPases, however, are known to be membrane-related enzymes and to play a role in cell-to-cell interactions. Thus, their activity may be interpreted as an expression of the attempt to compensate decreasing efficiency in aging Leydig cells.

REFERENCES

Goslar HG, Hilscher B, Haider SG, Hofmann N, Passia D & Hilscher W 1982 Enzyme histochemical studies on the pathological changes in human Sertoli cells. Journal of Histochemistry and Cytochemistry 30 (12): 1268–1274.

Holstein AF, Roosen-Runge EC & Schirren C 1988 Leydig cells during senescence. In Illustrated pathology of human spermatogenesis, pp. 237–238. Eds AF Holstein , EC Roosen-Runge & C Schirren Grosse Verlag Berlin.

Passia D, Behrendt H, Hilscher B, Hilscher W, Hofmann N & Haider SG 1983 Histological, enzyme-histochemical and electron microscopic study on the testis of infertile patients. Acta Histochemica Suppl XXVIII, 303–308.

Passia D, Haider SG, Hofmann N, Hilscher B & Hilscher W 1985 Enzymehistochemical studies on the disturbances of human spermatogenesis. Acta Histochemica Suppl XXXI, 135–138.

25

DIFFERENTIAL DISPLAY PCR CLONING OF W/W^V-MUTANT TESTIS SPECIFIC GENES

C. Hansis, D. Jähner, A.-N. Spiess, and R. Ivell

Institute for Hormone and Fertility Research
University of Hamburg
Grandweg 64, 22529 Hamburg, Germany

The W/W^V mouse is an important model with which to investigate azoospermia and its consequences in a mammalian system. The W locus encodes the transmembrane tyrosine kinase receptor c-kit (Chabot et al., 1988). Mutations in the c-kit gene affect the differentiation of hematopoetic stem cells, melanocyte progenitors and primordial germ cells. The latter results in infertility due to azoospermia.

Several mutations in the W locus are known (Nocka et al., 1990). The original W mutation consists of a 78 amino acid deletion in the transmembrane domain of the receptor which leads to the removal of the kinase activity. The W^V mutation substitutes threonine to methionine in the kinase domain of the protein. The heterozygeous W/W^V mutation renders the mouse infertile but viable, with surprisingly few physiological side-effects.

The "Differential Display Reverse Transcription Polymerase Chain Reaction" (DDRT-PCR) is an efficient technique to identify differentially expressed genes in two different tissues. By means of DDRT-PCR, known and unknown differentially expressed genes with a putative role in cell function can be identified (Liang & Pardee, 1992; Bauer et al., 1993).

In our study, we employed DDRT-PCR to identify genes, that are up- or downregulated in the testes of W/W^v mutant mice by comparison with wild type mice. It was our aim to characterize potential paracrine factors in the testis responding to the azoospermia. For this purpose, we reverse transcribed the mRNA of wildtype and w/w^v testes with one base-anchored oligo dT-primers, followed by a subsequent PCR reaction with the same 3′-primers and a combination of 26 different ^{33}P-labeled decamer primers at the 5′-end. The resulting pools of cDNA were electrophoresed on a conventional sequencing gel. After autoradiography, the most prominent differentially displayed bands were excised and reamplified with the corresponding primers. Those bands with an enhanced signal in the w/w^v mutant were classified as "upregulated", while bands with a stronger signal in the wildtype were designated as "downregulated". After gel-purification of the PCR fragments and cloning into T-vectors, the differential expression of the genes was verified using Northern Blot hybridization with probes from the most frequent clones initially

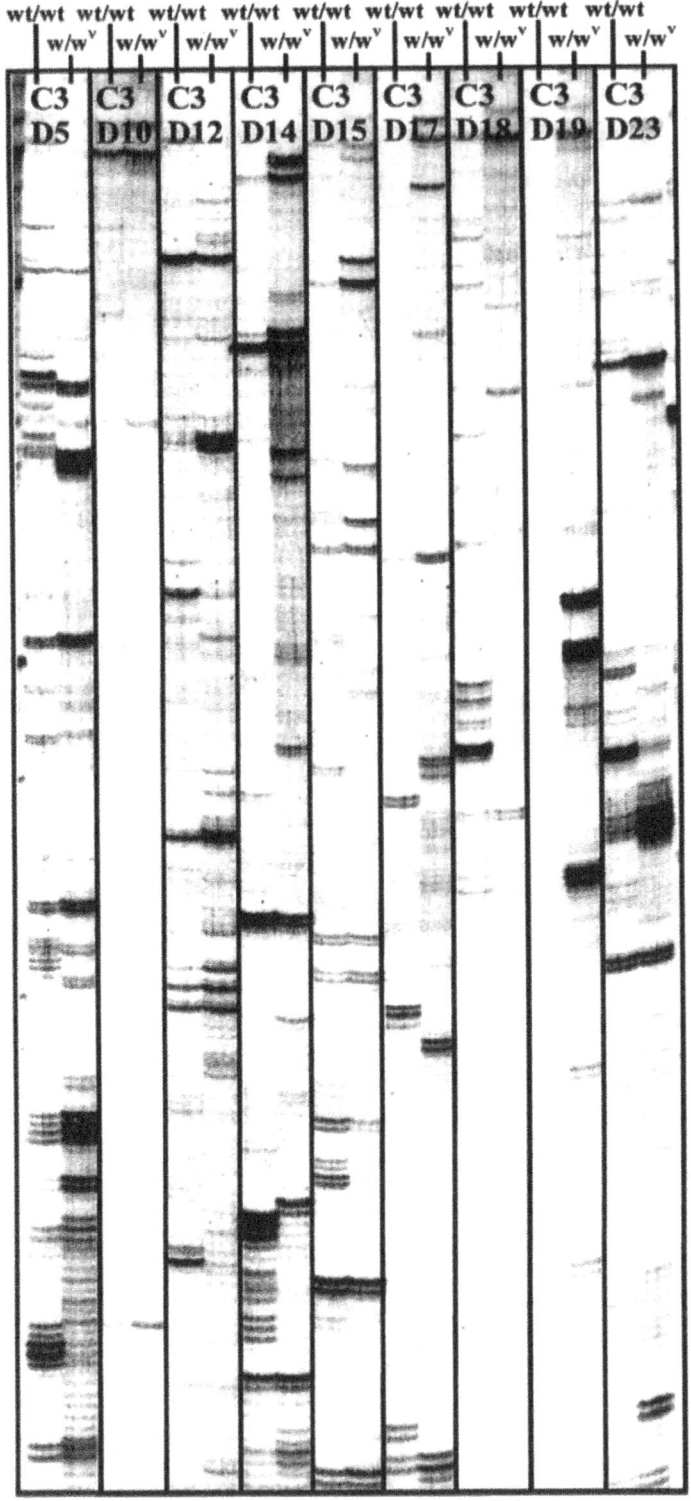

Figure 1. Differential display PCR comparing the effects of different primer combinations on the products of wild-type and W/W^V mouse testis RNA.

characterized by T-track sequencing. Finally, DNA databases were searched with the full sequences of the genes of interest.

The electrophoretic separation of DDRT-PCR fragments from the testes of wildtype (wt/wt) and w/w^v mutant mice resulting from a combination of C3 as the 3'-primer and a combination of different 5'-primers (D5-D23) is shown in Figure 1.

From a total of 432 PCR reactions, we were able to clone 25 upregulated and 9 downregulated gene products. After Northern Blot analysis, one wildtype-specific gene and four w/w^v mutant-specific genes were identified, of which three had homology to known genes and two are yet to be characterized.

Based on the ratio of the clones analysed so far and their differential expression (15%), we expect that at least a further 24 differentially expressed genes will be identified.

REFERENCES

Bauer D, Muller H, Reich J, Riedel H, Ahrenkiel V, Warthoe P & Strauss M 1993
Identification of differentially expressed mRNA species by an improved display technique (DDRT-PCR).Nucleic Acids Research 21: 4272–4280.
Chabot B, Stephenson D, Chapman V, Besmer P & Bernstein A 1988 The proto-oncogene c-kit encoding a transmembrane tyrosine kinase receptor maps to the mouse W locus. Nature 335: 88–89.
Liang P & Pardee AB 1992 Differential display of eukaryotic messenger RNA by means of the polymerase chain reaction. Science 257: 967–971.
Nocka K, Tan J, Chiu E, Chu T, Ray P, Traktman P & Besmer P 1990 Molecular bases of dominant negative and loss of function mutations at the murine c-kit/white spotting locus: W^{37}, W^V, W^{41}, W. EMBO Journal 9: 1805–1813.

26

IMMUNOREACTIVITY FOR GLIAL CELL MARKERS IN THE HUMAN TESTIS

M. S. Davidoff,[1] R. Middendorff,[1] D. Müller,[2] E. Köfüncü,[1] and A. F. Holstein[1]

[1]Institute of Anatomy
University of Hamburg (UKE)
Martinistraße 52, 20246 Hamburg, Germany
[2]Institute for Hormone and Fertility Research
University of Hamburg
Grandweg 64, 22529 Hamburg
Germany

Previous studies provide evidence for the neuroendocrine nature of Leydig cells of the human testis (Schulze et al., 1987; Davidoff et al., 1993). In a comparative study of the testes of different vertebrates and man, Maunoury et al. (1991) established immunoreactivity for glial fibrillary acidic protein (GFAP) in Leydig cells and adrenal cortex cells of the Syrian hamster. Holesh et al. (1993) found GFAP in rat Leydig cells situated in the vicinity of microvessels. The fact that Leydig cells of the testis may contain glial cell antigens encouraged us to search for GFAP (found in astrocytes and cell types of non-neural tissues), galactocerebroside (GalC; expressed in oligodendrocytes and peripheral Schwann cells), cyclic 2′,3′-nucleotide-3′-phosphodiesterase (CNPase; expressed in oligodendrocytes and peripheral Schwann cells), and A2B5-antigen (recognizing GT3 and other c-series gangliosides expressed by oligodendrocytes, type II astrocyte progenitors, some neurons and neuroendocrine cells) in structures of the human testis.

Testes from 14 patients aged 30–86 years, who underwent orchidectomy as primary treatment for prostatic carcinoma, were studied. The material was fixed by immersion for 24 h with Bouin's fixative and embedded in paraffin. 6-μm thick sections were stained immuncytochemically using a combination of the PAP and ABC method (Davidoff et al., 1993). The following monoclonal antibodies were applied: anti-GFAP (Sigma; diluted 1:1000), anti-GalC (Boehringer; 5μg/ml), anti-CNPase (Boehringer; 10μg/ml), and anti-A2B5 (Boehringer; 10μg/ml). Western blot analysis for GFAP and CNPase was performed with total homogenates from isolated Leydig cells as well as with cytosolic fractions from human testis, isolated seminiferous tubules, and rat brain homogenates.

Immunoreactivity for all glial markers was established in the Leydig cells of the human testis. The staining intensity varied between the cells of the testes of individual patients as well as between Leydig cell groups and individual Leydig cells of a histological section. In addition, immunoreactivity for GalC, CNPase and GFAP was seen within the cytoplasm of numerous Sertoli cells. Also, a different number of connective tissue cells, the so called "covering cells" (Holstein and Davidoff 1997), showed immunostaining for GalC, A2B5 and GFAP. Western blot studies confirmed the presence of GFAP (50 kD) and CNPase (45/48 kD).

The results of the present study show for the first time that Leydig cells, Sertoli cells and cells of the intertubular tissue possess immunoreactivity for antigens usually found in structures of neural origin. These results show that Leydig cells of the human testis, in addition to the well established neuroendocrine substances, exhibit also antigens characteristic for glial cells. This fact may be an additional argument toward the neuroectodermal origin of these cells. It is well known that the gonadal Anlage develops in close vicinity to the neural crest. Thus, the possibility exists for migrating pluripotent neural crest stem cells to invade the developing gonad. These stem cells may further develop to progenitor cells that differentiate to different cell lineages (Anderson, 1989, 1995; Lo and Anderson, 1995). It could be presumed that Leydig cells originating from neural crest cells remain in a differentiation stage of precursors that possess both neuronal and glial characteristics (Gard and Pfeiffer, 1990). The results obtained, together with already established similarities between Leydig and Sertoli cells (Davidoff et al., 1996), raise the possibility that Sertoli cells as well as some intertubular cells and probably some peritubular cells are also neural crest derivatives and may have common progenitors.

REFERENCES

Anderson DJ 1989 The neural crest cell. Lineage problem: neuropoiesis? Neuron 3:1–12.

Anderson DJ 1995 A molecular switch for the neuron-glia developmental decision. Neuron 15:1219–1222.

Davidoff MS, Schulze W, Middendorff R & Holstein AF 1993 The Leydig cell of the human testis - a new member of the diffuse neuroendocrine system. Cell and Tissue Research 271: 429–439.

Davidoff MS, Middendorff R & Holstein AF 1996 Dual nature of Leydig cells of the human testis. Bio Medical Reviews 6: (in press).

Holesh JA, Harik SI, Perry G & Stewart PA 1993 Barrier properties of testis microvessels. Proceedings of the National Academy of Sciences U.S.A. 90: 11069–11073.

Holstein AF & Davidoff MS 1997 Organization of the intertubular tissue of the human testis. Recent Advances in Microscopy of Cells, Tissues and Organs, Univerity of Roma "La Spienza" (in press).

Gard AL & Pfeiffer SE 1990 Two proliferative stages of the oligodendrocyte lineage (A2B5+O4- and O4+GalC-) under different mitogenic control. Neuron 5: 615–625.

Lo L & Anderson DJ 1995 Postmigratory neural crest cells expressing c-RET display restricted developmental and proliferative capacities. Neuron 15: 527–539.

Maunoury R, Portier M-M, Léonard N & McCormik D 1991 Glial fibrillary acidic protein immunoreactivity in adrenocortical and Leydig cells of the Syrian golden hamster (Mesocricetus auratus). Journal of Neuroimmunology 35: 119–129.

Schulze W, Davidoff MS & Holstein AF 1987 Are Leydig cells of neural origin? Substance-P-like immunoreactivity in human testicular tissue. Acta Endocrinologica 115: 373–377.

27

INSULIN-LIKE GROWTH FACTOR-BINDING PROTEIN (IGFBP)-5 IN HUMAN TESTICULAR TUBULES

B. Drescher,[1] W. Zumkeller,[2] H. Lauke,[1] M. Hartmann,[3] and M. S. Davidoff[1]

[1]Department of Anatomy
[2]Department of Paediatrics
University of Hamburg (UKE)
Martinistr. 52, 20246 Hamburg, Germany
[3]Armed Forces Hospital Hamburg
Department of Urology
Lesserstr. 180, 22049 Hamburg
Germany

Insulin-like growth factor-binding proteins constitute a protein family of six members (IGFBP-1 to -6), which bind insulin-like growth factor (IGF)-I and -II with high affinity and thereby regulate the mitogenic and metabolic activities of IGFs (Jones and Clemmons, 1995). Many tumours and neoplastic cell lines produce IGFBPs (Werner and LeRoith, 1996) and there is ample evidence for the presence of membrane-bound IGFBPs in tumour cell lines (Müller et al., 1996) indicating the existence of specific receptors for IGFBPs. Increased levels of IGFBPs, and in particular IGFBP-2, are found in the serum from patients with Wilms' tumours (Zumkeller et al., 1993). The presence of IGFBPs in germ cell tumours has not yet been investigated systematically and there are no data available on intratubular tumour cells presenting the early stages of germ cell tumours (carcinoma-in-situ (CIS)).

Testicular tissue adjacent to testicular germ cell tumours, exhibiting both tubules with histologically normal germinal epithelium and tubules with CIS cells, were investigated. Additionally, seminiferous tubules were isolated and cultured for four days as described elsewhere (Lauke et al., 1991).

Immunohistochemical staining was performed on tissue sections of paraffin-embedded material using a combination of peroxidase antiperoxidase (PAP)- and avidin-biotin-peroxidase complex (ABC)-techniques (Davidoff and Schulze, 1990). The primary antibody was a polyclonal antiserum raised in rabbits against HPLC-purified human IGFBP-5 (Upstate Biotechnology Inc., New York, U.S.A.). Controls were performed using antiserum pre-absorbed with the corresponding antigen, human recombinant IGFBP-5, purchased from Austral Biologicals, San Ramon, CA, U.S.A.

All CIS cells showed intense IGFBP-5 immunoreactivity seen as a peripheral ring and/or as paranuclear accumulation of the IGFBP-5-anti-IGFBP-5-complex. CIS cells maintained the IGFBP-5 immunoreactivity under culture conditions. In normal germ cells, IGFBP-5 immunoreactivity was found in the nuclei and/or cytoplasm of spermatogonia, in the cytoplasm of spermatocytes as well as in the cytoplasm and/or nuclei of round spermatids. Sertoli cells were not stained whereas Leydig cells showed a moderate staining intensity of cytoplasm and a stronger staining of Reinke crystals.

The IGFBP-5-immunoreactivity in all cell types of germ cells indicates a role for IGFBP-5 in spermatogenesis. The different location in the cytoplasm and nucleus may reflect a possible differential pathway for IGFBP-5 in the cell. IGFBP-5 immunoreactivity of both spermatogenetic cells and CIS cells provides further evidence for their relationship. Furthermore, the visualization of IGFBP-5 adds an additional marker for the characterization of CIS cells.

REFERENCES

Davidoff MS & Schulze W 1990 Combination of the peroxidase anti-peroxidase (PAP)- and avidin-biotin-peroxidase complex (ABC)-techniques: an amplification alternative in immunocytochemical staining. Histochemistry 93: 531–536.

Jones J & Clemmons DR 1995 Insulin-like growth factors and their binding proteins: biological actions. Endocrine Reviews 16: 3–34.

Lauke H, Seidl K, Hartmann M & Holstein AF 1991 Carcinoma-in-situ cells of cultured human seminiferous tubules. International Journal of Andrology 14: 33–43.

Müller D, Günther C, Biernoth R, Westphal M & Zumkeller W 1996 Demonstration of membrane-bound insulin-like growth factor-binding proteins in human glioma cell lines. International Journal of Oncology 9 [Suppl]: 144.

Werner H & LeRoith D 1996 The role of the insulin-like growth factor system in human cancer. Advances in Cancer Research 68: 183–223.

Zumkeller W, Schwander J, Mitchell CD, Morrell DJ, Schofield PN & Preece MA 1993 Insulin-like growth factor I, II, and IGF binding protein-2 in the plasma of children with Wilms' tumours. European Journal of Cancer 29A: 1973–1977.

28

A NOVEL ROLE FOR ATRIAL NATRIURETIC PEPTIDE (ANP) IN TESTIS

D. Müller[1] and R. Middendorff[2]

[1]Institute for Hormone and Fertility Research at the University of Hamburg
Grandweg 64, 22529 Hamburg, Germany
[2]Institute of Anatomy
University Hospital Eppendorf
Martinistr. 52, 20246 Hamburg
Germany

Previous studies have demonstrated that atrial natriuretic peptide (ANP) influences the testosterone production of testicular Leydig cells (Mukhopadhyay et al.,1986; Pandey et al., 1986; Schumacher et al., 1992). This effect was found to be elicited by binding of ANP to a guanylate cyclase (GC) activity-containing plasma membrane receptor, termed GC-A, leading to cellular increases in the second messenger, guanosine-3':5'-cyclic monophosphate (cGMP). The presence of GC-A in Leydig cells could be established by various molecular and biochemical methods (Pandey et al., 1986; Pandey and Singh, 1990).

Aim of the present study was to examine whether ANP receptors are also present in other cell types of the testis. Crude membranes were prepared from whole testes, isolated Leydig cells and seminiferous tubules, respectively, of adult Wistar rats. Equal amounts of membrane protein were incubated with ^{125}I-ANP (1 nM) for 20 min at 20 °C in 40 μl of 20 mM Hepes buffer, pH 7.5, containing 5 mM $MgCl_2$, 125 mM NaCl, and various protease inhibitors. To induce radioligand/receptor crosslinks, the samples were irradiated for 12 min with UV light (peak wave length: 302 nm). Reactions were terminated by chilling and immediate addition of 20 μl of 3x SDS-PAGE sample buffer consisting of 0.3 M Tris-HCl, pH 6.8, 200 mM DTT, 30 % (v/v) glycerin, 15 % (w/v) SDS and 0.06 % bromophenol blue, and the proteins were resolved by SDS-PAGE under reducing conditions in 7.0 % polyacrylamide separation gels. Autoradiographic analyses of the dried gels revealed a specifically labeled protein of 130 kDa, consistent with the size of GC-A, in all three membrane preparations. However, as indicated by different band intensities, reflecting different amounts of receptor protein, the GC-A concentration was found to be fivefold higher in tubular than in Leydig cell membranes. Thus, seminiferous tubules rather than Leydig cells represent the predominant sites of ANP receptor expression in the rat testis.

To investigate the functional activity of tubular GC-A, the effect of ANP on cGMP production by tubules was analyzed. Freshly dissected testes of 90-day-old animals were cut into two pieces and transferred to dishes containing Ham's F12/DMEM culture medium (Gibco, Eggenstein, Germany) supplemented with 15 mM NaHCO$_3$, 20 mM Hepes pH 7.4, 100 IU/ml penicillin, 100 µg/ml streptomycin, 2.5 µg/ml amphotericin B, 10 µg/ml transferrin, 5 µg/ml hydrocortisone and 2% fetal bovine serum. Apparently intact tubules were isolated and transferred to 24-well microtiter plates (approximately 10 pieces of tubular sections/well) containing the culture medium stated above. After an incubation for 24 h at 34 °C in an atmosphere of 5% CO$_2$ and 95% O$_2$, the medium was removed and tubules were washed twice in Locke's salt solution (154 mM NaCl, 5.6 mM KCl, 2.2 mM CaCl$_2$, 1 mM MgCl$_2$, 6 mM NaHCO$_3$, 10 mM glucose, 2 mM Hepes, pH 7.4) supplemented with 20 µg/ml bacitracin. To measure basal and ANP-dependent guanylate cyclase activities, the tubule preparations were incubated successively for 1 h at 34 °C in 250 µl Locke's solution, containing additionally 0.25 mM 3-isobutyl-1-methyl-xanthine (IBMX), in the absence and presence of 1 µM ANP. Solutions were removed after each incubation, and the cGMP content was determined by radioimmuno assays (IBL, Hamburg, Germany). ANP-induced increases in cGMP were measured in all experiments (n = 6) performed. Remarkably however, the extent of stimulation, calculated as ANP-dependent cGMP production versus basal guanylate cyclase activity, varied considerably between individual tubule preparations, ranging from 4- to 27-fold. These findings demonstrate that tubular ANP receptors are functionally active and in addition suggest a heterogeneous expression of GC-A among particular tubule segments.

Finally, the occurrence of ANP within testicular structures was analyzed by using immunohistocemical approaches. These studies were carried out on paraffin sections of adult rat testes. Experiments performed with two different anti-ANP antibody preparations (Peninsula; Affinity, Braunschweig, Germany) consistently revealed the presence of ANP in Leydig cells.

Taken together, this study demonstrates for the first time the presence and activity of ANP receptors in seminiferous tubules of the testis. The expression of GC-A in these functional units suggests a novel role for ANP related to germ cell development. The immunological detection of ANP in Leydig cells might indicate a local production and function of the peptide hormone. The interesting question whether circulating or Leydig cell-produced ANP predominates in the interaction with tubular ANP receptors remains to be established.

REFERENCES

Mukhopadhyay AK, Bohnet HG & Leidenberger FA 1986 Testosterone by mouse Leydig cells is stimulated in vitro by atrial natriuretic factor. FEBS Letters 202: 111–116.

Pandey KN, Pavlou SN, Kovacs WJ & Inagami T 1986 Atrial natriuretic factor regulates steroidogenic responsiveness and cyclic nucleotide levels in mouse Leydig cells in vitro. Biochemical and Biophysical Research Communications 138: 399–404.

Pandey KN, Inagami, T & Misono KS 1986 Atrial natriuretic factor receptor on cultured Leydig tumor cells: ligand binding and photoaffinity labeling. Biochemistry 25: 8467–8472.

Pandey KN & Singh S 1990 Molecular cloning and expression of murine guanylate cyclase/atrial natriuretic factor receptor cDNA. Journal of Biological Chemistry 265: 12342–12348.

Schumacher H, Müller D & Mukhopadhyay AK 1992 Stimulation of testosterone production by atrial natriuretic peptide in isolated mouse Leydig cells results from promiscuous activation of cyclic AMP-dependent protein kinase by cyclic GMP. Molecular and Cellular Endocrinology 90: 47–52.

29

NERVE GROWTH FACTOR (NGF) RECEPTORS IN MALE REPRODUCTIVE ORGANS

D. Müller,[1] H.-J. Paust,[1] R. Middendorff,[2] and M. S. Davidoff[2]

[1]Institute for Hormone and Fertility Research at the University of Hamburg
Grandweg 64, 22529 Hamburg
[2]Institute of Anatomy
University Hospital Eppendorf
Martinistr. 52, 20246 Hamburg
Germany

In addition to its essential role for the survival, development and differentiation of neurons of the central and peripheral nervous systems, nerve growth factor (NGF) seems to exert also a variety of nonneurotrophic activities. In particular, there is considerable evidence for crucial functions of NGF within the male reproductive system. For example, functional studies indicated the presence of 13 ng of biologically active NGF per g of mouse testis and even a tenfold higher amount of NGF per g of epididymis (Ayer-LeLievre et al., 1988). By in situ hybridization, NGF mRNA was localized to germ cells in seminiferous tubules and to epithelial cells of the ductus epididymis (Ayer-LeLievre et al., 1988). Similar levels of NGF were also ascertained in human testis (Seidl and Holstein, 1990).

Further studies demonstrated the presence of the two types of NGF receptors, designated as TrkA and p75. Gene expression of p75 was found in the Sertoli cells of rat testis, suggesting that NGF produced by male germ cells exerts an effect on spermatogenesis through the interaction with Sertoli cells (Persson et al., 1990; Parvinen et al., 1992). Immunoblot analyses (Djakiew et al., 1994) confirmed the presence of p75 in rat Sertoli cells and showed changes in testicular receptor levels during postnatal development. The same study demonstrated the presence of TrkA in testis. However, Western blot analyses of TrkA in cell lysates of isolated cell types remained inconclusive.

In this study, we used (a) RT-PCR analyses to examine the gene expression of the two NGF receptors in rat testes during different stages of postnatal development and (b) immunohistochemical approaches to reveal the local distribution of these receptors in adult rodent and human male reproductive organs.

To assess the development-dependent gene expression of NGF receptors, total RNA isolated from whole testes of 15-, 24-, 26-, 45- or 90-days-old rats was reverse transcribed into cDNA, and p75- and TrkA-specific sequences were amplified by PCR. Reaction products were size fractionated by agarose gel electrophoresis, blotted and hybridized to 32P-labeled

The Fate of the Male Germ Cell, edited by Ivell and Holstein
Plenum Press, New York, 1997

oligonucleotide probes. Autoradiography revealed different amounts of radioactivity associated with the TrkA-specific (546 bp) and p75-specific (634 bp) PCR products each. The TrkA gene was found to be expressed maximally around days 24 to 26, whereas TrkA transcript levels are markedly (by 50 %) lower at days 15 and 45 and are even more (by 90 %) reduced in the adult (at day 90). Unlike TrkA, p75 mRNA levels were found to be highest at the earliest stage examined (day 15), followed by a progressive decline with increasing age.

Immunohistochemical analyses were used to detect expression sites of Trk A in the rat testis. Paraffin sections (6 μm) of tissues from adult animals were exposed to a polyclonal anti-TrkA antibody (Santa Cruz Biotechnology, diluted 1:200), and the receptor was visualized by the PAP/ABC approach (Davidoff and Schulze, 1990). This study showed an intense staining of the Leydig cells. Considerable weaker immunoreactivity could be localized to late germ cells and Sertoli cells. No staining was observed on blood and lymphatic vessels.

TrkA-immunoreactivity was also established in the Leydig cells of the human testis. Apart from a relatively homogenous cytoplasmic staining, some cells showed a pronounced labeling at their plasma membranes. In addition, staining of Sertoli cells, located primarily to the basal compartment, was observed.

The expression of both, TrkA and p75, was investigated immunohistochemically in the mouse epididymis, using additionally a polyclonal anti-p75 antibody (Chemicon International; diluted 1 : 200). Immunoreactivity for both receptors was found to be confined to the epithelial layer of the ductus epididymis. In case of TrkA, immunoreactivity was localized strictly to plasma membranes connecting neighbouring epithelial cells. Remarkably, the luminal part (i.e., the free surface) of the cells was devoid of staining. In contrast, there was a granular, predominantly cytoplasmic staining for p75.

In conclusion, the developmental expression pattern of TrkA and p75 in testis suggests a particular role for NGF during the functional maturation of this organ. In the adult testis, Leydig cells appear to represent a major site of TrkA expression. Moreover, both types of NGF receptors were found to be present in the epithelial layer of the ductus epididymis. Together with previous evidence for NGF mRNA and protein (Ayer-LeLievre et al., 1988), these findings point to crucial auto- and/or paracrine activities of NGF in this part of the male reproductive tract.

REFERENCES

Ayer-LeLievre C, Olson L, Ebendal T, Hallböök F & Persson H 1988 Nerve growth factor mRNA and protein in the testis and epididymis of mouse and rat. Proceedings of the National Academy of Sciences of the United States of America 85: 2628–2632.

Davidoff MS & Schulze W 1990 Combination of the peroxidase anti-peroxidase (PAP)- and avidin-biotin-peroxidase complex (ABC)-techniques: an amplification alternative in immunocytochemical staining. Histochemistry 93: 531–536.

Djakiew D, Pflug B, Dionne C & Onoda M 1994 Postnatal Expression of Nerve Growth Factor Receptor in the Rat Testis. Biology of Reproduction 51: 214–221.

Parvinen M, Pelto-Huikko M, Söder O, Schultz R, Kaipia A, Mali P, Toppari J, Hakovirta H, Lönnerberg P, Ritzen EM, Ebendal T, Olson L, Hökfelt T & Persson H 1992 Expression of b-Nerve Growth Factor and Its Receptor in Rat Seminiferous Epithelium: Specific Function at the Onset of Meiosis. The Journal of Cell Biology 117: 629–641.

Persson H, Ayer-LeLievre C, Söder O, Villar MJ, Metsis M, Olson L, Ritzen M & Hökfelt T 1990 Expression of b-Nerve Growth Factor Receptor mRNA in Sertoli Cells Downregulated by Testosterone. Science 247: 704–707.

Seidl K & Holstein AF 1990 Evidence for the presence of nerve growth factor (NGF) and NGF receptors in human testis. Cell and Tissue Research 261: 549–554.

30

DELAYED ONSET OF SPERMATID ELONGATION IN THE PUBERTAL GOLDEN HAMSTER TESTIS DEPENDS ON A DEVELOPMENTAL DEFICIENCY OF LEYDIG CELL-11ß-HSD

A. Miething

Institute of Anatomy
University of Bonn
Nussallee 10, 53115 Bonn
Germany

RESULTS

In the seminiferous epithelium of the pubertal golden hamster, the most advanced germ cells reach step 6/7 of spermatid maturation (which is immediately before commencement of the process of elongation) on the 26th dpp (day post partum). However, this level of spermatid development is not exceeded until beyond the 30th dpp, when for the first time elongating spermatids of steps 8 and onwards appear. This temporary developmental disturbance exclusively affects the pre-elongational round spermatids, while the other types of germ cells, i.e. spermatogonia, spermatocytes and younger spermatids, regularly continue to differentiate. The areas of the seminiferous epithelium concerned therefore show a progressive desynchronization of the local germ-cell associations, which no longer fit into one of the regular stages of the seminiferous epithelium (Clermont, 1954). In addition, many of the affected spermatids are subjected to degeneration.

The semiquantitative histochemical evaluation of the 11ß-HSD (11ß-hydroxysteroid dehydrogenase) activity within the pubertal Leydig cells of the golden hamster reveals a steep increase up to an almost adult-like level between the 28th and 32nd dpp.

DISCUSSION

The transformation of round to elongated spermatids is known to be highly testosterone-dependent and cannot be maintained under testosterone-suppressed conditions (Bartlett et al., 1989; O'Donnell et al., 1994). In addition, recent studies have indicated that the enzyme 11ß-HSD, by oxidatively inactivating intracellular glucocorticoids, promotes a considerable enhancement of Leydig cell testosterone production (Monder et al., 1994).

Taken together, these constellations and the present results make it most probable that it is actually the delayed rise of the 11ß-HSD which, by limiting the testosterone output of the Leydig cells, is responsible for the observed temporary arrest of spermatid differentiation and the resulting partial desynchronization of the pubertal seminiferous epithelium. Not before the 11ß-HSD activity of the Leydig cells substantially rises around the 30th dpp, is this specific developmental block of spermiogenesis overcome.

These results are in accord both with a reported increase of serum testosterone in the male golden hamster around the 30th dpp (Vomachka & Greenwald, 1979) and with the temporal coincidence of the appearance of elongated spermatids and of the demonstration of Leydig cell-11ß-HSD in the pubertal testis of the rat (Neumann et al., 1993). It is, however, not clear whether species other than the golden hamster display a comparable delay of the onset of pubertal spermatid elongation; on the whole, the biological significance of the described temporal discrepancy between the maturation of the seminiferous epithelium and that of the Leydig cells is as yet obscure.

REFERENCES

Bartlett JMS, Weinbauer GF & Nieschlag E 1989 Differential effects of FSH and testosterone on the maintenance of spermatogenesis in the adult hypophysectomized rat. Journal of Endocrinology 121: 49–58.

Clermont Y 1954 Cycle de l'epithelium seminal et mode de renouvellement des spermatogonies chez le hamster. Revue Canadienne de Biologie 13: 208–245.

Monder C, Miroff Y, Marandici A & Hardy MP 1994 11ß-hydroxysteroid dehydrogenase alleviates glucocorticoid-mediated inhibition of steroidogenesis in rat Leydig cells. Endocrinology 134: 1199–1204.

Neumann A, Haider SG & Hilscher B 1993 Temporal coincidence of the appearance of elongated spermatids and of histochemical reaction of 11ß-hydroxysteroid dehydrogenase in rat Leydig cells. Andrologia 25: 263–269.

O'Donnell L, McLachlan RI, Wreford NG & Robertson DM 1994 Testosterone promotes the conversion of round spermatids between stages VII and VIII of the rat spermatogenic cycle. Endocrinology 135: 2608–2614.

Vomachka AJ & Greenwald GS 1979 The development of gonadotropin and steroid hormone patterns in male and female hamsters from birth to puberty. Endocrinology 105: 960–966.

31

COMPARTMENTALIZATION OF THE INTERTUBULAR SPACE IN THE HUMAN TESTIS

A. F. Holstein and M. Davidoff

Institute of Anatomy
Department of Microscopical Anatomy
University of Hamburg
Martinistrasse 52, 20246 Hamburg
Germany

During the last decades the interest of most testicular research was focussed on seminiferous tubules, germ cells, Sertoli cells and Leydig cells. The intertubular space was neglected. That is not surprising, because the tissue in the intertubular space can be preserved by fixation in adequate structure only with difficulty. During the removal of testicular tissue the seminiferous tubules are shifted against each other so that the intertubular tissue disrupts.

Structural constitutents of the intertubular space are the microvasculature, the Leydig cells, macrophages and sporadically a few mast cells. In the testicular tissue of the human, lymph vessels predominantly are found in the septula testis. They are not a constituent of the intertubular space.

In detailed and very carefully performed studies of human testicular tissue it has now been possible to elaborate a further cell system which contributes to a characteristic organization of the intertubular space. Large flat cells were identified which separate tissue compartments in the intertubular space. These cells resemble connective tissue cells and have been designated Co-cells (connective tissue cells, compartmentalizing cells, covering cells). Together with a thin layer of extracellular matrix the Co-cells form locally a sheath around the vessels of the microvasculature, around groups of Leydig cells and around parts of seminiferous tubules (fig.1). Serial sections of human testicular tissue showed that Co-cells unite these structures within a common compartment. This novel finding opens up new possibilities in the explanation of locally acting regulatory mechanisms for spermatogenesis by paracrine factors.

The Co-cells are obviously also involved in the production of the extracellular matrix of the intertubular space. Thus, mRNA for the proteoglycan decorin has been found in the Co-cells around blood vessels, Leydig cells and seminiferous tubules (Ungefroren et al., 1995).

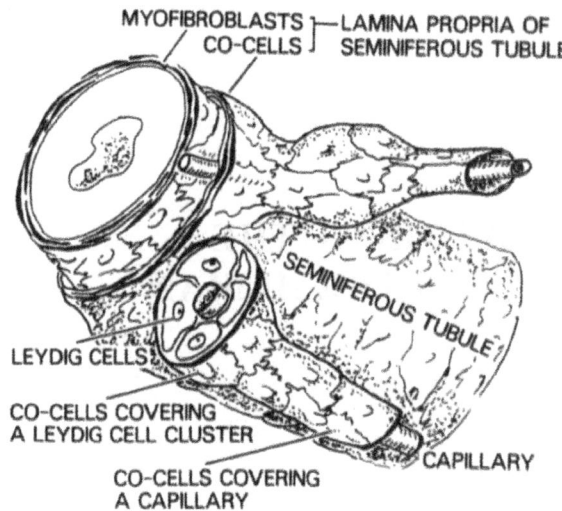

MYOFIBROBLASTS ⎤—LAMINA PROPRIA OF
CO-CELLS ⎦ SEMINIFEROUS TUBULE

SEMINIFEROUS TUBULE

LEYDIG CELLS

CO-CELLS COVERING
A LEYDIG CELL CLUSTER

CAPILLARY

CO-CELLS COVERING
A CAPILLARY

Figure 1. Concept of the compartmentalization of the intertubular space: Co-cells surround capillaries of the microvasculature, Leydig cell clusters and seminiferous tubules constituting a common compartment.

Immunocytochemical investigations of the Co-cells of the intertubular space showed that these cells, dependent upon their contacts to blood vessels, Leydig cells or seminiferous tubules, express different antigens. An interesting result is the occurence of glia-cell antigens (GFAP and CNPase) predominantly in the Co-cells surrounding the Leydig cells. In Co-cells surrounding blood vessels and seminiferous tubules, components of the NO/cGMP-system could be demonstrated. Antigens of the connective tissue (vimentin, fibroblast surface protein) are found everywhere. The functional meaning of this Co-cell heterogeneity still remains to be clarified.

On histological diagnosis of fertility disturbances, it is often possible to observe in the testicular tissue considerable changes within the intertubular space, which are related to a pronounced disturbance of spermatogenesis. Especially striking is an enormous increase of the extracellular matrix. This is recognizable by the fact that the individual Leydig cells are also surrounded by more or less broad sheaths of extracellular matrix, whereas the Co-cells remained very well preserved. These changes reflect a fibrosing process of the intertubular space. Presumably they are irreparable and represent barriers that only can be penetrated with difficulty by nutritive substances and hormones. Mostly in such testes an arrest of spermatogenesis is found at the level of primary spermatocytes or even at the level of spermatogonia. Since the extracellular matrix in the intertubular space can be neither phagocytosed by macrophages nor dissolved by other mechanisms, this histological picture points towards an unfavourable prognosis concerning the recovery of a man's fertility.

ACKNOWLEDGMENTS

This study was supported by a grant from the Bundesministerium für Bildung, Wissenschaft, Forschung und Technologie (BMBF) as a part of a larger concerted project "Fertilitätsstörungen" (01 KY 9502).

REFERENCE

Ungefroren H, Ergün S, Krull NB & Holstein AF 1995 Expression of the small proteoglycans biglycan and decorin in the adult human testis. Biology of Reproduction 52: 1095–1105.

32

MICROCIRCULATION AND THE VASCULAR CONTROL OF THE TESTIS

S. Ergün,[1] N. Kilic,[1] S. Harneit,[1] H. J. Paust,[2] H. Ungefroren,[1]
A. Mukhopadhyay,[2] M. Davidoff,[1] and A. F. Holstein[1]

[1]Institute of Anatomy
University of Hamburg (UKE)
Martinistr. 52, 20246 Hamburg, Germany
[2]Institute for Hormone and Fertility Research
University of Hamburg
Grandweg 64, 22529 Hamburg
Germany

1. SUMMARY

The arterial blood supply of the human testicular parenchyma within a testicular lobule demonstrates a segmental organization. In correlation to Leydig cells and seminiferous tubules within such a testicular tissue segment the capillary pathway was subdivided into three parts, namely arterial side inter-Leydig cell capillaries, intramural capillaries and venous side inter-Leydig cell capillaries. The organization, ultrastructure and functional aspects of the human testicular microvasculature were studied in detail. The casting preparation demonstrated a tightly organized vascular network. Computer aided 3-D reconstructions revealed that the capillary pathway in the human testis interconnects Leydig cells and seminiferous tubules. Electron microscopically the endothelial cells (EC) of arterial- and venous side inter-Leydig cell capillaries as well as of intertubular capillaries free of Leydig cells were of the continuous type without fenestrations. The intramural capillaries consisted of non-fenestrated and fenestrated sections. In all cases studied, the fenestrations faced the germinal epithelium. Transcytotic vesicles were numerous in the EC of inter-Leydig cell capillaries and in the non-fenestrated part of intramural capillaries. Leydig cells and Sertoli cells demonstrated immunostaining for vascular endothelial growth factor (VEGF) and its receptors flt-1 and KDR. In agreement with data obtained by RT-PCR analyses, human testicular capillaries were negative for VEGF but positive for its receptors. Immunohistochemically Leydig cells and Sertoli cells were also positive for Big-Endothelin and Endothelin-1. Endothelin receptors ET-A and ET-B were localized in Leydig cells while Sertoli cells showed only ET-A immunostaining. Testicular blood vessels and peritubular cells were only positive for ET-B. Androgen receptors could be localized on the arterial side of the human testicular microvasculature and on intramural capillaries, whereas the intralobular veins were negative. In contrast, estrogen receptors were found in all parts of testicular vasculature. From these results we conclude: 1) In the human testis, capillaries interconnect Leydig cell clusters and seminiferous tubules in a serial manner. A part of the capillaries runs in between the layers of the lamina propria and therefore represents the capillarization of human seminiferous tubules. It was presumed that the multilayered lamina propria requires its own capillary supply which may allow a rapid exchange of hormones and other nutritious substances between the microvasculature and the germinal epithelium. 2) VEGF produced and released from Leydig and Sertoli cells could act as a paracrine factor on the testicular microvasculature which possesses VEGF receptors and could modulate the permeability of capillaries in the adult testis, e.g. by fenestration of intramural capillaries. 3) The presence of androgen receptors (AR) and estrogen receptors (ER) on the testicular vasculature indicate that the testicular microvasculature is involved in the local endocrine and paracrine regulation of spermatogenesis. Particularly the presence of AR on the arterial side and on intramural capillaries leads to the assumption that androgens could influence the blood supply to Leydig cells and seminiferous tubules. 4) Leydig cells and Sertoli cells could also influence the blood flow via the potent vasoconstrictor ET-1.

2. INTRODUCTION

The microvasculature of an organ in general consists of arterioles, metarterioles, capillaries and venules. This part of the vascular system is responsible for the exchange of gases, nutritional substances and hormones. In contrast to the human testis, the organization of the testicular microvasculature in some animals like rat or mouse have been well studied, and these studies have shown a "rope-ladder"-like organization of the capillaries of rat and mouse testis (Müller 1956; Hundeiker & Keller 1963; Hundeiker & Mullert 1966; Kormano 1967a; Suzuki 1982; Weerasooriya & Yamamoto 1985). There is a degree of controversy concerning the organization of the microvasculature of the human testis (Hundeiker 1971; Kormano & Suoranta 1971a; Suzuki & Nagano 1986). A capillary organization was also reported as "rope-ladder"-like for the capillary bed of the human testis by Kormano and Suoranta (1971b) and Hundeiker (1971). However, Suzuki and Nagano (1986) found no evidence for the existence of such a "rope-ladder"-like capillary organization in the human testis. Once again it isevident that even within the same organ system, for example the testis, results cannot easily be transferred from one species to another.

Using serial sections and computer aided 3-D reconstructions, Ergün et al. (1994a) described a segmental vascular organization in the human testis. Detailed studies of the capillary organization of the human testis provided evidence for the first time that capillaries connect Leydig cell clusters with adjacent seminiferous tubules in a serial manner (Ergün et al. 1994b). Another interesting aspect which has been neglected for a long time is the functional role of the testicular vasculature. During the past decade the testicular microvasculature has attracted attention because of its assumed modulation by endocrine and paracrine mechanisms which are important for the regulation of spermatogenesis (Setchell & Rommerts 1986; Bergh et al. 1988; Damber et al. 1987; 1989; Sharpe 1990; Setchell 1990; Damber & Bergh, 1992). The literature concerning the ultrastructural features of testicular capillaries is very sparse. In this respect, Hundeiker (1971) found that, in contrast to other endocrine organs, the endothelium of the capillaries in the human testis was not fenestrated, showed a low degree of transcytosis and belonged to the A-1-α type capillaries (classification according to Bennet et al. 1959), similar to those found in muscle (Fawcett et al. 1969). There is also evidence that in the testis of some laboratory animals the intertubular and peritubular capillaries are non-fenestrated (Wolf & Merker, 1966; Fawcett et al. 1970; Meyerhoffer et al., 1989; Meyerhoffer & Bartke 1990). In contrast, it was recently demonstrated that capillaries in the lamina propria of normal human seminiferous tubules are partly fenestrated (Ergün et al. 1996b).

This article sets out to provide 1) results concerning the organization and the ultrastructure of the human testicular microvasculature in general compared with the testicular vasculature of other species such as the rat or mouse, and 2) results of a detailed study on the functional aspects of the human testicular microcirculation.

3. ORGANIZATION AND ULTRASTRUCTURE OF TESTICULAR MICROVASCULATURE

Earlier studies concerning the organization of the microvasculature of rat or mouse testis showed a "rope-ladder"-like organization of the capillaries (Figure 1).

In this form of capillary organization there are intertubular capillaries running along the seminiferous tubules and localized in a triangular space between three adjacent seminiferous tubules. These intertubular capillaries are connected together via the peritubular capillaries which run around the seminiferous tubules and partly contact the lamina propria of seminiferous tubules without penetration into this (Figure 2).

A detailed review about this capillary organization is given by Setchell and Brooks (1994). Casting preparations of human testicular vasculature revealed a tightly organized vascular network but no indication of the existence of a capillary organization in form of a "rope-ladder"-like system.

Computer aided 3-D reconstructions based on large semi-thin cross sections from human testicular tissue showed also no indications for a "rope-ladder"-like capillary organization in the human testis, but demonstrated a segmental arterial supply of the testicular parenchym (Figure 4).

The basis for such a segmental vascular organization was formed by segmental arteries which arised from the recurrent artery in regular distances at about 300 µm. The tissue segments were orientated at right angles to the longitudinal course of seminiferous tubules and contained parts of different seminiferous tubules located within a testicular lobule. This vascular organization of the human testis has been demonstrated in detail by Ergün et al. (1994a) and could provide an explanation for the focal atrophy of testicular parenchyma reported by Holstein (1988). Within such a tissue segment, the capillary showed a close association to Leydig cells and seminiferous tubules. A spatial view of the vascular organization in correlation to Leydig cells and seminiferous tubules could be obtained by means of fluorescence microscopic studies which showed interstitial and peritubular capillaries running first in a peritubular fashion, then penetrating into the lamina propria of seminiferous tubules. The light microscopic investigations of serial semi-thin sections of the human testis showed that interstitial capillaries, which arise from arterioles in the interstitium, branch repeatedly into multiple capillaries mostly situated among Leydig cells (Ergün et al. 1994b). The relationship between the testicular capillaries and Leydig cells as well as seminiferous tubules is visualized by means of the computer-aided 3-D-reconstruction (Figure 5).

Note, that after the bifurcation within the Leydig cell cluster (arrow) one capillary penetrates into the lamina propria (arrowhead) of a seminiferous tubule which shows the

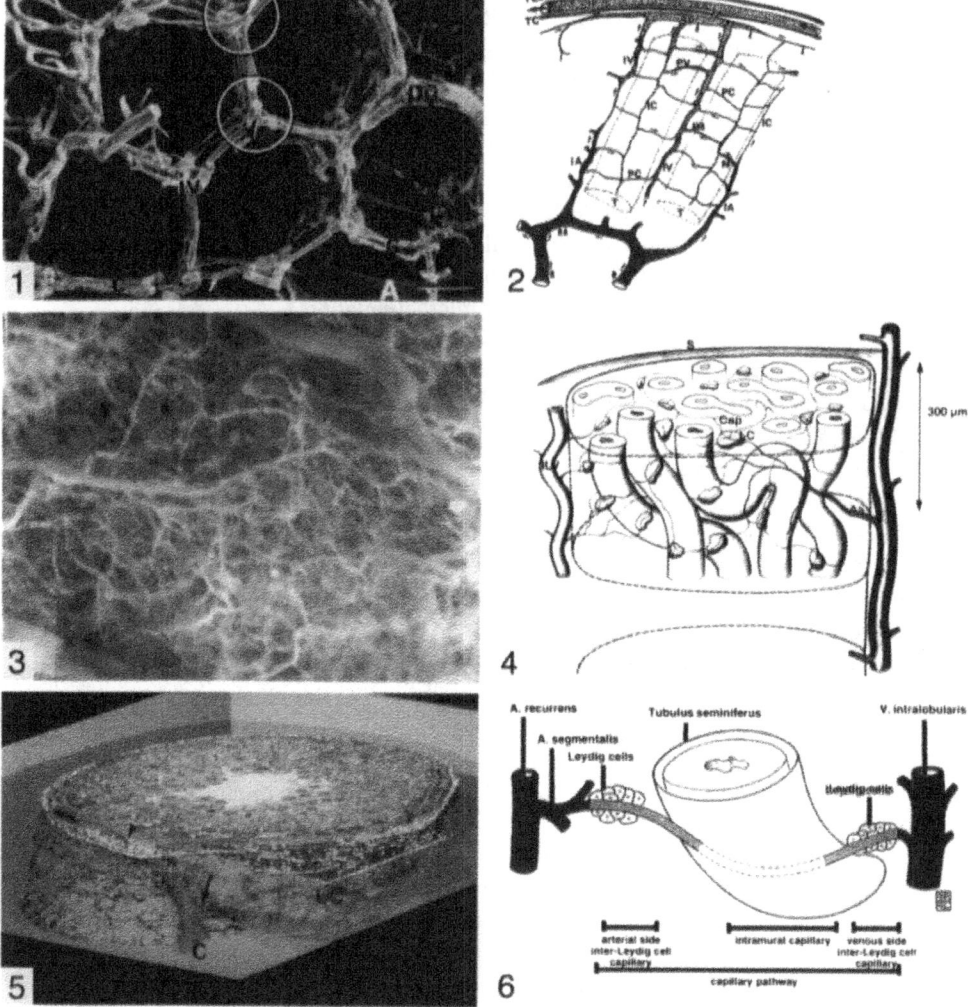

Figure 1. Cast of the blood vessels inside the rat testis sectioned at right angles to the seminiferous tubules, showing the three-dimensional arrangement of the intertubular and peritubular vessels. Note the clear hexagonal pattern of the vessels with only one intertubular vessel (arrowhead) at each interstitial space (circled). iv, intertubular venule; pv, peritubular venule; c, peritubular capillaries; A, artery; Bar = 100μm).

Figure 2. Diagrammatic representation of the intratesticular blood vessels in a plane longitudinal to seminiferous tubules (T). R, radiate artery; AA, arterio-arterial anastomotic arcade; IA, intertubular arteriole; IC, intertubular capillary; IV, intertubular venule; PA, peritubular arteriole; PC, peritubular capillary; PV, peritubular venule; AVA, arterio-venous anastomosis; IP, intra-albugincal venous plexus within the testicular capsule (TC); SP, sub-albugineal venous plexus and (arrowheads) its component vein, venule, and capillary, respectively, from left to right. Veno-venous anastomotic channels in SP are not shown. Blood flow direction is indicated by small and large arrows.

Figure 3. Casting preparation of human testicular vasculature showing an intralobular artery (A) (X150)).

Figure 4. Graphical representation of the segmental angioarchitecture of the human testis (Ergün et al., 1994a).

Figure 5. Computer-aided 3-D-reconstruction shows the organization of the testicular capillaries (C) in relation to the Leydig cell cluster (LC) and the seminiferous tubule.

Figure 6. Graphical demonstration of the subdivision of the capillary pathway of human testis in relation to Leydig cells and seminiferous tubules. Arterial side inter-Leydig cell capillaries are passing through Leydig cell clusters on the arterial side of the testicular microvasculature. Intramural capillaries are localized within the lamina propria of seminiferous tubules. Venous side inter-Leydig cell capillaries are surrounded by Leydig cells on the venous side of the testicular microvasculature.

branching of an interstitial capillary into further capillaries within a Leydig cell cluster. One of them penetrated into the lamina propria of a seminiferous tubule (Figure 5). After a distance of variable length within the lamina propria the capillaries leave the wall of the tubules (Ergün et al. 1996b). In their further course they are again surrounded by interstitial Leydig cells before they terminate in interstitial venules or intralobular small veins. Figure 6 graphically displays the vascular organization in the human testis and summarizes the results above about the fundamental subdivision of the capillary pathway in connection to Leydig cells and seminiferous tubules, into *inter-Leydig cell capillaries* on the arterial side, *intramural capillaries* (localized within the lamina propria of seminiferous tubules) and *inter-Leydig-cell-capillaries* on the venous side (Figure 6).

This means, that Leydig cells and seminiferous tubules in the human testis are connected serially, Leydig cells - seminiferous tubules - Leydig cells, by the blood stream via the microvasculature.

These results indicate that this capillary organization interconnects Leydig cells within the interstitium which represents the hormonal compartment, and the seminiferous tubules which is the compartment for spermatogenesis. The reason for this capillarization of the lamina propria of human seminiferous tubules may lie in the multilayered architecture of the human lamina propria (Davidoff et al., 1990) which needs its own blood supply. Therefore, this capillary organization may play an important role in the distribution and transport of hormones and other factors from the interstitium to seminiferous tubules and vice versa, also because of the poorly developed lymphatic network in the human testis (Holstein et al. 1979). In correlation to seminiferous tubules of the human testis this capillary pathway was subdivided into efferent and afferent parts of capillaries. Additionally, the presence of Leydig cell clusters on both the arterial and the venous side of the capillary pathway leads to the assumption that these cells may influence the capillary function via vasoactive substances on both sides of the capillary pathway. This presumably could play a key role in the regulation of testicular microcirculation. In this context the permeability of capillaries is of importance for exchange of the necessary factors between the blood and the testicular tissue. Indications for the status of the capillary permeability in an organ could be obtained by enzyme-histochemical analyses and by electron microscopical studies on the ultrastructure of capillaries.

A marker for transcytotic activity of EC is alkaline phosphatase was could be observed in arterioles, capillaries and venules (Ergün et al. 1996b). In contrast, large vessels such as intralobular arteries and veins, did not show any activity. Alkaline phosphatase activity was found in all interstitial sections of the capillaries, surrounded or not by Leydig cells, as well as in the intramural capillaries. In contrast to the results obtained in rat testis (Kormano, 1967b), the peritubular cells of the lamina propria of human seminiferous tubules were devoid of alkaline phosphatase activity, while intratubular spermatogonia and Sertoli cells were clearly positive (Ergün et a. 1996b). Electron microscopic examination of human testicular capillaries showed that the individual sections of these vessels were of different endothelial structure (Ergün et al. 1996b). Arterial-side inter-Leydig cell capillaries were surrounded by Leydig cells and exihibited a continuous endothelial lining, a continuous basal lamina in which pericytes were included, and belong to the A-1-α type of capillaries. Electron microscopic magnification demonstrated numerous endocytotic vesicles of variable size located on both sides of the endothelium, directed to the capillary lumen and to the basal lamina (Ergün et al. 1996b). Intramural capillaries ran within the lamina propria and were surrounded by processes of myofibroblasts. They comprise capillary segments characterized by marked differences of the endothelial ultrastructure. Figure 7 shows a cross section through a capillary within the lamina propria with a continuous non-fenestrated endothelium and continuous basal lamina.

Accordingly, this segment of the intramural capillaries resembles the A-1-α type. Figure 8 illustrates a section of an intramural capillary that contains two different endothelial segments, fenestrated and non-fenestrated (figure 8).

The fenestrated endothelial segment possesses a continuous basal lamina in which pericytes are included (A-2-α type). The fenestrated side of the endothelium is generally oriented towards the germinal epithelium. Higher electron microscopic magnification of this capillary shows that the fenestrations are closed by typical diaphragms (Ergün et al. 1996b). In this fenestrated capillary section only few intracytoplasmatic vesicles can be observed in the opposite non-fenestrated segment of the endothelium. Within the intramural segment tight junctions were observed at the contact zones between the EC. Venous side inter-Leydig cell capillaries possessed in all investigated cases a continuous non-fenestrated endothelium with continuous basal lamina (A-1-α type). In this segment again numerous and large intracytoplasmatic vesicles which are partly surrounded by a double membrane were reported (Ergün et al. 1996b).

Summarizing these results concerning the course and the ultrastructure of intramural capillaries, it could be demonstrated that the intramural capillary pathway of the human testis consists of non-fenestrated and fenestrated parts. In relation to the lamina propria of the seminiferous tubules and the blood flow it is of significance that the fenestrated side of the capillary wall was facing always toward the germinal epithelium. There is no evidence in the literature for similarly organized capillary sections in the testis of other species (Hundeiker 1971; Setchell 1994). The functional significance of these different types of capillaries in the human testis are not known. The fact that the fenestrated segments of the intramural capillaries faced the germinal epithelium allows the presumption that they are responsible for an enhanced selective exchange of substances. This selectivity of exchange may depend on the proteoglycan composition of the diaphragms of the endothelial fenestrations (Simionescu & Simionescu 1988). Presumably the fenestrated segments of the intramural capillaries fulfill special functions regarding the exchange of metabolites and hormones between the seminiferous tubules and the blood circulation. In this respect the fenestrated segments of the intramural capillaries differ significantly from the organization of the remaining segments of the microvasculature of human testis. In the arterial inter-Leydig cell capillaries and in the venous inter-Leydig cell capillaries a closed endothelial lining is established. It must be noted that the EC of the capillary sections that are surrounded by Leydig cells at the arterial and the venous side contain many more transcytotic vesicles and channels than the remaining interstitial capillary segments which are free of Leydig cells. The EC of these different capillary segments showed variable transcytotic activity which probably plays a crucial role in the permeability of various solutes and specific substances (Schnittler et al. 1990). The numerous transcytotic channels in these capillary segments reflect receptor-mediated transport of proteins (Simionescu 1987). Concerning the transcytotic activity, the non-fenestrated intramural capillary segments resemble the intra-Leydig cell segments. The numerous transcytotic structures in the EC of the testicular capillaries reflect an active exchange rate between the interstitial Leydig cells, the periphery of the seminiferous tubules and the blood circulation.

In comparison to the rat or mouse testis the above demonstrated differences in the organization of human testicular microvasculature, for example the capillarization of the lamina propria of seminiferous tubules, lead to the assumption that angiogenetic mechanisms and vasoactive substances could be responsible. In this context we focused our studies on the expression of growth factors, vasoactive substances and steroid receptors.

4. GROWTH FACTORS IN THE TESTICULAR VASCULATURE

Most growth factors have been described in the testis, but there are few data about the expression and the role of these factors in testicular vascular biology. The IGFs and their receptors, FGF (bFGF) and its receptors, TGFß and its receptors, NGF and NGF receptors, IL and IL receptors, as well as EGF and EGF receptors have all been found within different cell types of testis. The localization of these factors and their receptors as well as their functional role in the testis is reported in detail in the review given by Benahmed M. (1996). One growth factor responsible for both angiogenesis and vascular permeability is vascular endothelial growth factor (VEGF).

5. VEGF AND ITS RECEPTORS IN THE HUMAN TESTIS

VEGF was first described as a protein secreted by tumors and capable of causing increased vascular leakage (Senger et al. 1993). It was isolated from pituitary follicular cells (Ferrara & Henzel 1989) and from pituitary-derived folliculo-stellate cells (Gospodarowicz et al. 1989). It is a ~46 kD homodimeric glycopeptide with a specific mitogenic effect on vascular EC and is structurally related to the platelet-derived growth factor (Conn et al. 1990). Four isoforms of VEGF of 121, 165, 189 and 206 amino acids have been characterized (Tischer et al. 1991). The angiogenetic potency (Connolly et al. 1989; Keck et al. 1989; Leung et al. 1989) is the main effect of VEGF during the embryonic development, while in adults VEGF acts on the permeability of EC which are mainly resting cells (Brown et al. 1992; Breier et al. 1992). The permeability potency of VEGF/VPF is 50000 times higher than that of histamine (Connolly et al. 1989; Keck et al. 1989; Connolly 1991). An intradermal injection or topical application of VEGF according to the Miles Test (Miles & Miles 1952) shows an increasing permeability of postcapillary venules, muscular venules and capillaries within minutes (Roberts & Palade 1995). This correlates well with fenestrations in the capillaries and small venules (Roberts & Palade 1995). Receptors of VEGF belong to the tyrosine kinase receptor family and constitute a subgroup within the class III tyrosine kinases with seven immunoglobulin-like loops in their extracellular domain (Shibuya et al. 1990; Terman et al. 1991; Terman et al. 1992; Quinn et al. 1993). Two related high affinity receptors for VEGF, fms-like tyrosine kinase (Flt-1) and fetal liver kinase (Flk-1), designated as KDR in human, have been identified (Shibuya et al. 1990; Yamaguchi et al. 1993). Consistent with the fact that VEGF does not increase mitogenesis in any other cell type (Jakeman et al. 1992; Plate et al. 1992), both receptors are known to be expressed primarily on EC. The expression of these receptors at the highest levels in tissues with proliferating vascular EC (both in pathology and in normal development) has been well documented (Plate et al. 1993; Takagi et al. 1996). On the other hand, there are a few recent reports demonstrating the presence of VEGF receptor Flt-1 also in non-EC like bovine retinal pericytes (Takagi et al. 1996) and renal mesangial cells (Takahashi et al. 1995). VEGF receptor KDR has been shown to be expressed on ovarian tumor cells (Boocock et al. 1995). Recently, VEGF-B (Olofsson et al., 1996), VEGF-C (Joukov et al., 1996) and VEGF-related protein (VRP) (Lee et al., 1996) have been described. A detailed survey about structural, molecular and biological properties of VEGF and its receptors has been recently given by Klagsbrun and D'Amore (1996). VEGF-B is also mitogenic for EC while VEGF-C stimulates EC migration. VEGF-C is a ligand for the receptors Flt-4 and KDR while the receptor for VEGF-B has not yet been identified yet. The presence of VEGF mRNA has been demonstrated in fetal

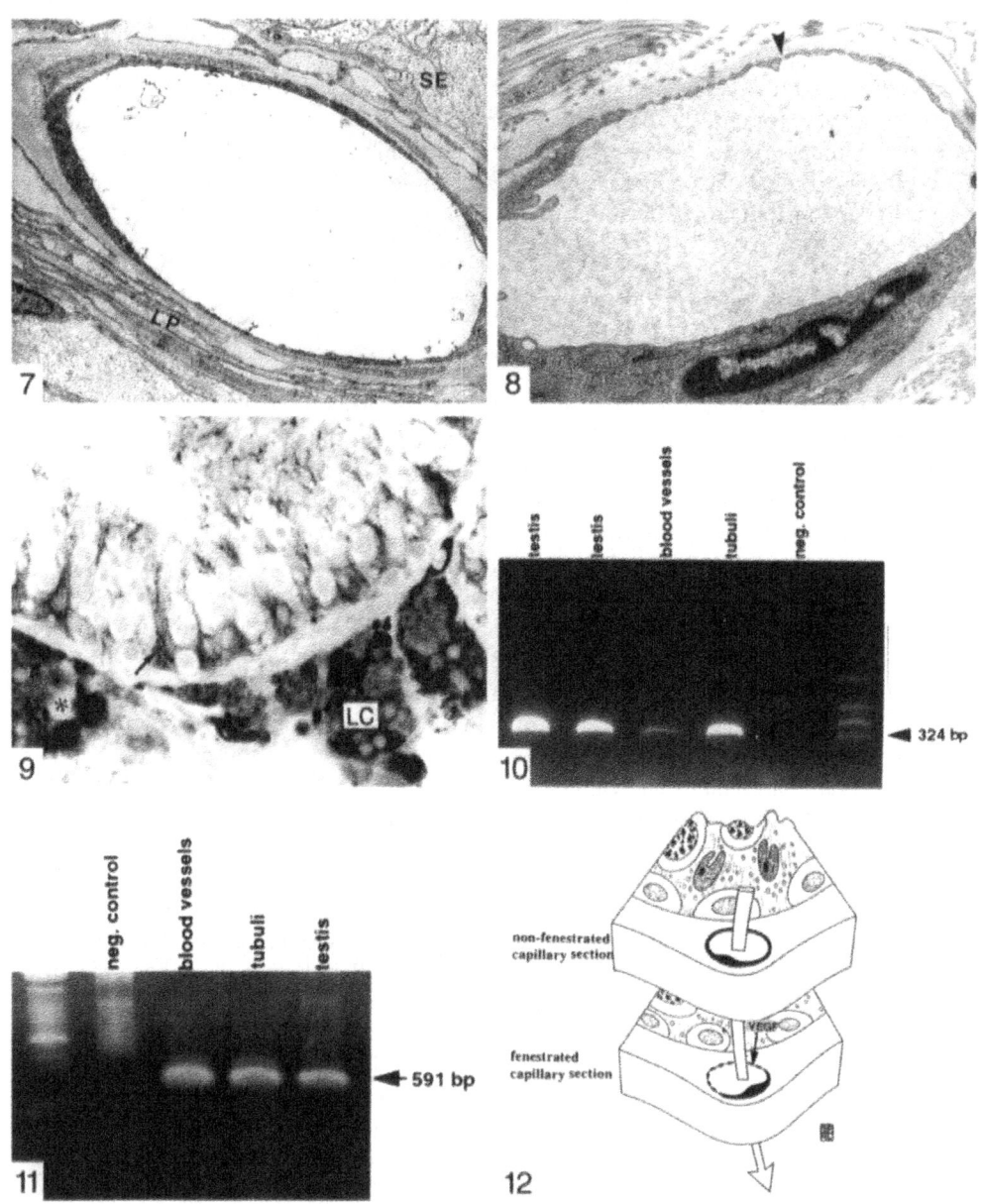

testicular tissue (Shifren et al. 1994) by means of a ribonuclase protection assay and northern blot analysis, and also in testicular germ cell tumors mRNA (Viglietto et al. 1996). However, expression of VEGF and its receptors in normal adult testes have not so far been examined in detail. In keeping with our interest in understanding the role of VEGF in the development of testicular microvasculature and in maintaining its function, we have studied the intratesticular sites of VEGF production, the sites of its actions and, we present additional evidence for a role of VEGF in human testicular physiology.

The expression of VEGF mRNA was studied in whole normal human testicular tissue, isolated fragments of seminiferous tubules and isolated fragments of testicular microvessels by means of RT-PCR using human VEGF specific primers. Agarose gel analysis by ethidium bromide staining of PCR products showed a double band migrating at positions corresponding to 648 and 516 bp for total testicular tissue and for isolated seminiferous tubules, whereas for isolated blood vessels from the testis no amplification was achieved (Ergün et al., 1996a; Ergün et al., 1997). In agreement with these data from RT-PCR, two protein bands corresponding to ~24 and ~49 kDa could be detected in Western blotting analyses for whole testicular tissue and seminiferous tubules, but no similar staining was observed for testicular blood vessels. Immunohistochemically, VEGF was present in both Leydig cells and Sertoli cells while blood vessels, germ cells and peritubular cells did not stain for VEGF (Figure 9, Ergün et al., 1997, MCE).

Furthermore, by means of RT-PCR we could show that mRNA for the VEGF receptor flt-1 is expressed in whole human testicular tissue, in isolated fragments of testicular microvessels and in seminiferous tubules (Figure 10, Ergün et al., 1997)

In agreement with the RT-PCR data, we observed strong immunohistochemical staining for flt-1 protein in Leydig and Sertoli cells as well as in the EC of the interstitial capillaries whereas peritubular and germ cells were negative. Flt-1 immunoreactivity could be detected also in the smooth muscle cells and EC of large interstitial arteries and arterioles. Also, the other specific VEGF receptor, KDR (human homolog for murine receptor flk-1), was found to be expressed in whole testicular tissue, isolated seminiferous tubules and in testicular microvessels based on RT-PCR data shown in Figure 11 (Ergün et al., 1997)

KDR protein was immunohistochemically localized in EC and pericytes of the capillaries in the interstitium as well as within the lamina propria of seminiferous tubules. In addition, Leydig and Sertoli cells were positive.

From these results we conclude that a) within the testis VEGF is probably produced in Sertoli and in Leydig cells, b) VEGF can act as a paracrine factor on the inter-Leydig

---

Figure 7. Cross-section of a non-fenestrated capillary localized within the lamina propria. SE: seminiferous epithelium, LP: lamina propria.

Figure 8. Another cross-section of a capillary within the lamina propria. This capillary part shows endothelial fenestrations which are closed by diaphragms (arrow head).

Figure 9. VEGF immunoreactivity is localized in interstitial Leydig cells (LC) and in Sertoli cells (arrow). Interstitial capillaries are negative (*) (X 740).

Figure 10. Flt-1 gene expression in human testes as detected by RT-PCR. A 324 bp fragment was amplified from RNA isolated from two testes of two different individuals and from TF-1 cell line which acted as positive control.

Figure 11. KDR gene expression in human testicular tissue as detected by RT-PCR. A 591 bp fragment corresponding to KDR was amplified using RNA from testicular tissue, isolated seminiferous tubules and isolated testicular blood vessels. The mRNA of the cell line TF-1 was used as positive control.

Figure 12. The graphic represents the hypothetic assumption of the influence of VEGF on the intramural capillary part of the human testicular capillary pathway produced and secreted by Sertoli cells which probably causes the fenestration of this capillary section.

cell capillaries via its receptors flt-1 and KDR and influence the permeability of these capillaries perhaps via the increase of transcytotic activity of the endothelium, c) at least VEGF produced and secreted by Sertoli cells may act as a paracrine factor via KDR on the intramural capillaries and may in part mediate the fenestration of these capillaries as described above (Figure 12, Ergün et al., 1997).

In addition, it can be assumed that VEGF may act as an autocrine factor on Leydig cells via the VEGF receptors KDR and flt-1 which were present in these cells also. An another interesting aspect ist the absence of VEGF staining in endothelial cells of interstitial blood vessels and of capillaries within the lamina propria, confirming that these endothelial cells are non-proliferative. But according to the recently published "balance hypothesis for the angiogenic switch" by Hanahan & Folkman (1996) the normally quiescent vasculature can be activated to sprout new capillaries. This process has been called angiogenesis (Folkman, 1986; Folkman & Shing 1992) which is probably controlled by changes in the relative balance of inducers and inhibitors of angiogenesis. This balance could also be of importance for the neovascularization of testicular tumors.

6. STEROID RECEPTORS ON THE HUMAN TESTICULAR VASCULATURE

Androgens in the form of testosterone or dihydrotestosterone are necessary for spermatogenesis and for development of the male reproductive organs. It was earlier reported that testosterone influences the composition of the interstitial fluid in the rat testis (Maddocks & Sharp, 1989). At the same time, other hormones like gonadotropins can have a similar effect (Sharp, 1979; Damber et al. 1981; Setchell & Sharp, 1981 Widmark et al., 1986). Like all other steroid hormones the androgens unfold their effects via the androgen receptor that belongs to the family of steroid/thyroid nuclear hormone receptors (Chang et al. 1988; Tan et al. 1988; Lubahn et al. 1990). The androgen bound to its receptor forms a ligand-receptor complex which translocates from the cytoplasm to the nucleus and there regulates the gene transcription. In the testis, testosterone is produced in Leydig cells and is necessary for the support of spermatogenesis. In this context, based particularly on the results obtained from rat testis, it has been suggested that testosterone acts as a paracrine factor on the peritubular and Sertoli cells (Sharpe 1986). In addition to the earlier ligand binding studies, the cloning of AR-cDNA (Chang et al. 1988) and the synthesis of the corresponding receptor peptide (Tan et al. 1988) promoted a number of studies concerning the androgen receptor during the past decade. In the rat testis, it has already been shown that smooth musclecells of the small arteries contain androgen receptors (Bergh & Damber 1992). These authors demonstrated also that testosterone could influence the vasomotion of rat testicular blood vessels.

Considering the different vascular organization in the human testis reported by Ergün et al. (1994a,b; see above) it was the aim of our studies to show the expression and localization of steroid receptors on the human testicular microvasculature. AR mRNA could be shown using a ribonuclease protection assay (RPA) in isolated capillaries and seminiferous tubules of the human testis (Ergün et al. 1995a). Figure 13 shows the detection of AR mRNA in testicular capillaries and in testicular arteries.

In further studies, we were able to show that isolated fragments of veins from human testis were negative for AR mRNA. Corresponding to these results by RPA we found AR immunoreactivity within the nuclei of the smooth muscle cells of the arterial wall, as well as of the endothelial cells and the pericytes of the arterial side and the intramural capillaries. No similar immunoreactivity was present in venous side capillaries, in venules and in the intralobular veins (Ergün et al. 1995a). Nuclear localization of the androgen receptor was also found in Leydig, peritubular and Sertoli cells. The immunohistochemical localization of the androgen receptor was confirmed by in situ hybridization. Estrogen receptor mRNA was detected by means of RT-PCR (figure 14).

and Southern hybridization in isolated human testicular blood vessels and in isolated human seminiferous tubules (Ergün et al. 1995a). Immunohistochemically, the estrogen receptor was found in the nuclei of the muscular layer of arteries and veins as well as of endothelial cells of all capillaries and Sertoli cells, whereas the peritubular cells were negative. Figure 15 summarizes the distribution of AR and ER in the normal human testicular tissue, particularly in the vasculature.

AR could be localized in arteries, arterioles, arterial-side inter-Leydig cell capillaries and intramural capillaries whereas venous-side capillaries, venules and veins were negative. ER, on the other hand, was found in all testicular blood vessels, in Leydig and Sertoli cells. With these results, we have been able to demonstrate for the first time the existence of androgen and oestrogen receptors in capillaries of the human testis. The results obtained allow the assumption that blood flow and permeability of the different parts of the capillaries may be influenced by androgens and estrogen. In the rat testis, it could be demonstrated that the blood flow is reduced by Leydig cell depletion but that after treatment with testosterone blood flow and vasomotion were normalized (Damber et al. 1992). The demonstration of AR in the smooth musculature of arteries was interpreted to suggest a possible influence of testosterone on precapillary arteries and subsequently the blood flow (Bergh and Damber 1992). Our results suggest that the arterial side capillaries of the human testis may be controlled by testosterone and that the capillaries are involved in local endocrine and paracrine regulatory mechanisms of spermatogenesis. The fact that ER are present in all human testicular blood vessels studied leads to the assumption that estrogen may be the basal steroid acting on all testicular blood vessels, whereas because of the presence of AR on the arterial side of the testicular vasculature androgens may play a role in the arterial supply of Leydig cells and seminiferous tubules.

7. VASOACTIVE SUBSTANCES IN THE TESTIS

Testicular blood flow and the permeability of testicular capillaries are the main parameters which influence the transport and exchange of hormones, nutritious and secreted substances from or to the testicular tissue. Therefore the regulation of these parameters is essential for the function of Leydig cells and for the maintenance of spermatogenesis. For the control of both blood flow and capillary permeability vasoactive substances are of importance. The functional role of some vasoactive substances for testicular microcirculation, like catecholamines and acetycholine, serotonin, histamine, prostaglandins, kallikrein, adenosine, arginine-vasopressin, the renin-angiotensin system, oxytocin and substance P has been reviewed by Bergh and Damber (1993) and by Setchell and Brooks (1994). We have concentrated our studies on the expression and the functional role of endothelin and its receptors in the human testis, particularly in the human testicular microvasculature.

8. ENDOTHELIN AND ITS RECEPTORS IN THE HUMAN TESTIS

Endothelin-1 (ET-1) was first described as a potent vasoconstrictor substance derived from porcine aortic endothelial cells (Yanagisawa et al. 1988). Since then it has been shown to be a multifunctional peptide composed of 21 amino acids. It belongs to a family of structurally related peptides, consisting of ET-1, ET-2 and ET-3. These peptides bind specifically to two types of receptors, endothelin receptor subtypes A and B (ET-A, ET-B) which belong to the superfamily of G protein-coupled receptors. ET-A and ET-B differ in their affinities for the endothelins: ET-B binds all members of the endothelin family of peptides with similar affinity whereas ET-A binds preferentially to ET-1 (Arai et al 1990). ET-1 is synthesized from preproendothelin 1, a precursor peptide consisting of 203 amino acids. Preproendothelin-1 is proteolytically cleaved to produce the 38 amino acid intermediate proendothelin-1 or big endothelin-1 (Big ET-1). Big ET-1 is subsequently converted into mature ET-1 by endothelin converting enzyme 1 (ECE-1) (Xu et al. 1994). ET-1 mediates various biological actions. It is able to cause contractions of vascular and nonvascular smooth muscle and possibly plays a role in both the local and systemic regulation of blood flow (Simonson & Dunn 1990). It has also been shown to be co-mitogenic for vascular smooth muscle cells (Weissberg et al. 1990). ET-1 acts as a neuropeptide (Koseki et al. 1989) and enhances steroid secretion by bovine adrenal glomerulosa cells (Cozza et al. 1992). In the human reproductive system, ET-1 or its receptors have been found in the uterus, testis, epididymis and seminal fluid (Davenport et al. 1991; Hammami et al. 1994; Maggi et al. 1995). Studies on the localization and function of ET-1 in the testis have mainly focussed on the rat, where it has been localized immunohistochemically in Sertoli cells (Fantoni et al. 1993). Within the testis, receptors for ET have been shown by means of autoradiography to be present in peritubular myoid cells and Leydig cells (Sakaguchi et al. 1992). ET-1 has been shown to regulate positively the steroidogenesis in isolated rat Leydig cells (Conte et al. 1993) and also to influence the intracellular calcium level of cultured rat peritubular myoid cells and Sertoli cells (Filippini et al. 1993, Sharma et al. 1994).

The aim of our study was to demonstrate the localization of Big ET-1, ET-1, ET-A and ET-B in different human testicular tissue components, particularly on the testicular vasculature, and to propose a functional role for ET-1 in the testicular microcirculation.

RT-PCR analyses and Southern hybridization showed that mRNA for ET-1 is expressed in both isolated fragments of human seminiferous tubules and in whole testicular tissue (Figures 16a, 16b), while the isolated fragments of testicular capillaries were negative in both methods. Expression of mRNA for ET receptor subtype A (ET-A) was detected in whole testicular tissue and in isolated fragments of seminiferous tubules by RT-PCR and Southern hybridization (Figures 16c, 16d) while with neither method could a corresponding signal be seen in isolated fragments of testicular capillaries. The mRNA for ET-receptor subtype B (ET-B) was found by both RT-PCR and Southern hybridization in whole testicular tissue, and in isolated fragments of seminiferous tubules (Figures 16e, 16f) while in isolated fragments of testicular capillaries a corresponding signal could be observed only by Southern hybridization (Figure 16f). Big ET-1 (not shown) and ET-1 immunoreactivity was found to be present in Leydig cells and Sertoli cells, while the endothelial cells of capillaries within Leydig cell clusters and within the lamina propria of seminiferous tubules were negative (Figure 17).

Note that early and mature spermatids displayed ET-1 immunoreactivity also while peritubular cells and germ cells were negative (Figure 17). ET-A immunoreactivity could be localized in Leydig cells, Sertoli cells and in a granular reaction in the adluminal compartment of the tubuli seminiferi (not shown). Light microscopical examination of sections counterstained with Calcium red revealed this intratubular granular reaction to be associated with the heads of early and mature spermatids (not shown). Large blood vessels, interstitial and intramural capillaries (capillaries localized within the lamina propria) as well

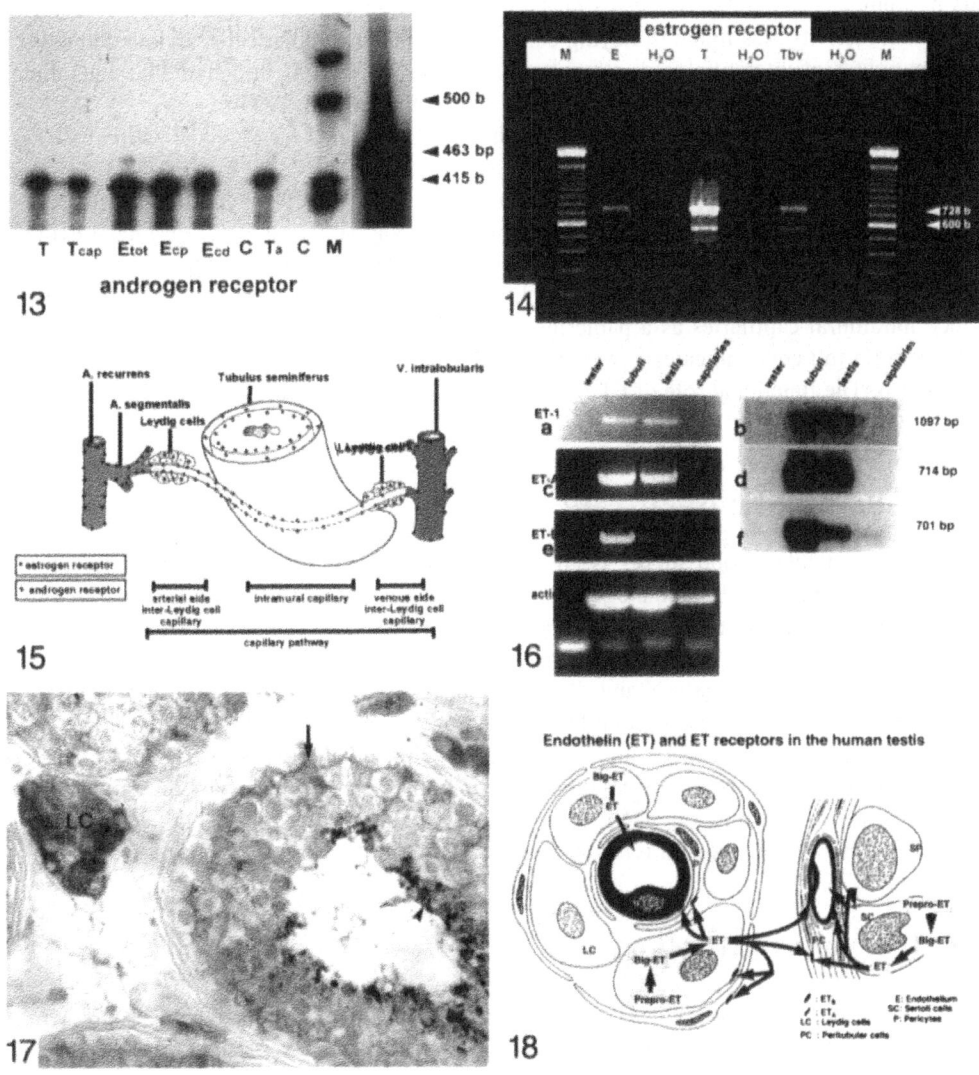

Figure 13. Detection of AR mRNA at 415 bp in testicular arteris (Ta) and capillaries (Tcap) by means of ribonuclease protection assay. T: whole testicular tissue, Etot: whole epididymal tissue, Ecp: caput epididymidis, Ecd: cauda epididymidis, C: negative control, M: marker.

Figure 14. Detection of ER mRNA in human testicular blood vessels (Tbv) at 728 bp by means of RT-PCR. The other band detected here could be arisen by alternative splicing. E: human epididymis, T: whole human testicular tissue.

Figure 15. This graphic shows in summary the distribution of AR and ERs in the human testis, particularly on the vasculature.

Figure 16 a,b. Detection of mRNA of ET-1 and ET receptors (ET-A and ET-B) in the human testicular tissue: In both RT-PCR and Southern hybridization, ET-1 and ET-A mRNAs are present in isolated seminiferous tubules and whole testicular tissue but not in isolated testicular capillaries while ET-B mRNA could be also found also in isolated human testicular capillaries by Southern hybridization.

Figure 17. ET-1 immunoreactivity could be localized in Leydig (LC) and Sertoli cells (arrow) as well as in mature spermatids (arrow head) (X 400).

Figure 18. This graphic represents the localization of ET-1 and the distribution of ET-A and ET-B in the human testis. It summarizes also our hypothesis of paracrine and autocrine action sites of ET-1 in the human testis.

as peritubular cells and germ cells showed no ET-A immunoreactivity. Like ET-A, ET-B immunoreactivity was found in Leydig cells, but also in peritubular cells, endothelial cells and pericytes of interstitial and intramural capillaries, as well as in endothelial cells and vascular smooth muscle cells of large blood vessels (Ergün et al. 1995b).

The graphic in Figure 18 summarizes the results obtained for the localization of ET-1 and its receptors in the human testis.

It can be concluded from our results that ET-1 is obviously produced in Sertoli and Leydig cells. ET-1 produced in Leydig cells can act as a paracrine factor on inter-Leydig cell capillaries. The presence of both ET receptors in Leydig cells indicates an autocrine effect of ET-1 on these cells. On the other hand ET-1 produced in Sertoli cells could influence intramural capillaries as a paracrine factor via receptor subtype B. The presence of ET-A in Sertoli cells indicates however that ET could act as an autocrine factor in Sertoli cells also. The functional role of ET receptor A in spermatids remains to be clarified.

9. CONCLUDING REMARKS

The testicular blood supply is an essential factor for the maintenance of testicular function, the support of spermatogenesis and hence for the fate of male germ cells. The blood supply is determined by blood flow, local microcirculation and capillary permeability. The results described here provide new perspectives for understanding the organization of the microvasculature and its functional significance in the performance of complicated regulatory processes in human spermatogenesis. We hope that with these results, particularly those concerning the human testicular microcirculation, have been able to emphasize the importance of vascular control in testicular function, and thus contribute a small part to the missing piece "vascular control of the testis" mentioned by Prof. Setchell in his plenary chapter (chapter 1).

10. ACKNOWLEDGMENTS

The authors are grateful to Mrs. S. Schwartz, Mrs. S. Verago, Mrs. M. Böge, Mrs. A. Salewski and Mrs. M. Schwartz for their excellent technical assistance. We are also very grateful to Mrs. M. Lück for the drawing of the graphics.

Supported by the Bundesminister für Forschung und Technologie, Bonn, Germany, as a part of a larger concerted project "Fertilitätsstörungen" (01 KY 9103) and in part supported by the Deutsche Forschungsgemeinschaft (Ho 388/6–1).

11. REFERENCES

Arai H, Hori S, Aramori I, Ohkubo H & Nakanishi S 1990 Cloning and expression of a cDNA encoding an endothelin receptor. Nature 348: 730–732.

Benahmed M 1996 Growth factors and cytokines in the testis. In: Male Infertility, pp 55–95. Ed Comhaire FH, Chapman & Hall, London.

Bennet HS, Luft JH & Hampton JC 1959 Morphological classification of vertebrate blood capillaries. American Journal of Physiology 196: 381–390.

Bergh A, Damber JE & Widmark A 1988 Hormonal control of testicular blood flow, microcirculation and vascular permeability. In: Molecular and Cellular Endocrinology of the Testis, pp 132–133. Eds Cooke BA & Sharpe RM, Raven Press, New York.

Bergh A & Damber JE 1992 Immunohistochemical demonstrations of androgen receptors on testicular blood vessels. International Journal of Andrology 15: 425–434.

Bergh A & Damber JE 1993 Vascular controls in testicular physiology. In: Molecular Biology of the Male Reproductive System, pp 439–468. Ed de Kretser D, Academic Press Inc.

Boocock CA, Charnock-Jones DS, Sharkey AM, McLaren J, Barker PJ, Wrigth KA, Twentyman PR, & Smith SK 1995 Expression of vascular endothelial growth factor and its receptors flt and KDR in ovarian carcinoma. Journal of the National Cancer Institute 87: 506–16.

Breier G, Albrecht U, Sterrer S & Rissau W 1992 Expression of vascular endothelial growth factor during embrionic angiogenesis and endothelial cell differentiation. Development 114:521–532.

Brown LF, Yeo KT, Berse B, Yeo TK, Senger DR, Dvorak HF & Van De Water L 1992 Expression of vascular permeability factor (vascular endothelial growth factor) by epidermal keratinocytes during wound healing. Journal of Experimental Medicine 176: 1375–1379.

Chang CS, Kokontis J & Liao ST 1988 Molecular cloning of human and rat complementary DNA encoding androgen receptors. Science 240:324–326.

Conn G, Soderman DD, Schaeffer MT, Wile M, Hatcher VB & Thomas KA 1990 Purification of a glycoprotein vascular endothelial cell mitogen from a rat glioma-derived cell line Proceedings of the National Academy of Sciences USA, 87: 1323–1327.

Connolly DT, Olander JV, Heuvelman D, Nelson R, Monsell R, Siegel N, Haymore BL, Leimgruber R, & Feder J 1989 Human permeability factor. Isolation from u937 cells. Journal of Biological Chemistry 264: 20017–20024.

Connolly DT 1991 Vascular permeability factor: A unique regulator of blood vessel function. Journal of Cellular Biochemistry 47: 219–223.

Conte D, Questino P, Fillo N, Isidori A & Romanelli F 1993 Endothelin stimulates testosterone secretion by rat Leydig cells. Journal of Endocrinology 136: R1-R4.

Cozza EN, Chiou S & Gomez-Sanchez CE 1992 Endothelin-1 potentiation of angiotensin II stimulation of aldosterone production. American Journal of Physiology 262: R85-R89.

Damber JE, Selstam G & Wang J 1981 Inhibitory effect on estradiol-17ß on human chorionic gonadotrophin-induced increment of testicular blood flow and plasma testosterone concentration in rats. Biology of Reproduction 25: 555–559.

Damber JE, Bergh A & Widmark A 1987 Effect of an LHRH-agonist on testicular microcirculation in hypophysectomized rats. International Journal of Andrology 10: 785–791.

Damber JE, Bergh A & Widmark A 1989 Effects of hormones on testicular microvasculature. In Perspectives in Andrology, vol 53: pp 97–109. Ed. M. Serio, Raven Press New, York.

Damber JE & Bergh A 1992 Testicular microcirculation-a forgotten essential in andrology? International Journal of Andrology 15: 285–292.

Damber JE, Madocks S, Widmark A & Bergh A 1992 Testicular blood flow and vasomotion can be maintained by testosterone in Leydig cell-depleted rats. International Journal of Andrology 15: 385–393.

Davenport AP, Cameron IT, Smith SK & Brown MJ 1991 Binding sites for iodinated endothelin-1, endothelin-2 and endothelin-3 demonstrated on human uterine glandular epithelial cells by quantitative high-resolution autoradiography. Journal of Endocrinology 129: 149–154. .

Davidoff MS, Breucker H, Holstein AF & Seidl K 1990 Cellular architecture of the lamina propria of human seminiferous tubules. Cell and Tissue Research 262: 253–261.

Ergün S, Stingl J & Holstein AF 1994a Segmental angioarchitecture of the testicular lobule in man. Andrologia 26: 143–150.

Ergün S, Stingl J & Holstein AF 1994b Microvasculature of the human testis in correlation to Leydig cells and seminiferous tubules. Andrologia 26: 235–262.

Ergün S & Ungefroren H 1995a Androgen and Estrogen receptors in the microvasculature of the human testis. Acta Anatomica 152: 278–279.

Ergün S, Ungefroren H & Holstein AF 1995b Vasoaktive Substanzen im menschlichen Hoden- Hinweise auf para- und autokrine Regulation der testikulären Mikrozirkulation. Verhandlungen der Anatomischen Gesellschaft Vol 177 (Suppl): 14–15.

Ergün S, Killic N, Fiedler W & Mukhopadhyay AK 1996a Vascular endothelial growth factor and its receptors in the human testicular tissue. Miniposter, 9th Eur. Workshop on Molecular and Cellular Endocrinology of the Testis, Geilo, Norway, c17.

Ergün S, Davidoff M & Holstein AF 1996b Capillaries in the lamina propria of human seminiferous tubules are partly fenestrated. Cell and Tissue Research, 286: 93–102.

Ergün S, Kilic N, Fiedler W & Mukhopadhyay AK 1997 Vascular endothelial growth factor (VEGF) and its receptors in the normal human testicular tissue. Molecular and Cellular Endocrinology (in press).

Fantoni G, Morris PL, Forti G, Vannelli GB, Orlando C, Barni T, Sestini R, Danza G & Maggi M 1993 Endothelin-1: a new autocrine/paracrine factor in rat testis. American Journal of Physiology 265: E267-E274.

Fawcett DW, Heidger PJ & Leak LV 1969 The lymph-vascular system of the interstitial tissue of the testis as revealed by electron microscopy. Journal of Reproduction and Fertility 19: 109–119.

Fawcett DW, Leak LV & Heidger PJ 1970 Electron microscopic observations on the structural components of the blood-testis barrier. Journal of Reproduction and Fertility, Suppl. 10: 105–119.

Ferrara N & Henzel WJ 1989 Pituitary follicular cells secrete a novel heparin-binding growth factor specific for vascular endothelial cells. Biochemical and Biophysical Research Communications 161: 850–858.

Filippini A, Tripiciano A, Palombi F, Teti A, Paniccia R, Stefanini M & Ziparo E 1993 Rat testicular myoid cells respond to endothelin: characterization of binding and signal transduction pathway. Endocrinology 133: 1789–1796.

Folkman J 1986 How is blood vessel growth regulated in normal and neoplasmatic tissue? G.H.A. Clowes memorial award lecture. Cancer Research 46: 467–473.

Folkman J & Shing Y 1992 Angiogenesis. Journal of Biological Chemistry 267: 10931–10934.

Gospodarowicz D, Abraham JA & Schilling J 1989 Isolation and characterization of a vascular endothelial cell mitogen produced by pituitary derived folliculo stelate cells. Proceedings of the National Academy of Sciences USA 86: 7311–7315.

Hammami MM, Haq A & AlSedairy S 1994 The level of endothelin-like immunoreactivitiy in seminal fluid correlates positively with semen volume and negatively with plasma gonadotrophin levels. Clinical Endocrinology 40: 361–366.

Hanahan D & Folkman J. 1996 Patterns and emerging mechanisms of the angiogenic switch during tumorigenesis. Cell 86: 353–364.

Holstein AF, Orlandini GE & Möller R 1979 Distribution and fine structure of the lymphatic system in the human testis. Cell and Tissue Research 200: 15–27.

Holstein AF 1988 In: Illustrated Pathology of Human Spermatogenesis, pp 76–116. Eds Holstein AF, Rosen-Runge EC, Schirren C, Grosse, Berlin.

Hundeiker M & Keller L 1963 Die Gefäßarchitektur des menschlichen Hodens. Morphologisches Jahrbuch 105:26–73.

Hundeiker M & Mullert LV 1966 Vermeidbare Risiken bei der Hodenbiopsie. Der Hautarzt 17:546–547.

Hundeiker M 1971 Die Kapillaren im Hodenparenchym. Archive für klinische und experimentelle Dermatologie 239:426–435.

Jakeman LB, Winer J, Bennet GL, Altar CA, & Ferrara N 1992. Binding sites for vascular endothelial growth factor are localized on endothelial cells in adult rat tissues. Journal of Clinical Investigation 89: 244–253.

Keck PJ, Hauser SD, Krivi G, Sanzo K, Warren T, Feder J & Connolly DT 1989 Vascular permeability factor, an endothelial cell mitogen related to PDGF. Science 246: 1309–1312.

Jukov V, Pajusola K, Kaipainen A, et al. 1996 A novel vascular endothelial growth factor, VEGF-C is a ligand for the Flt-4 (VEGFR-3) and KDR (VEGFR-2) receptor tyrosine kinases. EMBO Journal 15: 290–298

Klagsbrun M & D'Amore P 1996 Vascular endothelial growth factor and its receptors. Cytokine and Growth Factor Reviews 7(3): 259–270.

Kormano M 1967a An angiographic study of the testicular vasculature in the postnatal rat. Zeitschrift für Anatomie und Entwicklungsgeschichte 126:138–153.

Kormano M 1967b Dye permeability and alkaline phosphatase activity of testicular capillaries in the postnatal rat. Histochemie 9: 327–338.

Kormano M & Souranta H 1971a Microvascular organisation of the adult human testis. Anatomical Records 170:31–40.

Kormano M & Souranta H 1971b An angiographic study of the arterial pattern of the human testis. Anatomischer Anzeiger 128:69–76.

Koseki C 1989 Autoradiographic distribution in rat tissues of binding sites for endothelin: a neuropeptide? American Journal of Physiology 256: R858-R866.

Lee J, Gray A, Yuan J, Louh SM, Avraham H & Wood W 1996 Vascular endothelial growth factor-related protein: A ligand and specific activator of the tyrosine kinase receptor Flt-4. Proceedings of the National Academy of Sciences USA 93: 1988–1992.

Leung DW, Cachianes G, Kuang WJ, Goeddel DV & Ferrara N 1989 Vascular endothelial growth factor is a secreted angiogenic mitogen. Science 246: 1306–1309.

Maddocks S & Sharpe RM 1989 Dynamics of testosterone secretion by the rat testis: Implications for measurements of intratesticular levels of testosterone. Journal of Endocrinology 120: 323–329.

Maggi M, Barni T, Orlando C, Fantoni G, Finetti G, Vannelli GB, Mancina R, Gloria L, Bonaccorsi L, Yanagisawa M & Forti G 1995 Endothelin-1 and its receptors in human testis. Journal of Andrology 16: 213–224.

Meyerhoffer A, Sinha Hikim AP, Bartke A & Russel LD 1989 Changes in the testicular microvasculature during photoperiod-related seasonal transition from reproductive quiescence to reproductive activity in the adult golden hamster. Anatomical Record 224: 495–507.

Meyerhoffer A & Bartke A 1990 Developing testicular microvasculature in the golden hamster, Mesocricetus auratus: a model for angiogenesis under physiological conditions. Acta Anatomica 139: 78–85.

Miles AA & Miles EM 1952 Vascular reactions to histamine, histamin liberators, or leukotoxins in the skin of guinea pigs. Journal of Physiology 118: 228–257.

Müller I 1956 Kanälchen und Capillararchitektonik des Rattenhodens. Zeitschrift für Zellforschung 45:522- 537.

Olofsson B, Pajusola K, Kaipainen A, et al. 1996 Vascular endothelial growth factor B, a novel growth factor for endothelial cells. Proceedings of the National Academy of Sciences USA 93: 2576–2581.

Plate KH, Breier G, Weich HA & Risau W 1992 Vascular endothelial growth factor is a potential tumour angiogenesis factor in human gliomas in vivo. Nature 359: 845–848.

Plate KH, Breier G, Millauer B, Ullrich A & Risau W 1993 Up-regulation of vascular endothelial growth factor nad its cognate receptors in a rat glioma model of tumour angiogenesis. Cancer Research 53: 5822–5827.

Qu-Hong, Nagy JA, Senger DR, Dvorak HF & Dvorak AM 1995 Ultrastructural localization of vascular permeability / vascular endothelial growth factor (VPF / VEGF) to the abluminal plasma membrane and vesiculovacuolar organelles of tumor microvascular endothelium. Journal of Histochemistry and Cytochemistry 43: 381–389.

Quinn TP, Peters KG, De VC, Ferrara N & Williams LT 1993 Fetal liver kinase I is a receptor for vascular endothelial growth factor and is selectively expressed in vascular endothelium. Proceedings of the National Academy of Sciences USA 90: 7533–7537.

Roberts WG & Palade GE 1995 Increased microvascular permeability and endothelial fenestration induced by vascular endothelial growth factor. Journal of Cell Science 108: 2369–2379.

Sakaguchi H, Kozuka M, Hirose S, Ito T & Hagiwara H 1992 Properties and localization of endothelin-1-specific receptors in rat testicles. American Journal of Physiology 263: R15-R18.

Schnittler HJ, Wilke A, Gress TH, Suttorp N & Drenckhahn D 1990 Role of actin and myosin in the control of paracellular permeability in pig, rat and human vascular endothelium. Journal of Physiology 431: 379–401.

Senger DR, Van De Water L, Brown L, Nagy J, Yeo KT, Yeo TK, Berse B, Jackman R, Dvorak A & Dvorak H 1993 Vascular permeability factor (VPF, VEGF) in tumor biology. Cancer Metastasis Reviews 12: 303–324.

Setchell BP 1990 Local controls of testicular fluids. Reproduction, Fertility and Development 2: 291–309.

Setchell BP 1994 Anatomy, vasculature, innervation, and fluids of the male reproductive tract. In:) The Physiology of Reproduction, pp 1065–1175. Eds Knobil E, Neill J et al. Raven Press, New York

Setchell BP & Sharpe RM 1981 The effect of human chorionic gonadotrophin on capillary permeability, extracellular fluid volume and flow of lymph and blood in the testis of rats. Journal of Endocrinology 91: 245–254.

Setchell BP & Rommerts FFG 1986 The importance of Leydig cells in the vascular response to hCG in rats. International Journal of Andrology 8:436–440.

Sharma OP, Flores JA, Leong DA & Veldhuis JD 1994 Mechanisms by which endothelin-1 stimulates increased cytosolic free calcium ion concentrations in single rat Sertoli cells. Endocrinology 135:127–134.

Sharpe RM 1986 Paracrine control of the testis. Clinics in Endocrinology and Metabolism, 15:185–207.

Sharpe RM 1990 Intratesticular control of steroidogenesis. Clinical Endocrinology 33: 787–807.

Shibuya M, Yamaguchi S, Yamane A, Ikeda T, Tojo A, Matsushime H & Sato M 1990 Nucleotide sequence and expression of a novel human receptor type tyrosin kinase gene (flt) closly related to the fms family. Oncogene 5: 519–524.

Shifren JL, Doldi N, Ferrara N, Mesiano S & Jaffe RB 1994 In the human fetus, vascular endothelial growth factor is expressed in epithelial cells and myocytes, but not vascular endothelium: implication for mode of action. Journal of Clinical Endocrinology and Metabolism 79: 316–22.

Simonson MS & Dunn MJ 1990 Cellular signaling by peptides of the endothelin gene family. FASEB Journal 4: 2989–3000.

Simionescu N & Simionescu M 1987 Receptor-mediated transcytosis of albumin: Identification of albumin binding proteins in the plasma membrane of capillary endothelium. In: Proceedings of the IV World Congress on Microcirculation, Elsevier, Amsterdam and New York.

Simionescu N & Simionescu M 1988 The cardiovascular system. In: Cell and Tissue Biology, pp. 355–398. Ed Leon Weiss, Urban und Schwarzenberg Baltimore-Munich.

Suzuki F 1982 Microvasculature of the mouse testis and excurrent duct system. American Journal of Anatomy 163: 309–325.

Suzuki F & Nagano T 1986 Microvasculature of the human testis and excurrent duct system. Cell and Tissue Research 243: 79–89.

Tan JA, Joseph DR, Quarmby VE, Lubahn DB, Sar M, French FS & Wilson EM 1988 The rat androgen receptor: primary structure, autoregulation of its mesenger ribonucleic acid and immunocytochemical localization of the receptor protein. Molecular Endocrinology 2: 1276–1285.

Takagi H, King GL & Aiello LP 1996 Identification and characterization of vascular endothelial growth factor receptor (Flt) in bovine retinal pericytes. Diabetes 45: 1016–1023.

Terman BI, Carrion ME, Kovacs E, Rasmusen BA, Eddy RL & Shows TB 1991 Identification of new endothelial cell growth factor receptor tyrosin kinase. Oncogene 6: 1677–1683.

Terman BI, Dougher Vermazen M, Carrion ME, Dimitrov D, Armellino DC, Gospodarowicz D & Bohlen P 1992 Identification of the KDR tyrosin kinase as a receptor for vascular endothelial cell growth factor. Biochemical and Biophysical Research Communications 187: 1579–1586.

Tischer E, Mitchell R, Hartman T, Silva M, Gospodarowicz D, Fiddes JC & Abraham JA 1991 The human gene for vascular endothelial growth factor. Multiple protein forms are encoded through alternative exon splicing. Journal of Biological Chemistry 266: 11947–11954.

Viglietto G, Romano A, Maglione D, Rambaldi M, Paoletti I, Lago CT, Califano D, Monaco C, Mineo A, Santelli G, Manzo G, Botti G, Chiappetta G & Persico MG 1996 Neovascularization in human germ cell tumors correlates with a marked increase in the expression of the vascular endothelial growth factor but not the placenta-derived growth factor. Oncogene 13: 577–87.

Weerasooriya TR & Yamamoto T 1985 Three-dimensional organisation of the vasculature of the rat spermatic cord and testis. Cell and Tissue Research 241: 317–323.

Weissberg PL, Witchell C, Davenport AP, Hesketh TR & Metcalfe JC 1990 The endothelin peptide ET-1, ET-2, ET-3 and sarafotoxin S6b are co-mitogenic with platelet-derived growth factor for vascular smooth muscle cells. Arteriosclerosis 85: 257–262.

Widmark A, Damber JE & Bergh A. 1986 The relationship between human chorionic gonadotrophin-induced changes in testicular microcirculation and the formation of testicular interstitial fluid. Journal of Endocrinology 109: 419–425.

Wolff J & Merker HJ 1966 Ultrastruktur und Bildung von Poren im Endothel von porösen und geschlossenen Kapillaren. Zeitschrift für Zellforschung 73: 174–191.

Xu D, Emoto N & Giaid A 1994 A membrane-bound metalloprotease that catalyzes the proteolytic activation of big endothelin-1. Cell 78: 473–485.

Yamaguchi TP, Dumont DJ, Conlon RA, Breitman ML & Rossant J 1993 flk-1, an flt-related receptor tyrosine kinase is an early marker for endothelial cell precursors. Development 118: 489–498.

Yanagisawa M, Kurihara H, Kimura S, Tomobe Y, Kobayashi M, Mitsui Y, Yazaki Y, Goto K & Masaki T 1988 A novel potent vasoconstrictor peptide produced by vascular endothelial cells. Nature 332: 411–415.

33

EXPRESSION OF VEGF AND ITS RECEPTORS AND CAPILLARY DENSITY IN LEYDIG CELL TUMORS OF THE HUMAN TESTIS

N. Kilic,[1] W. Fiedler,[2] A. F. Holstein,[1] and S. Ergün[1]

[1]Institute of Anatomy
[2]Department of Hematology and Oncology
University of Hamburg (UKE)
Martinistr. 52, 20246 Hamburg
Germany

INTRODUCTION

A close association can be observed between the Leydig cells and the capillaries of normal human testis (Ergün et al., 1996). A prerequisite for tumor growth is neoangiogenesis (Folkman, 1982). One of the angiogenetic factors responsible is vascular endothelial growth factor (VEGF) which is a very potent mitogenic factor for endothelial cells (Conn et al., 1990) and increases the capillary permeability by the induction of fenestrations (Roberts and Palade, 1995). The role of VEGF in the growth of Leydig cell tumors and their capillarization has not so far been studied.

MATERIALS AND METHODS

Densitometric studies were carried out for analysis of the capillarization of Leydig cell tumors of the human testis. ELISA was used to detect VEGF peptide in serum obtained from human pampiniform plexus. Immunohistochemistry for VEGF and its receptors flt-1, flt-4 and KDR was performed on 5 cases of Leydig cell tumors. The ultrastructure of tumor blood vessels was studied by means of electron microscopic examination.

RESULTS

There is a marked increase of capillary density from the center to the periphery of Leydig cell tumors. In pampiniform serum of one patient with a Leydig cell tumor, VEGF peptide, detected by ELISA, was considerably higher than in normal serum. Endothelial cells of capillaries in the periphery of Leydig cell tumors stained more strongly than in capillaries in the center while capillaries within Leydig cell clusters of normal testis stained negatively. Leydig cells were positive in both normal testis and in Leydig cell tumors. Flt-1 immunoreactivity can be found in endothelial cells of capillaries in the periphery of, and adjacent to Leydig cell tumors. A strong flt-4 immunoreactivity was observed in endothelial cells of peripheral blood vessels in these tumors as well as in some tumor Leydig cells. KDR immunostaining was found in some Leydig cells and faintly also in endothelial cells of all tumor blood vessels. Electron microscopic examination of tumor capillaries revealed fenestrations and gaps of the capillary wall.

DISCUSSION

The positive VEGF immunoreactivity in endothelial cells of tumor capillaries and the higher capillary density at the periphery of these tumors lead to the assumption that the endothelial cells are probably proliferative in this area. The presence of all studied VEGF receptors, particularly of KDR and flt-4 in the endothelial cells of blood vessels in the same area supports the assumption that VEGF plays an important role in tumor neovascularization. The markedly higher level of VEGF in serum of pampiniform plexus from a tumor patient than in normal serum indicates that VEGF could be produced and secreted by tumor Leydig cells. The functional role of VEGF and its receptors present in some tumor Leydig cells remains to be clarified. In contrast to normal inter-Leydig cell capillaries, the open gaps and absent basal lamina of some tumor capillaries indicate that these capillaries could be newly formed and therefore represent tumor neovascularization.

REFERENCES

Conn G, Soderman DD, Schaeffer MT, Wile M, Hatcher VB & Thomas KA 1990 Purification of a glycoprotein vascular endothelial cell mitogen from a rat glioma-derived cell line. Proceedings of the National Academy of Sciences USA 87: 1323–1327.

Ergün S, Davidoff M & Holstein AF 1996 Capillaries in the lamina propria of human seminiferous tubules are partly fenestrated. Cell and Tissue Research 286: 93–102.

Folkman J 1982 Angiogenesis: initiation and control. Annals of the New York Academy of Sciences 401: 212–227.

Roberts WG & Palade GE 1995 Increased microvascular permeability and endothelial fenestration induced by vascular endothelial growth factor. Journal of Cell Science 108: 2369–2379.

34

ANGIOARCHITECTURE OF THE HUMAN SPERMATIC CORD

S. Ergün,[1] T. Bruns,[2] and R. Tauber[2]

[1]Institute of Anatomy
University of Hamburg
Martinistr.52, 20246 Hamburg, Germany
[2]General Hospital Barmbek
Department of Urology
Rübenkamp 148, 22291 Hamburg
Germany

INTRODUCTION

In the spermatic cord of all species investigated a tightly organized venous network around the testicular artery has been described, with certain species-specific differences (Hess et al., 1984; Noordhuizen-Stassen et al., 1985; Rerkamnuaychoke et al., 1991a). Several functions have been assigned to this intimate association between the testicular artery and the pampiniform plexus. These include as the regulation of temperature and arterial pulse in the testis as well as exchange of gas and hormones between the testicular artery and the pampiniform plexus (Waites and Moule, 1961; Jacks and Setchell, 1973; Hess et al., 1984). However, our knowledge about the vascular organization of the human pampiniform plexus is not sufficient to explain the mechanistic aspects of hormonal exchange between the testicular artery and pampiniform plexus.

MATERIAL AND METHODS

40 human spermatic cords were investigated by means of casting preparations, light microscopic examinations, radiographic analyses and computer aided 3-D-reconstructions on the basis of serial sections of paraffin-embedded material.

RESULTS

After leaving the testis, the testicular veins form two principal groups coexisting side by side. Numerous veno-venous anastomoses could be observed within each individual group, whereas only a few mutual intergroup anastomoses were found. The testicular artery runs within one group without showing close topographic relationship to the other group: The vein group without close topographic relationship to the testicular artery runs for several centimeters embedded within fatty tissue. Computer-aided 3-D-reconstructions provided a spatial picture of the vascular organization. These results permit the following classification of the veins of the pampiniform plexus: Group I: Veins which by means of veno-venous anastomoses form a tight plexus around the testicular artery. Group II: Veins with veno-venous anastomoses between each other and running at a sizeable distance embedded in fatty tissue, but without close topographical relationship to the testicular artery. Group III: Veins forming veno-venous anastomoses between groups I and II. Group IV: Veins forming arterio-venous anastomoses with the testicular artery.

DISCUSSION

This organization of the veins of the human pampiniform plexus reveals that smaller veins near to the testicular artery formed arteriovenous anastomoses on the one hand and venovenous anastomoses with the larger veins on the other hand. This could be of importance for further understanding of physiological processes like transfer of hormones and other substances from the veins to the testicular artery and vice versa. It should also facilitate tracing of the veins during antegrade sclerosing.

REFERENCES

Hess H, Leiser R, Kohler T & Wrobel KH 1984 Vascular morphology of the bovine spermatic cord and testis. Cell and Tissue Research 237: 31–38.

Jacks F & Setchell BP 1973 A technique for studying the transfer of substances from venous to arterial blood in the spermatic cord of wallabies and rams. Journal of Physiology 233: 17P-18P.

Noordhuizen-Stassen EN, Charbon GA, de Jong FH & Wensing CJG 1985 Functional arterio-venous anastomoses between the testicular artery and the pampiniform plexus in the spermatic cord of rams. Journal of Reproduction and Fertility 75: 193–201.

Rerkamnuaychoke W, Nishida T, Kurohmaru M & Hayashi Y 1991a Morphological studies on the vascular architecture in the boar spermatic cord. Journal of Veterinary Medical Sciences 53: 233–239.

Waites GMH & Moule GR 1961 Relation of vascular heat exchange to temperature regulation in the testis of the ram. Journal of Reproduction and Fertility 2: 213–224.

35

MORPHOLOGICAL AND FUNCTIONAL ASPECTS OF THE HUMAN SPERMATIC CORD VEINS

T. Bruns,[1] S. Ergün,[2] and R. Tauber[1]

[1]Department of Urology
Barmbek General Hospital
Rübenkamp 148, 22291 Hamburg, Germany
[2]Institute of Anatomy
University of Hamburg (UKE)
Martinistr. 52, 20246 Hamburg
Germany

INTRODUCTION

It is well known that varicocele of the testis may lead to a damage of testicular tissue and therefore is one etiologic factor for male infertility. Successful varicocele treatment may result in preserving or restoring fertility. Antegrade sclerotherapy has proved to be safe, economical and effective (Bruns et al., 1996). The major component of this procedure is the preparation of the pampiniform plexus. After insertion of a canula into a selected vein of the plexus and fluoroscopic control showing drainage by the internal spermatic vein, a sclerosing agent (Äthoxysklerol[R]) is injected in an antegrade direction. By analysis of the vascular organization of the pampiniform plexus (Ergün et al., 1994) it could be demonstrated that the testicular veins are organized in two main groups and form two vein plexus. One of these vein groups forms a tight plexus around the testicular artery, the other vein group is embedded in a large macroscopically visible section of fatty tissue. To save the testicular artery from injury during dissection, the vein group in this lateroventral compartment should be dissected for antegrade sclerotherapy.

The objective of this study was to assess whether preparation of a vein in the fatty embedded vein group leads to drainage by the internal spermatic vein.

PATIENTS AND METHODS

256 patients were prepared for antegrade sclerotherapy because of left-sided varicocele. None of the idiopathic varicoceles had been treated previously. The dissection of a vein of the pampiniform plexus was performed as described previously (Tauber et al., 1993, 1994, Mottrie et al., 1995).The intention is to select a dilated and straight vein of the

fatty embedded vein group. The selected vein is exposed and distally ligated. After incision, a 24 gauge thin-walled canula is inserted into the vein in an antegrade fashion and secured with a single ligature. Phlebography is performed by injecting 5ml of nonionic contrast medium under fluoroscopic control. After documentation of contrast drainage through the internal spermatic vein subsequent sclerosing procedere begins.

RESULTS

Canulating a vein in the fatty embedded group was possible in 255 of 256 patients (99,6%). The venous drainage through the internal spermatic vein could be demonstrated by phlebography with its typical vessel variations in all of these 255 cases. None of the phlebographies showed simultaneous drainage through the external spermatic vein and/or vein of the vas deferens. In 5 cases (2%) the correct venous group could not be identified on the first examination. A second dissection was necessary and successful in these cases. Phlebographic examination on the first occasion showed drainage through the external spermatic vein or ductal vein. None of these phlebographies showed simultaneous drainage also through the internal spermatic vein. In 1 case with a subclinical varicocele no canulation of a vein was possible (0,4% technical failure). In none of the 256 operations had the testicular artery been dissected or canulated.

DISCUSSION

In the literature, numerous references are made to anastomoses between the pampiniform plexus, or the distal internal spermatic vein, and the external (cremasteric) spermatic vein (Harrison, 1966). Antegrade sclerotherapy may only be done, when no drainage is visible to the external spermatic vein or ductal vein. We were unable to see any anastomoses to these veins in 255 fluoroscopic controls. There was no visualization, neither of the cremasteric nor the ductal vein, or of their path of drainage via the iliac vessels when the contrast medium is drained by the internal spermatic vein. The internal spermatic vein is usually visualized during canulation of a vein in the lateroventral compartment with the typical yellowish fat. Therefore this part of pampiniform plexus appears to offer a safe and sufficient approach for antegrade scrotal sclerotherapy of varicocele and saves the spermatic artery from injury. Phlebographic studies showed that this approach seems to be selective for the internal spermatic vein. Nevertheless antegrade phlebography is essential to prevent the external spermatic or ductal vein from being sclerosized.

REFERENCES

Bruns T & Tauber R. 1996 Antegrade Sklerosierung der Vena testicularis. In: Moderne Aspekte der Diagnostik und Therapie der Varicocele testis. Podium Urologie Bd.2, pp 113–122. Eds. D Fahlenkamp, S Lenk, W Weidner. Blackwell Wissenschafts-Verlag, Berlin Wien.

Ergün S, Bruns T, Holstein AF & Tauber R 1994 Die Angioarchitektur des menschlichen Samenstranges. Der Urologe A Suppl. 1/94: S76.

Harrison RG 1966 The Anatomy of Varicocele. Proceedings of the Royal Society of Medicine 59: 763–765.

Mottrie AM, Bürger RA, Voges GE, Baert L & Hohenfellner R 1994 Die antegrade skrotale Sklerotherapie der Varicocele testis. AktuelleUrologie OperativeTechniken 25: I-VI.

Tauber R & Johnsen N 1993 Die antegrade skrotale Verödung zur Behandlung der Testisvarikozele. Der Urologe A 32: 320–326.

Tauber R & Johnsen N 1994 Antegrade scrotal sclerotherapy for the treatment of varicocele: technique and late results. Journal of Urology151: 386–390.

36

SEMEN ANALYSIS AFTER TREATMENT OF VARICOCELE BY ANTEGRADE SCROTAL SCLEROTHERAPY

N. Johnsen,[1] I. Johnsen,[2] and R. Tauber[1]

[1]Department of Urology
Barmbek General Hospital
Rübenkamp 148, 22291 Hamburg, Germany
[2]Department of Internal Medicine
Eilbek General Hospital
Friedrichsberger Str. 60, 22081 Hamburg
Germany

INTRODUCTION

Varicocele is a major cause of male infertility. Different studies have shown that varicocele is associated with a progredient deterioration of testicular function. This fact constitutes the indication for treatment of varicocele, which has been underscored by the results of studies demonstrating that the correction of varicocele can arrest testicular damage and can lead to improved spermiogenesis (Hadziselimovic et al., 1989; Kass & Belman et al., 1987). We have established antegrade scrotal sclerotherapy (ASS) for treatment of varicocele (Tauber & Johnsen, 1994). This study wants to examine the influence of antegrade sclerotherapy of varicocele on spermiogenesis and generative power.

METHODS

Patients

From February 1992 to June 1994, 986 patients were treated using ASS. 234 of them presented with two semen analyses preoperatively. Semen analysis of 103 successful treated patients have been controlled 6 months postoperatively. These were aged between 17 and 47 years old, average age being 27.9 years. Indications for operation have been: Varicocele and pathological semen analysis in 35 patients (34%). Heaviness or pain in the scrotum in 28 patients (27%). Infertility in 38 patients (37%). Persistence after sclerotherapy in 2 patients (2%).

Parameters

For a uniform definition of varicocele we use the classification of Dubin and Amelar. 93 patients had unilateral varicocele, 10 a bilateral. 12 varicoceles have been subclinical, 15 grade I, 51 grade II and 35 grade III. Sperm density in mill/ml, morphology and class A and B motility in percent have been evaluated. Furthermore, we surveyed if conception occured postoperatively.

RESULTS

The table shows the results of semen analysis. In 16 of 38 patients operated because of infertility the partners became pregnant, giving a pregnancy rate of 42%.

n=103	preop	postop	p in %
density mill/ml	39.4+/-47.4	57.2+/-50.1	2.5
motility in %	36.5+/-20.7	51.7+/-20	0.01
normal morphology in %	45.6+/-23.6	53.7+/-20.9	2.5

DISCUSSION AND CONCLUSION

After treatment of varicocele by ASS sperm density, morphology und class A and B motility improved significantly. Our results correspond to findings of other treatment forms like high ligation, retrograde sclerotherapy or microsurgical methods (Mamar et al., 1994; Mordel et al., 1990). Since we do not have a control group we cannot assume that treatment of varicocele causes improvement of fertility parameters.

Our results show that antegrade scrotal sclerotherapy is equivalent to other kinds of operative treatment for varicocele with regard to improvement of fertility parameters.

REFERENCES

Hadziselimovic F 1989 Testicular and vascular changes in children and adults with varicocele. Journal of Urology 142: 583.

Kass EJ & Belman AB 1987 Reversal of testicular growth failure by varicocele ligation. Journal of Urology 137: 475.

Mamar JL & Kim Y 1994 Subinguinal microsurgial varicocelectomy: a technical critique and statistical analysis of semen and pregnancy data. Journal of Urology 152: 1127.

Mordel 1990 Spermatic vein ligation as treatment for male infertility. Journal of Reproductive Medicine 35: 123

Tauber R & Johnsen N 1994 Antegrade srotal sclerotherapy for treatment of varicocele: Technique and late results. Journal of Urology 151: 386

37

VEGF MODULATES THE CAPILLARIES OF THE HUMAN EPIDIDYMIS

S. Ergün,[1] W. Empen,[1] and W. Fiedler[2]

[1]Institute of Anatomy
[2]Department of Hematology and Oncology
University of Hamburg (UKE)
Martinistr. 52, 20246 Hamburg, Germany

INTRODUCTION

Vascular endothelial growth factor (VEGF), a high specific endothelial cell mitogen (Conn et al., 1990), stimulates angiogenesis and enhances endothelial permeability by fenestration of venular and capillary endothelium (Roberts & Palade, 1995), through the opening of endothelial cell tight junctions. The expression and function of VEGF in the epididymis which is divided into caput, corpus and cauda epididymidis (Holstein, 1969, Yeung et al. 1991) are not yet sufficiently clarified.

MATERIAL AND METHODS

Expression and function of VEGF and its receptors (flt-1 and KDR) in the human epididymis were analysed by means of RT-PCR, immunohistochemistry and by *in vitro* topical application of $VEGF_{165}$.

RESULTS

VEGF mRNA was detected in different parts of the human epididymis by RT-PCR. VEGF protein could be localized in peritubular myoid cells of ductuli efferentes and of the ductus epididymidis, in basal cells of the ductus epididymidis, as well as in ciliated cells of the ductuli efferentes. Endothelial cells of blood vessels in the interstitium and in the lamina propria of ductuli efferentes as well as of ductus epididymidis were negative. The mRNA of VEGF receptors flt-1 and KDR was demonstrated by RT-PCR. Flt-1 protein was immunohistochemically detected in epithelial cells of certain regions of ductuli efferentes and in interstitial lymph vessels but not in endothelial cells of blood vessels. In

The Fate of the Male Germ Cell, edited by Ivell and Holstein
Plenum Press, New York, 1997

contrast KDR protein could be located in endothelial cells of capillaries and of large interstitial blood vessels. After treating epididymal tissue with $VEGF_{165}$ fenestration of endothelial cells, opening of tight junctions between the endothelial cells of capillaries and transendothelial gaps could be induced.

DISCUSSION

These results permit us to conclude, that VEGF acts in a paracrine manner via its receptor flt-1 on the interstitial lymph vessels, and via the receptor KDR on the interstitial blood vessels, as well as on the capillaries localized within the lamina propria of the ductuli efferentes and the ductus epididymidis. The induction of transendothelial gaps by application of $VEGF_{165}$ may be of importance for the capillary permeability for blood cells. The localization of VEGF in peritubular myoid cells signifies, that VEGF is able to modulate vascularization of the lamina propria of the ductuli efferentes and of the ductus epididymidis.

REFERENCES

Conn G, Soderman DD, Schaeffer MT, Wile M, Hatcher VB & Thomas KA 1990 Purification of a glycoprotein vascular endothelial cell mitogen from a rat glioma-derived cell line. Proceedings of the National Academy of Sciences USA 87: 1323–1327.

Holstein AF 1969 Morphologische Studien am Nebenhoden des Menschen. In Zwanglose Abhandlungen aus dem Gebiet der normalen und pathologischen Anatomie, pp 1–91. Eds W Bargmann & W Doerr. Georg Thieme Verlag, Stuttgart.

Roberts WG & Palade GE 1995 Increased microvascular permeability and endothelial fenestration induced by vascular endothelial growth factor. Journal of Cell Science 108: 2369–2379.

Yeung CH, Cooper TG, Bergmann M & Schulze H 1991 Organization of tubules in the human caput epididymidis and the ultrastructure of their epithelia. American Journal of Anatomy, 191: 261–279.

38

ENDOTHELIN-1 AND ITS RECEPTORS IN THE HUMAN EPIDIDYMIS

S. Harneit,[1] S. Ergün,[1] H. -J. Paust,[2] A. K. Mukhopadhyay,[2] and
A. F. Holstein[1]

[1]Institute of Anatomy
University of Hamburg
Martinistr. 52, 20246 Hamburg, Germany
[2]Institute of Hormone and Fertility Research
Grandweg 64, 22529 Hamburg
Germany

INTRODUCTION

Endothelin-1 (ET-1) is a potent vasoconstrictor localized for the first time in porcine aortic endothelial cells (Yanagisawa et al., 1988). Endothelin receptors can be classified as subtype A (ET-A) and subtype B (ET-B). The aim of this study was to examine the expression of ET-1 and its receptors in the human epididymis. Morphologically, the human epididymis is commonly divided into three major parts: caput, corpus and cauda epididymidis (Holstein, 1969). The caput epididymidis contains mainly parts of efferent ducts (Yeung et al., 1991) while the corpus and cauda epididymidis comprise epididymal duct.

MATERIAL AND METHODS

Expression of ET-1 and its receptors ET-A and ET-B was assessed by means of RT-PCR. Immunohistochemistry was carried out on longitudinal sections through 10 whole normal human epididymides.

RESULTS

ET-1 mRNA was detected in caput, corpus and cauda epididymidis. Immunohistochemically, ET-1 protein was localized mainly in ciliated cells of efferent ducts and in some principal cells of the epididymal duct. Whereas no ET-1 immunoreactivity was detected in epididymal microvasculature, larger arteries stained positively. ET-A and ET-B mRNA were detected in the caput, corpus and cauda epididymidis. In the efferent ducts, ET-A immunoreactivity was localized in ciliated cells, in the proximal part of the epididymal duct this was mainly in basal cells whereas the distal part was devoid of any ET-A immunoreactivity. Throughout the epididymis, blood vessels stained positively for ET-B. Furthermore, ET-B immunoreactivity was also found in ciliated cells of the efferent ducts and in basal cells of the distal part of the epididymal duct.

DISCUSSION

In this study, we demonstrate the presence of ET-1 and its receptors, and their distribution in different tissue compartments within the human epididymis. The fact that ciliated cells of the efferent ducts contain ET-1 as well as its receptors ET-A and ET-B indicates that ET-1 could act as an autocrine factor in these cells. ET-1 produced by efferent ducts and the epididymal duct could be expected to influence the epididymal blood-flow in a paracrine manner via ET-B receptors located in the epididymal blood vessels. A further important aspect is the region-specific distribution of ET-1 and its receptors in the human epididymis. Similar regional differences have been described earlier (Pera et al., 1994). A correlation of the region-specific ET-1 distribution with the regional differences in the secretive and resorptive activities of the epididymal epithelium remains to be established.

REFERENCES

Holstein AF 1969 Morphologische Studien am Nebenhoden des Menschen. In Zwanglose Abhandlungen aus dem Gebiet der normalen und pathologischen Anatomie, pp 1–91. Eds W Bargmann & W Doerr. Georg Thieme Verlag, Stuttgart.
Pera I, Ivell R & Kirchhoff C 1994 Regional variation of specific gene expression in the dog epididymis as revealed by in situ transcript hybridization. International Journal of Andrology 17: 324–330.
Yanagisawa M, Kurihara H, Kimura S, Tomobe Y, Kobayashi M, Mitsui Y, Yazaki Y, Goto K & Masaki T 1988 Nature 332: 411–415.
Yeung CH, Cooper TG, Bergmann M & Schulze H 1991 Organization of tubules in the human caput epididymidis and the ultrastructure of their epithelia. American Journal of Anatomy, 191: 261–279.

39

THE ROLE OF APOCRINE RELEASED PROTEINS IN THE POST-TESTICULAR REGULATION OF HUMAN SPERM FUNCTION

G. Aumüller, H. Renneberg, P.-J. Schiemann, B. Wilhelm, J. Seitz,
L. Konrad, and G. Wennemuth

Department of Anatomy and Cell Biology
Philipps-Universität
Robert-Koch-Str. 6, 35033 Marburg
Germany

1. SUMMARY

A unifying hypothesis is presented postulating an apocrine release of several seminal proteins which mix and reaggregate in seminal fluid, thereby eventually forming particles designated either as "prostasomes", "vesiculosomes" or "seminosomes". The term "aposomes" should be restricted to the blebs released from secretory cells in the rat dorsal prostate and coagulating gland. Three different proteins present in human seminosomes along with the respective antibodies have been used to identify the localization, function and hypothetical interaction with spermatozoa. The proteins were (1) seminal vesicle-derived fibronectin, (2) prostate-derived 5'-nucleotidase and (3) a hitherto unidentified 100 kD membrane protein from epididymis, seminal vesicle and prostate. 1. Fibronectin is an extracellular matrix protein which is also secreted from the seminal vesicles participating in the formation of the seminal clot. Immunofluorescence and immunoelectron microscopy revealed a relatively broad distribution pattern of fibronectin immunoreactivity on spermatozoa from different donors. Adding a fibronectin antiserum at a moderate dilution to vital spermatozoa in vitro resulted in a significant increase in sperm motility. Purified plasma fibronectin added at various concentrations to a vital sperm preparation was found to inhibit sperm motility in a dose-dependent manner. Measurement of calcium fluxes in individual sperm in the presence of fibronectin showed a significant increase. These findings point to a possible post-testicular regulatory function of seminal fibronectin. 2. 5'-Nucleotidase (5'-NT) is an enzyme that hydrolyzes nucleotides such as AMP or IMP into inorganic phosphate and the respective nucleoside. The highest amount and activity of 5'-nucleotidase was present in glandular cells of the prostate; much less was detected in seminal vesicles and epididymis. On spermatozoa, the enzyme was localized on the outer leaflet of the plasma membrane covering the acrosomal region. Addition of purified en-

The Fate of the Male Germ Cell, edited by Ivell and Holstein
Plenum Press, New York, 1997

zyme to an in vitro incubation system of spermatozoa had no effect on sperm motility. A slight reduction of overall motility, however, was observed after addition of 5'-NT antibody to the spermatozoa. When 5'-nucleotidase inhibitors and adenosine channel antagonists were added to the sperm incubation system, a clear-cut inhibition of sperm motility occurred in a dose-dependent manner. This result is interpreted as indicating a significant role of ecto-5'-nucleotidase in the regulation of sperm motility. 3. A polyvalent antiserum against native human prostasomes recognized antigens in the range of 10–14 kD and of approximately 100 kD, respectively, in seminal fluid and prostate homogenates. Immunohistochemical studies revealed the presence of respective antigens in the epididymis, seminal vesicles and the prostate. Immunoelectron microscopy of ultracryo-sections showed labeling both of the apical plasma membrane in the prostate, as well as intraluminal secretory particles indicating the apocrine i.e. plasma-membrane bounded release of these particles. The secretory elements are termed "seminosomes". An affinity-purified fraction within the antiserum recognizes a 100 kD protein which is present both in the apical plasma membrane of the male genital glands, but also in the sperm head and principal piece of human spermatozoa. Incubation of spermatozoa with seminosomes and the respective purified antiserum had no effect on sperm motility. This is in contradistinction to former reports on motility increase induced by the so-called prostasomes.

2. INTRODUCTION

2.1. Sperm-Semen Interaction

Posttesticular sperm maturation and capacitation are still incompletely understood events in reproductive biology, although considerable progress has been made in elucidating the interaction between spermatozoa and secretions of the male accessory sex glands (for review, see Kirchhoff and Hale, 1996; Kirchhoff and Ivell, 1995; Lilja 1990; Aumüller and Seitz, 1990).

2.1.1. Seminal Protein Interaction. One of the most significant proteins in human semen is PSA, the prostate-derived antigen, a kallikrein-type protease of 33 kD (Lilja et al., 1987), which is released from the human prostate. It acts on semenogelin (for review, see Lilja 1990), a protein released from the seminal vesicle and forming the major component of the seminal coagulum which has recently been suggested to represent a sperm motility inhibitor (Robert and Gagnon 1996). Fibronectin is an additional secretion product of the seminal vesicles, which is integrated into the micromeshwork of the seminal coagulum (Lilja et al., 1987). During liquefaction, the seminal coagulum, essentially the semenogelins, are hydrolysed by the action of PSA into small fragments which bind to spermatozoa and induce hypermotility in the spermatozoa. We have focussed our interest on two seminal proteins, the secretion properties and functional significance of which were not well documented in the male genital system, namely 5'-nucleotidase and fibronectin.

2.1.2. Fibronectin, a Seminal Extracellular Matrix Protein. FN is a dimeric glycoprotein with an apparent molecular weight of 440 kD. It is a filament forming protein consisting of two nearly identical subunits. Three different forms of FN exist: 1. plasma FN, which is the soluble form, 2. cell bound FN, an oligomer, and 3. filament forming matrix FN (polymer). Binding sites for heparin, collagen and cell-cell interaction have been identified on the FN molecule. FN is bound to specific receptors and has been detected on the cell surface. The FN-receptor family is known as integrins. A tripeptide sequence—the RGD (Arg-Gly-Asp) sequence—has been identified in most of the integrins as the essential binding site. Integrins are mediators between the intracellular and extracellular matrix, because they are integrated in the cell membrane and have contact both to extracellular matrix proteins like FN and to intracellular binding sites of the cytoskeleton.

FN has been shown to play a role during fertilization. Hoshi et al. (1994) demonstrated that FN is an essential factor in sperm-egg-adhesion. The integrins involved in this mechanism were described as "spermadhesins", i.e. molecules adherent to the sperm membrane (Calvete et al., 1993). FN has been identified immunohistochemically on the surface of spermatozoa as the so-called equatorial FN band (EFB) (Glander et al., 1987). A few years later, Fusi et al. (1992) localized the RGD-sequence of the FN receptor binding site on human oocytes, too.

2.1.3. 5'-Nucleotidase, a Marker Enzyme of Prostasomes. 5'-Nucleotidase is a widely distributed enzyme in pro- and eukaryotic organisms (Zimmermann, 1992). One of its functions is purine salvage i.e. supply of cells with nucleosides during the synthesis of nucleotides. Cells are usually impermeable for 5'-mononucleotides. The latter are hence hydrolyzed by the ecto-form of the enzyme (bound to the outer face of the plasma membrane by a GPI-anchor) producing inorganic phosphate and nucleosides, e.g. adenosine. Adenosine can easily be taken up by the cells without energy through adenosine channels. It can be involved in ATP-generation and thereby provide the energy required to elicit specific cell functions such as movement, relaxation or contraction. The enzyme, therefore, is capable of allowing cells a great variety of tissue- or organ-specific actions, including neurotransmission and regulation of coronary, cerebral and muscular blood flow (Sundermann Jr., 1990).

The high amount of enzyme present in the human placenta, for instance, was suggested to modulate placental blood flow and energy supply for the fetus (Matsubara et al., 1987). If the ultimate function of the enzyme is to secure energy supply to cells, its presence in the male genital system, the prime source of spermatozoa with high fertilizing capacity, would allow further insight into the regulation of sperm functions within the male genital tract.

In the bovine male genital system the seminal vesicles and the ampulla of the deferent duct produce a high amount of secretory 5'-NT, which is synthesized in an androgen-dependent manner (Fini et al., 1983). The soluble enzyme present in the vesicular secretion adheres in a non-covalent manner to the surface of spermatozoa, thereby contributing to the overall activity of the endogenously present ecto-form of the enzyme located on the anterior tip of the acrosome (Schiemann et al., 1994). In the human male genital system a glycosylated form of the enzyme has been described (Fini et al., 1991). Here, we report on the localization of the enzyme in the male genital system and on human spermatozoa. Our results are in favour of a significant role in the regulation of sperm motility through 5'-NT added to the sperm surface from the secretions of the male accessory sex glands.

2.2. Secretion Modes in the Male Genital System

Two different modes of the release of secretory material from glandular cells have been shown in male accessory sex glands from different mammalian species: merocrine and apocrine secretion (for review see: Aumüller et al., 1994).

2.2.1. Merocrine Secretion. Synthesis of merocrine secreted proteins starts with translation of the mRNA on the ribosomes of the rough endoplasmic reticulum. A characteristic hydrophobic N-terminal amino acid sequence is bound to signal recognition parti-

Table 1. Survey on the cell biology of
apocrine secretion

BULK SECRETION

INCLUDING

- PLASMA MEMBRANE PROTEINS

  • BLEBS (RAT DORSAL PROSTATE)

  • STEREOCILIA (EPIDIDYMIS)

  • CELL DOMES (HUMAN PROSTATE)

- CYTOPLASMIC PROTEINS

- CYTOSKELETAL PROTEINS

- TRANSCYTOTIC PROTEINS

  • SERUM PROTEINS

  • INTERSTITIAL FLUID PROTEINS

OPTIONAL (HUMAN MALE ACCESSORY GLANDS)

- SECRETORY GRANULES

  • GRANULE PROTEINS

  • VACUOLE PROTEINS

  • MEMBRANE PROTEINS

cles supporting the translocation of the nascent peptide into the lumen of the rER. Imported proteins are further processed, followed by transport into the Golgi apparatus where posttranslational modifications and transfer into secretory granules take place. Secretory granules are translocated to the apical plasma membrane, where they release their contents after fusion of the granule membrane with the apical plasma membrane.

2.2.2. Apocrine Secretion. During the last few years, several secretory proteins were identified which lack the typical N-terminal sequence, and are synthesized on free ribosomes inside the cytoplasm (Muesch et al., 1990). Little is known about the intracellular processing and modification of these proteins. Apocrine secretion has been discussed as a feasible release mechanism for this type of protein in the rat dorsal prostate and coagulating gland, i.e. the regulated release of secretory material through apical protrusions or blebs into the glandular lumen (Seitz et al., 1990; Steinhoff et al., 1994). The particular form of blebs present in these glands has been termed "aposomes" (Aumüller and Seitz, 1990).

2.2.3. Diacytosis. In a series of papers Ronquist, Brody and collaborators (Brody et al., 1983; Ronquist and Brody, 1985; Ronquist 1987; Ronquist et al., 1988, 1990) described the ultrastructure of secretory granules and vesicles in human prostate epithelial cells, prostatic fluid and seminal plasma which they have termed "prostasomes". These particles show a tri- or multilamellar membrane architecture and occur freely or surrounded by a unit membrane, forming so called "storage vesicles".

In electron microscopic studies, Brody et al. (1983) and Ronquist and Brody (1985) described two different modes of release of prostasome-containing vacuoles from the interior of the prostatic cells into secretory fluid. The first mechanism is preceded by fusion of adjacent membranes belonging to the storage vesicle and the secretory cell. Alternatively, the intact storage vesicle on the whole is translocated from the interior of the cell into the

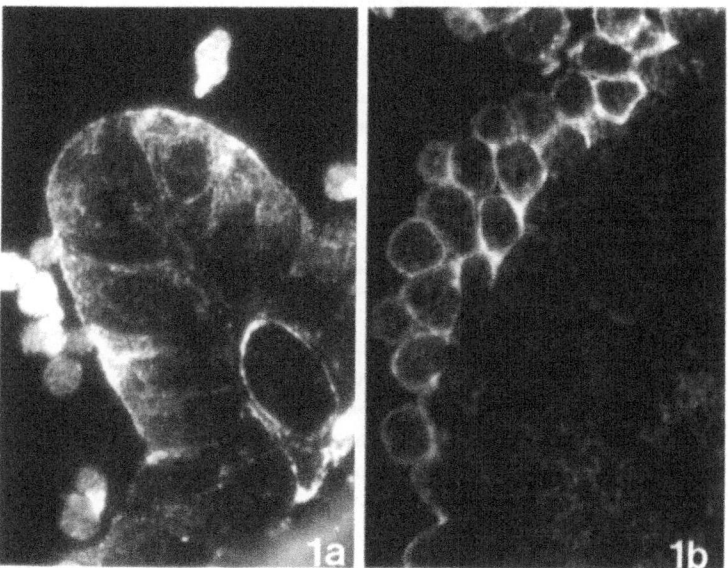

Figure 1. Morphology of apocrine secretion in the rat dorsal prostate (1a, x 1200) and the human prostate (1b, x 1200). Aposomes in rat dorsal prostate epithelium were visualized by immunofluorescence of secretory transglutaminase. Apocrine protrusions of human prostatic secretory cells were labeled with an antibody against a 100 kD membrane protein (see below).

acinar lumen. Brody et al. (1983) have termed this mode of secretion as "diacytosis". The question is, whether the aforementioned release mechanisms function separately or alternately in the male genital system, or if there is a common underlying process of release eventually forming seminal secretory particles. If so, such particles should be termed "seminosomes" rather than "prostasomes", indicating their release from different sites within the seminal pathways.

2.3. A Unifying Hypothesis of Secretion in the Male Genital System

Apocrine secretion has been described by Schiefferdecker (1917) in the mammary and axillary sweat glands and has later also been discussed as a conceivable mechanism of release in the human prostate (for review, see Aumüller et al., 1994). A peculiarity of adluminal cells in organs such as the prostate, the seminal vesicles and the epididymis is the presence of extensions or protrusions from the apical membranes of these cells. Dome-shaped protrusions or apical blebs are present on secretory cells of the prostate and the seminal vesicles, whereas the principal cells of the epididymis contain elongated stereocilia. As of yet, the secretory nature of these structures was controversial. Apocrine secretion in a narrow sense is suggested to occur exclusively in the rat dorsal prostate and coagulating gland, where apical blebs ("aposomes") are frequently encountered and shown to contain specific secretory proteins, e.g. transglutaminase (Seitz et al., 1990).

Release of the apical structures described in the human male accessory sex glands would suggest the presence of plasma membrane proteins, organ-specific secretory granule proteins, cytoplasmic proteins and cytoskeletal proteins, as well as blood-borne transcytotic proteins in human semen and perhaps in seminosomes. If detachment of the apical

Table 2. Survey on the terminology used in the present paper for the different structures involved in secretion

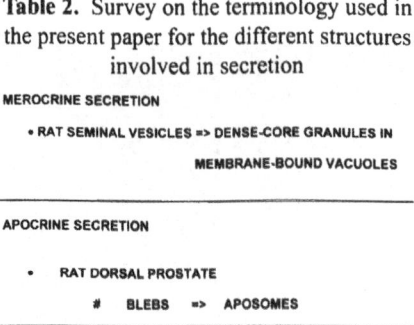

- MEROCRINE SECRETION

    • RAT SEMINAL VESICLES => DENSE-CORE GRANULES IN

                    MEMBRANE-BOUND VACUOLES

- APOCRINE SECRETION

    •   RAT DORSAL PROSTATE

        #   BLEBS => APOSOMES

    •   HUMAN PROSTATE

        #   CELL DOMES

        #   SECRETORY VACUOLES WITH

               GRANULES => PROSTASOMES

    •   BULL / HUMAN SEMINAL VESICLE

        #   SECRETORY GRANULES/VACUOLES

        #   CELL DOMES => VESICULOSOMES

    •   HUMAN EPIDIDYMIS

        #   STEREOCILIA

SEMINOSOMES

cell portion represents a physiological release mechanism, several repair mechanisms allowing the immediate closure of the apical cell pole would be necessary. Also, the release would require a well coordinated membrane flow or transport of lipids and proteins to the apical plasma membrane, a reshaping of the cells including a restructuring of the cytoskeleton, a coordinated synthesis of secretory proteins and a highly specialized organization for the release of the material.

The present paper has, therefore, used three different proteins typical in the male genital system and derived from seminal fluid/prostasomes to scrutinize the origin of these proteins, their hypothetical apocrine release from the different secretory sites and their interaction with spermatozoa.

We show that these proteins can indeed be related to apocrine release and are important in the post-testicular regulation of sperm function.

3. MATERIALS AND METHODS

3.1. Preparation of Spermatozoa

Fresh ejaculates were obtained from voluntary donors with proven infertility (n=5) and in addition from 16 patients with OAT-syndrome, attending the andrology unit. Samples were kept for 30 min at 37° C for liquefaction, subsequently washed in BWW-medium (Biggers et al., 1979), centrifuged for 10 min at 37° C at 300 x g. The pellet formed was used either for (i) swim-up and subsequent motility analysis, (ii) preparation of sperm smears, (iii) pre-embedding immunogold labeling or (iv) Western blotting. For swim up, an aliquot of the sample was covered with fresh BWW-medium and incubated for 1 h at

37° C in a CO_2 incubator. The supernatant was then used for motility analysis using the CASA system (Krause, 1995) or in vitro incubation (see below). For immunofluorescence, a small drop (50 µl) was placed on a chromalum-gelatin-coated glass slide which was slightly kinked and twisted to distribute the drop over an area of about 1 cm^2, air dried in horizontal position and then fixed in ice cold 40% methanol (1 min) and subsequently in 4% phosphate-buffered paraformaldehyde (10–30 min) at room temperature, then carefully washed in phosphate-buffered saline (PBS) containing 1% normal swine serum and stored in a refrigerator for further treatment. Another sample obtained after swim-up was fixed in a mixture containing 2.5 % paraformaldehyde and 0.1 % glutaraldehyde in PBS for 20 min at room temperature, washed 3 times in PBS and used for pre-embedding immunogold labeling.

3.2. Immunohistochemistry and Immunoelectron Microscopy

Smears of washed spermatozoa resuspended in PBS were spread on glass slides subbed with chromalum gelatine. Spermatozoa were fixed in cold acetone (-20° C) for 5 min, air dried and stored in a refrigerator until use. In order to prevent the smears from drying out, the subsequent incubation steps took place in a moist chamber. Incubation with the respective primary antibody was carried out at a dilution of 1:100 at room temperature or diluted 1:500 overnight at 4° C. The smears were washed three times for 5 minutes in PBS, pH 7.4 and incubated with either the secondary FITC-labeled anti-mouse IgG (Dako, Hamburg, Germany) or with a Cy3- labeled anti-mouse IgG (Dianova, Hamburg, Germany). These antibodies were diluted 1:50 in PBS (pH 7.4) and the smears were incubated for 30 min at room temperature. As the fluorescence of the Cy3 - and FITC - fluorochromes fades rapidly in daylight, the final incubation steps had to take place in the dark. The samples were photographed in a Zeiss fluorescence microscope (Oberkochen, Germany) using a Kodak TRI-X-PAN 400 film (Braunschweig, Germany).

For immunoelectron microscopy the pre-embedding staining method for immunoelectron microscopy was used. Sperm samples were incubated with the respective primary antibody (diluted in PBS 1:100) for 1 hour at room temperature. The secondary antibody was a gold-labeled anti-mouse IgG (BioCell, Plano, Marburg, Germany). The gold particles had a size of about 5 nm in diameter. Gold-labeled IgG was used at a dilution of 1:50 in PBS for 30 min at room temperature. Subsequent to the incubation procedure, specimen were either osmium-treated or immediately dehydrated with graded alcohol and embedded in Epon.

3.3. Sperm Motility Measurement

The computer-aided motility measurement was carried out at the Andrology Unit, Department of Dermatology, University of Marburg. The system for morphology and movement analysis consists of a microscope connected to a video camera (CASA, Computer Assisted Sperm Analysis). The analog pictures are transformed into digital signals by an A/D converter and are analysed by a computer, equipped with specialized software (Cell Motion Analysis, Medical GmbH, Germany). This particular software allows classification of cells according to various characteristics. The analysis includes many parameters such as number of spermatozoa per milliliter, viability (dead or alive), percentage of motility (immotile or moving on one spot), number/percentage of motile spermatozoa, spermatozoa running in circles, non-linear moving spermatozoa, velocity of motile spermatozoa and deformation of spermatozoa (head or tail). For statistics, a minimum of 200 spermatozoa has to be measured.

Sperm motility changes upon dilution of the ejaculate, e.g. with PBS. For control incubations run in parallel with antibody incubations, an equivalent amount of PBS was added to the sperm suspensions, resulting in identical dilutions and protein concentrations of experimental and control incubations.

3.4. Biochemical Studies

Antibodies against purified human seminal 5'-nucleotidase and a 100 kD peptide from human semen, respectively, were raised in rabbits. Microspectrophotometric determinations of calcium fluxes, enzyme activity determinations and RT-PCR studies of 5'-nucleotidase in the human genital tract, reported herein are published in detail elsewhere (Renneberg et al., in press; Schiemann et al., in press; Wennemuth et al., in press).

4. RESULTS

4.1. Apocrine Bulk Secretion of Seminosomes in the Male Genital System

Characteristic apical blebs, stereocilia or dome shaped protrusions of the apical cell pole are present on the secretory cells in the human prostate, seminal vesicles and epididymis (cf. Figs. 1a,b; 5a).

These apical protrusions are characterized immunohistochemically by the presence of numerous different secretory proteins, e.g. acid phosphatase and PSA in the prostate

Table 3. Characteristics of human seminal 5'-nucleotidase

MOLECULAR WEIGHT

- PROTEIN CORE OF 46 kD (TETRAMER)
- 70 kD (ECTO-FORM)
- 45 kD (SOLUBLE FORM)

STRUCTURE

- MEMBRANE-BOUND ECTO-FORM
- SOLUBLE SECRETORY FORM

POSTTRANSLATIONAL MODIFICATION

- GPI - ANCHOR (ECTO-FORM)
- GLYCOSYLATION

SYNONYMS

- CD 73 (LYMPHOCYTES)

DISTRIBUTION

- PROSTATE > SEMINAL VESICLES > EPIDIDYMIS
- SPERM HEAD

FUNCTIONS

- PURINE SALVAGE, ENERGY METABOLISM (?)

and semenogelin in the seminal vesicles, respectively. Fibronectin is present at higher concentrations only in the seminal vesicle epithelium, but lacking in prostatic and epididymal secretory cells. It is apparently released in an apocrine fashion from the glandular cell of the seminal vesicle. The apical blebs surrounded by plasma membrane and present on prostatic and seminal vesicle secretory cells, as well as the stereocilia observed in the epididymis are released into the lumen of the glands. Particles released in an apocrine fashion and containing a bulk of different secretory proteins are identified intraluminally in prostatic acini, seminal vesicle lumen and cross-sections of the epididymal duct. Apparently, the defect in the apical cell pole formed through the detachment of the secretion-filled blebs is immediately closed and a new apical pole develops. To identify the nature and significance of the apocrine proteins in the regulation of post-testicular sperm functions, the salient proteins (100 kD-membrane protein, 5'-nucleotidase, fibronectin) and the distribution of immunoreactive sites of these proteins on spermatozoa and their interaction with spermatozoa in vitro were studied.

4.2. Distribution of Seminosomal Proteins and Antibody Preparations

4.2.1. 5'-Nucleotidase. Enzyme activity determinations: Activity measurements indicated a very strong level of 5'-nucleotidase activity in preparations of the prostate, relative to preparations prepared from placenta (about 10 times higher). No enzyme activity was however found in testis. In the epididymis, it was in the same range as in the placenta.

Western blotting: The specificity of an antibody directed against a highly purified human seminal 5'-nucleotidase was studied in Western blotting experiments. Separation and immunostaining of highly purified seminal 5'-nucleotidase using our antibody resulted in one single band at a molecular mass of 69 kDa. A similar, though slightly less intense immunoreaction was obtained with seminal fluid as well as extracts from the prostate, seminal vesicle, epididymis and testis (decreasing intensity). Our antibody there-

Table 4. Characteristics of human seminal fibronectin (data adopted from Lilja et al., 1987)

MOLECULAR WEIGHT

- 225 - 250 kD

STRUCTURE

- 2 SUBUNITS

CONCENTRATION

- 1 mg/ml SEMINAL FLUID

SOURCE

- SEMINAL VESICLE

DISTRIBUTION

- SPERM HEAD (PEB)
- SPERM TAIL

FUNCTION

- STRUCTURAL PROTEIN (SEMINAL GEL)
- SPERM ATTACHMENT
- CALCIUM FLUX REGULATION

Table 5. Characteristics of human
seminal 100 kD protein

MOLECULAR WEIGHT

- • AROUND 100 kD

ISOELECTRIC POINT

- • pH 5.5

DISTRIBUTION

- • PROSTATE SECRETORY CELLS (FETAL, ADULT)
- • APICAL PLASMA MEMBRANE
- • SEMINAL VESICLE
- • EPIDIDYMIS
- • SEMINAL FLUID (SEMINOSOMES)
- • EXTRAGENITAL SITES, E.G. SALIVARY GLANDS

FUNCTIONS

- • SPERM BINDING (ACROSOME)
- • SPERM REGULATION (?)

fore was regarded specific for human seminal 5'-nucleotidase and was further used for immunohistochemical studies.

RT-PCR: One single positive signal at 296 pb was obtained with RNA preparations from placenta (which served as positive control). It was present in preparations from human prostate, seminal vesicle, epididymis and testis (decreasing signal intensity).

Taking together the findings on the protein and RNA-levels, the strongest signal for 5'-nucleotidase was present in the human prostate, exceeding that of the placenta.

4.2.2. Fibronectin. Western blot: A commercially available antiserum against human plasma FN was used for Western blotting and immunohistochemistry. To check the presence of FN in seminal plasma and the accessory sex glands, a Western blot analysis of seminal plasma was performed using purified fibronectin as a standard. One single immunoreactive band appeared both in the sample and the FN standard at about 220 kDa. No FN was found in extracts from the prostate and epididymis, respectively.

4.2.3. Prostate Membrane Specific Protein.

4.2.3.1. Antigen Preparation from Seminosomes. Electron microscopy of thin sections prepared from epon-embedded specimens of (i) fresh human ejaculate, (ii) a pellet obtained after centrifugation at 1,000 x g and (iii) a pellet obtained after ultracentrifugation at 105,000 x g and subsequent gel filtration on a Sephacryl S-500 HR column demonstrated a high degree of purity obtained in samples taken from the latter preparation. Small particles of slightly varying size and electron density were seen showing the ultrastructural features of seminosomes as previously described by Ronquist and Brody (1985) for prostasomes. The enrichment of prostasomes was checked by measuring the ATPase activity, a known marker enzyme for these particles (Ronquist, 1987). Concomitant with the different steps of purification, an increase in specific activity of ATPase activity and a relative decrease in protein concentration was obtained (results not shown).

The different fractions were further subjected to SDS-Page. The seminosomes exhibit several dominant protein bands, one at 14 kD and three bands at approximately 100

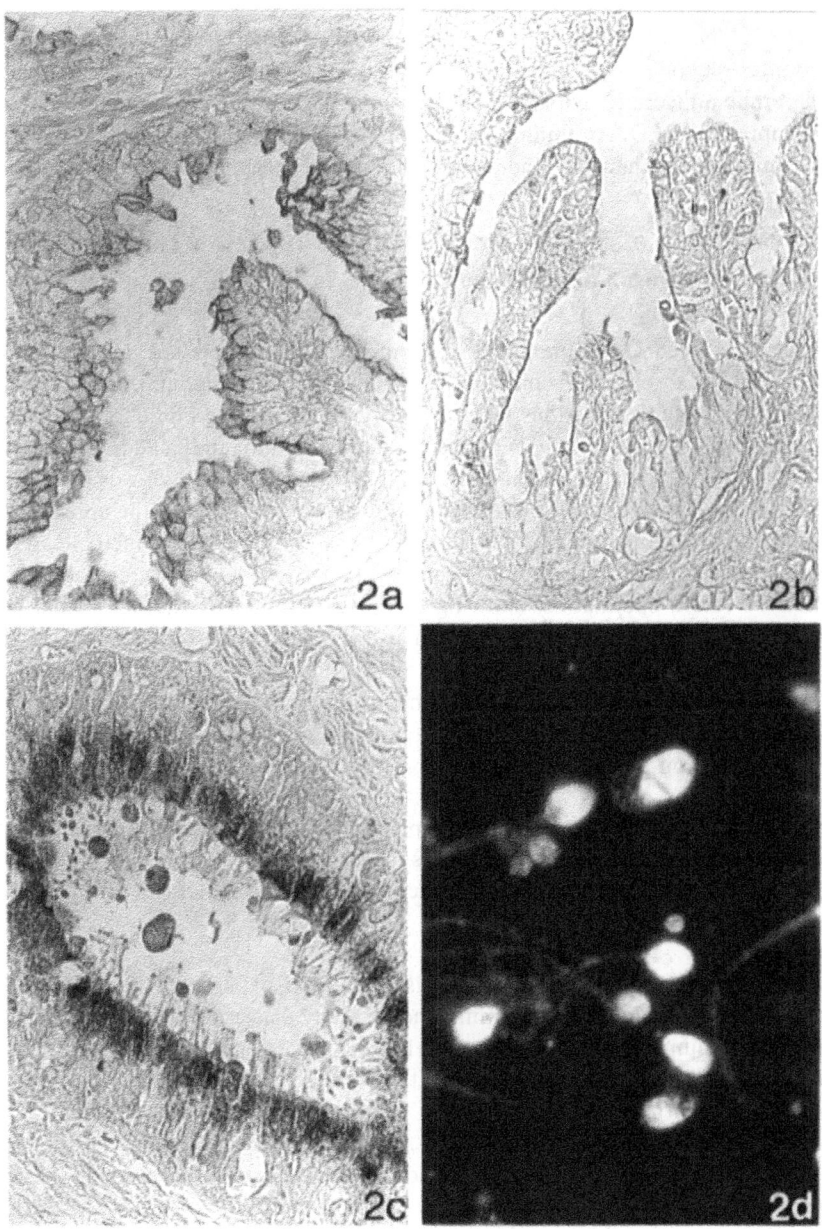

Figure 2. Immunohistochemical distribution of seminal 5′-nucleotidase in human prostate (a: apical blebs), seminal vesicle (b: apical plasma membrane), epididymis (c: stereocilia, secretions) and spermatozoa (d). Magn. a-c x 750; d x 1200.

kD. In Western blot analysis of seminosome fractions we were able to detect PSA and a secretory form of acid phosphatase, indicating that known secretory proteins are associated with the seminosomes.

4.2.3.2. Characterization of the Seminosome Antibody. Western blot analyses of seminosomes separated by reduced SDS-PAGE displayed different immunoreactive bands with the crude antiserum against the seminosome preparation ranging between 10 to 14 kD and approx. 100 kD. An immuno-affinity purified fraction of the 100 kD-protein was used for immunohistochemical studies and incubation experiments in spermatozoa.

4.3. Localization of Seminosomal Proteins in Male Accessory Sex Glands and on Spermatozoa

4.3.1. 5'-Nucleotidase. Immunohistochemistry of human tissues: In paraffin sections of the prostate the secretory epithelial cells displayed a rather uniform moderate immunoreaction. Using cryosections processed for 5'-nucleotidase immunofluorescence, the immunoreaction was present all over the prostatic epithelium and in addition in some stromal smooth muscle and endothelial cells. Seminal vesicle epithelium was of intermediate immunoreactivity and most of the reaction product was confined to the apical rim of the secretory cells. In epididymis, the supranuclear Golgi region of the principal cells and the stereocilia on these cells were immunoreactive. In testis, most cells were negative with the exception of round spermatids, showing a strong reaction of the acrosomal cap (not shown). No immunoreaction was observed in spermatozoa present in the rete testis or epididymal duct.

Localization of the enzyme on spermatozoa: Immunoreaction of washed human spermatozoa with the anti-5'-NT antibody resulted in a strong immunofluorescence of the anterior head portion and a slight reaction on the tail of the spermatozoa. No major variations in the staining pattern were observed. In control incubations where the primary antiserum had been omitted or had been replaced by an irrelevant antiserum (anti acid phosphatase), no immunofluorescence was observed. The distribution pattern of immunofluorescence with anti-5'-NT was rather similar to that obtained with the anti-seminosome (100 kD) antibody.

4.3.2. Fibronectin. Immunohistochemistry of human tissues: The seminal vesicles were the only site, where fibronectin immunoreactivity was associated with secretory epithelium and intraluminal secretory material. In the prostate and the epididymis, the epithelium was non-reactive. An only faint reaction was seen in stroma of the prostate and subepithelial capillaries in the epididymis.

Immunofluorescence of spermatozoa: Apart from the equatorial FN band (EFB) already described by Glander et al. (1987), additional immunoreactive sites on the neck and midpiece of the spermatozoa were frequently encountered. Throughout all specimens, a rather heterogenous distribution of FN immunoreactivity was observed. The postacrosomal region and the principal piece of the spermatozoa were also stained in some instances.

The immunohistochemical findings were confirmed by immunoelectron microscopy revealing an identical labeling pattern. In addition to the EFB-immunoreaction found by Glander et al. (1987), different distribution patterns of FN were observed on spermatozoa with intact acrosome relative to those devoid of the acrosome, where no labeling was achieved.

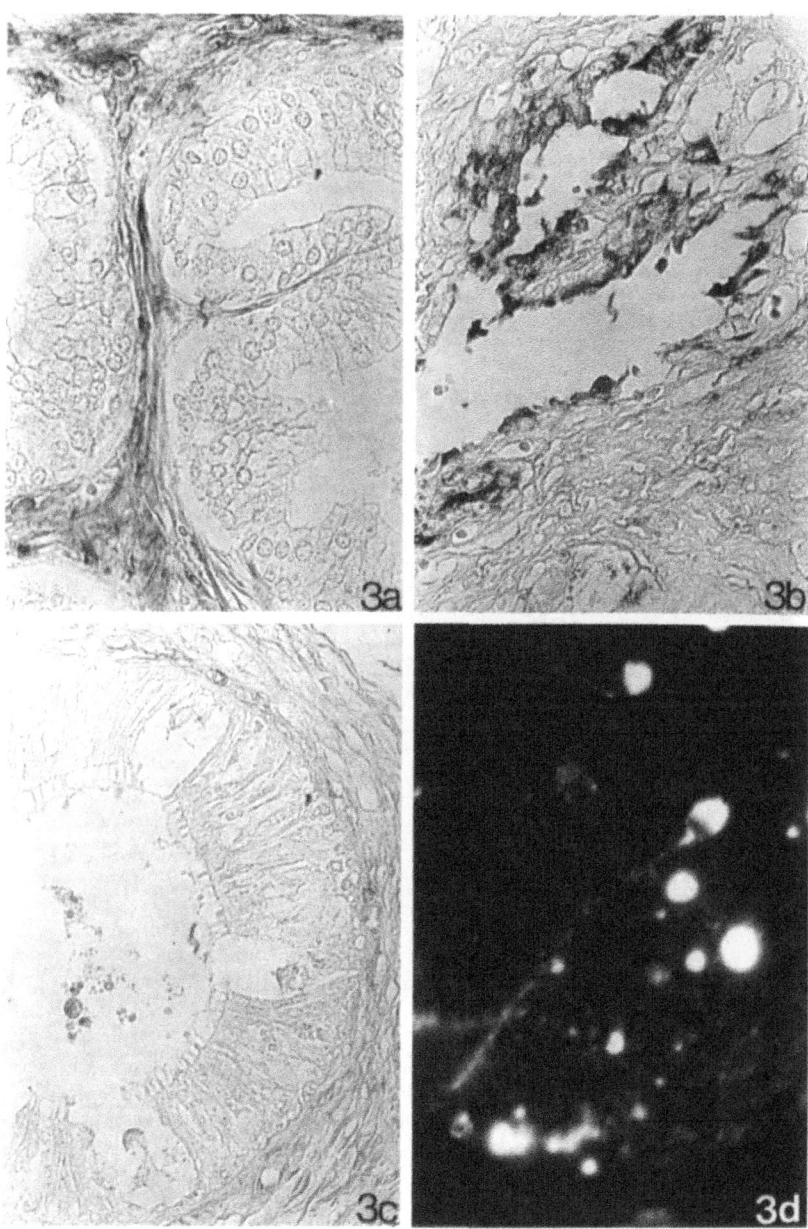

Figure 3. Immunohistochemical distribution of fibronectin in human prostate (a: only weak stromal reaction), seminal vesicle (b: secretory material labeled), epididymis (c: faint subepithelial reaction) and spermatozoa (d: differentially labeled sperm heads). Magn. a-c x 750, d x 1200.

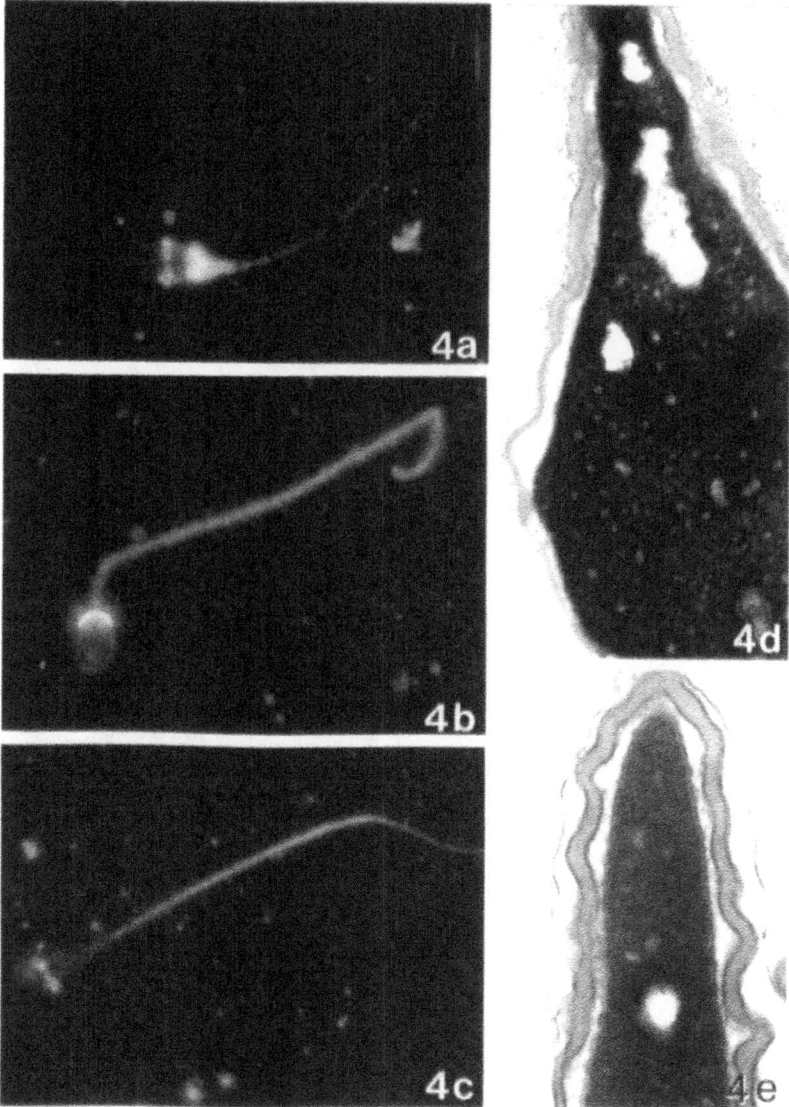

Figure 4. Immunofluorescence and immuno electron microscopy of fibronectin in human spermatozoa. a - d Variability of the immunofluorescence pattern. Magn. x 1200 e After the acrosomal reaction has been performed the anterior head portion is strongly labeled. Magn. x 22 000. f Control incubation with intact spermatozoon. No labeling is seen. Magn. x 20 000.

A regular staining on nearly all spermatozoa was observed on the head region. Gold particles were present in a restricted area on the surface of the membrane.

4.3.3. Seminosomal 100 kD Antigen. Immunohistochemistry of human tissues: The antibody reacted in the supranuclear and apical region of prostate epithelial secretory cells. A strong reaction was found at the apical plasma membrane and in the intraluminal

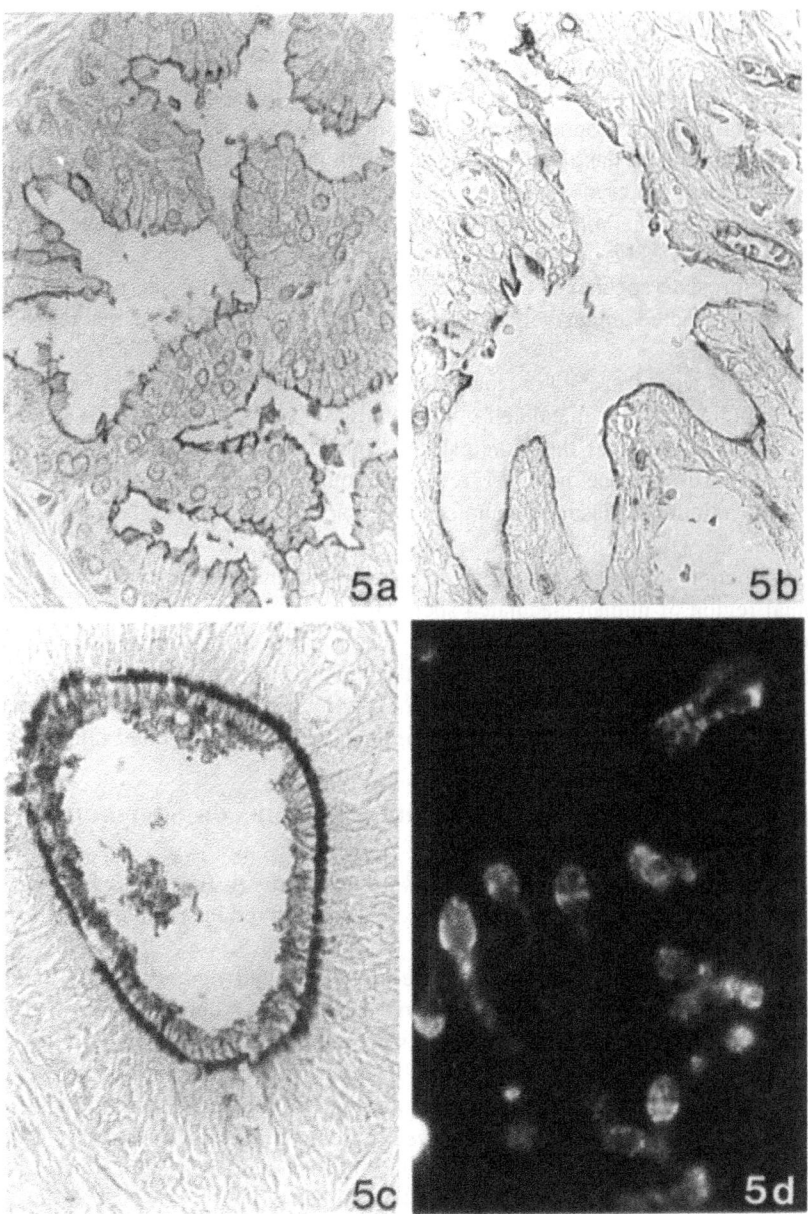

Figure 5. Immunohistochemical distribution of the seminal 100 kD membrane protein in the human prostate (a: apical bleb membrane), seminal vesicle (b: apical plasma membrane protrusions), epididymis (c: apical plasma membrane, stereocilia) and spermatozoa (d: sperm heads). Magn. a - c x 750, d x 1200.

particles of the acini. Depending on the intensity of fixation, cytoplasmic staining was also seen, which, however, was restricted to the apical portion of the glandular·cells. Basolateral plasma membranes or membranes and cytoplasm of the basal cells were never stained. Immunostaining of the adluminal plasma membrane of the secretory cells was particularly distinct on semithin cryosections, where a narrow rim of the adluminal plasma membrane was selectively labeled. Semithin cryo-sections incubated with an antibody against prostatic acid phosphatase and Prostate Specific Antigen, showed a more or less generalized staining of the apical cell portion with numerous intensely reacting secretory granules. On ultrathin cryosections incubated with the prostasome antibody, both intraluminal secretion and the apical plasma membrane were gold labeled, whereas labeling of the secretory vacuoles was in the background range.

Some prostate specimens contained a number of stromal cells, presumably macrophages which reacted intensively, too. Contrary to this, no labeled macrophages were found in spleen.

In seminal vesicles, a slight reaction occurred in the apical portion of secretory cells. In human epididymis, the apical cell portion and both kinocilia of the cells in the efferent ductules and stereocilia of the principal cells in the epididymal duct gave a strong immunoreaction. Testis tissue, however, remained completely unstained. In infantile prostate samples, the adluminal plasma membranes of the acinar precursor cells were also strongly immunoreactive with the prostasome antibody.

Distribution on Spermatozoa: A moderate immunoreaction was present over the acrosomal region of human ejaculated spermatozoa. No major differences were observed between different spermatozoa. Treatment of spermatozoa with a calcium ionophore or with Triton X-100 induced no increase in the intensity of the immunoreaction. Immuno electron microscopy showed labeling both of the outer plasma membrane, as well as the outer acrosomal membrane. Redistribution of the labeling could not be excluded.

4.4. Functional Effects of Seminosomal Proteins on Spermatozoa

4.4.1. Motility Measurements with FN and FN Antibody. Ejaculates from healthy donors: A significant increase from 56 % to 66 % was observed in sperm motility subsequent to adding FN antibody to ejaculate.

An increasing FN concentration added to a constant number of spermatozoa, resulted in a dose-dependent decrease of sperm motility. At a concentration of 0.18 to 0.5 mg FN/ml ejaculate, no motile spermatozoa were measured.

Ejaculates from patients assumed to be infertile (FN quantification): 58 samples from patients assumed to be infertile were submitted to FN quantification. Significant differences in FN concentrations were observed, ranging from 0.0008 mg/ml to 1.00 mg/ml plasma FN and 0.00 mg/ml to 2.082 mg/ml cellular bound FN, respectively. Measured sperm motilities were in the range between 0 % and 84 %.

A regression analysis was performed including the following parameters: plasma FN (ng/ml x 2500), cellular bound FN (ng/ml x 500), motile spermatozoa (%), linear motility (%), velocity (μm/s), circular running spermatozoa (%) and head deformation (%).

This procedure revealed a highly significant ($p=0.001$) inverse correlation (regression coefficient: -0.423) of sperm motility and plasma FN concentration (Fig.6), and a very highly significant ($p<0.0009$) inverse correlation ($r=-0.479$) of sperm motility and cellular bound FN: the higher the FN concentration, the lower the number of motile spermatozoa in ejaculate.

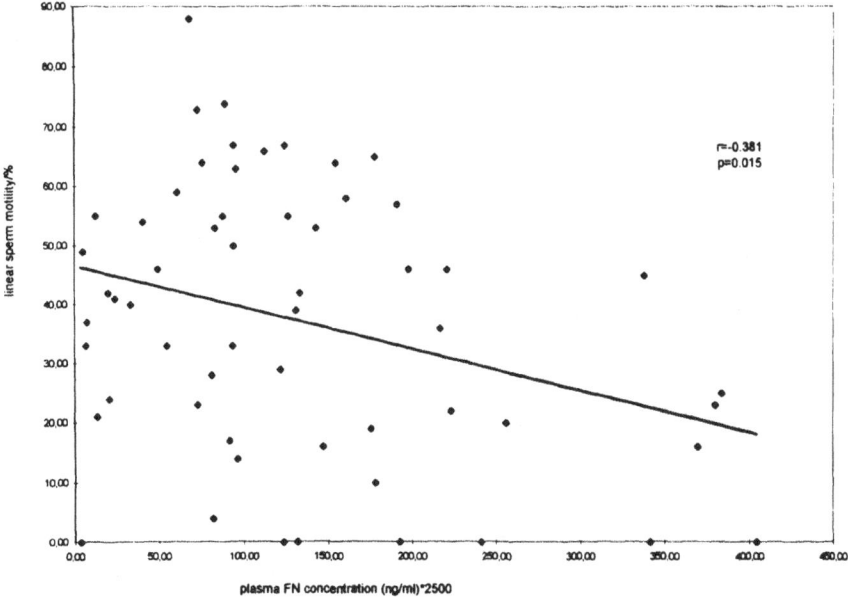

Figure 6. Inverse correlation between fibronectin concentration in seminal fluid and sperm motility.

The correlation between plasma FN and linear sperm motility was r=-0.318 at a p=0.015 (significant). Between cellular bound FN and linear sperm motility it was r=-0.360 at a p=0.006 (highly significant). There is obviously a clear-cut inverse correlation between FN content and linear sperm motility.

No significant correlation could be found between head deformation, circular running spermatozoa and plasma FN content in ejaculate or cellular bound FN on spermatozoa.

A different result was obtained by comparison of FN content and sperm velocity. A significant (p=0.017) correlation (r=-0.311) was found between cellular bound FN and

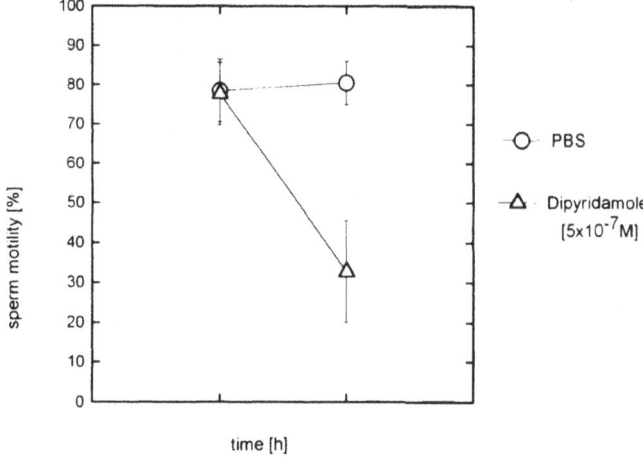

Figure 7. Influence of dipyridamole on sperm motility.

sperm velocity, but no significant (p=0.061) correlation (r=-0.248) between plasma FN concentration and sperm velocity.

4.4.2. Motility Measurements with 5'-Nucleotidase, 5'-Nucleotidase Inhibitors, 5'-Nucleotidase Antibody and Adenosine Channel Blockers. Whereas only marginal changes were observed in sperm motility after incubation of spermatozoa with purified 5'-nucleotidase (0.5 µg/ml) or the respective antibody (5 µl of the protein solution), a clear cut reduction of sperm motility (by about 50%) was observed within 1 h after incubation with either the 5'-nucleotidase inhibitor methylene-α,β-adenosine diphosphate (5×10^{-7} M) or the adenosine channel blocker dipyridamole (5×10^{-7} M).

4.4.3. Motility Measurements with Seminosomes and Seminosome Antibody. Contrary to reports by Fabiani (1994) and Fabiani et al. (1994, 1995) we were unable to detect any promotive effect of seminosomes on normal human spermatozoa. Using the respective antibody, no inhibition of spermatozoa could either be recorded.

4.4.4. Determination of Calcium Fluxes. Isolated live human spermatozoa were attached with their heads to poly-L-lysine coated glass slides (Wennemuth et al., in press) and centered in a computer-assisted microscpe microspectrophotometer equipped with a calcium quantification device (FURA-method-based). The experiments were calibrated by addition of progesterone (diluted in ethanol at concentrations ranging between 0.1–2.0 µg/ml). Time of measurement ranged between 20 and 300 seconds. Addition of 1µg/ml of fibronectin to the incubation medium resulted in an increase of intracellular calcium within 60 sec. The effect could be repeated 3 to 4 times during the subsequent incubation period.

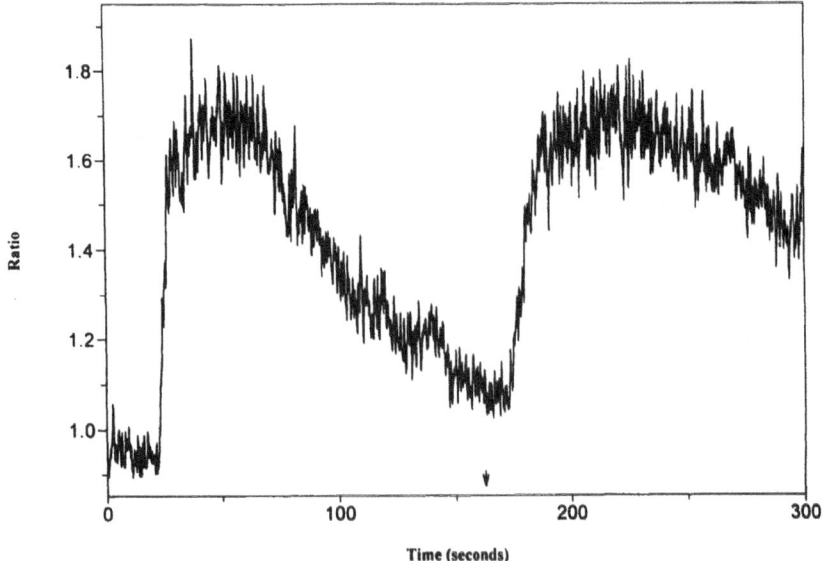

Figure 8. Influence of fibronectin (0.1 µg/ml) on calcium influxes in isolated human spermatozoa.

Table 6. Synopsis on the structure and
function of prostasome and
vesiculosomes (including data
of Ronquist and Brody, 1985,
and Agrawa and
Vanha-Perttula, 1987)

CHARACTERISTICS		
OF		
(PROSTASOMES)		(VESICULOSOMES)
SEMINOSOMES		
MORPHOLOGY		
SIZE RANGE		
20 - 150 nm		UP TO 200 nm
STRUCTURE		
MULTILAMELLATED, DENSE CORE		BLEBS, DENSE CORE
BIOCHEMISTRY		
PHYSICO-CHEMICAL PROPERTIES		
HEAT STABILITY		
HIGH CALCIUM CONTENT		
20 BANDS ON SDS - PAGE		20 BANDS ON SDS-PAGE
MAJOR ENZYMES:		MAJOR ENZYMES:
• MG^{++},CA^{++}-ATPase		• MG^{++}, CA^{++}-ATPase
• 5'-NUCLEOTIDASE		• 5'-NUCLEOTIDASE
• PROTEINKINASE		• PHOSPHOLIPASE A2
+ SYNAPTOPHYSIN		
(<=PROSTATIC ENDOCRINE CELLS)		

5. DISCUSSION

5.1. Origin of Seminosomal Proteins

5.1.1. Properties of Prostasomes. Our Western blotting results have clearly indicated that a few bands are prominent in prostate extracts which are in the range of 10–14 kD and of 100 kD. These may represent a rather heterogeneous group of proteins resulting in more or less incidental immunoreactions in different organs and cells. The significance and relevance of the immunoreactions has to be discussed in more detail, considering both the nature of prostasomes and the respective antigens present in immunoreactive sites.

Biochemical studies of isolated human prostasomes using one and two dimensional polyacrylamide gel electrophoresis (Lindahl et al., 1987) showed their complex protein composition. Arvidson et al. (1989) found high-ordered molecular membrane structures with very high cholesterol to phospholipid ratios in isolated prostasome fractions. These granules of human seminal fluid exhibited different associated enzyme activities such as ATPase (Ronquist, 1987), gamma-glutamyl-transferase (Lilja and Weiber, 1983; Ronquist et al., 1988), fucosyl transferase (Ronquist and Stegmayr, 1984), different peptidases (Vanha-Perttula, 1984; Ronquist et al., 1988), a protein kinase (Stegmayr et al., 1982), phospholipase A2 (Lindahl et al., 1987), lipoxygenase (Oliw and Sprecher, 1989), three isoforms of lactate dehydrogenase (Olsson and Ronquist, 1990) and 5'-nucleotidase (Fabiani and Ronquist, 1993). It is, therefore, conceivable that the use of native, freshly isolated prostasomes as antigen(s) for rabbit immunization would result in polyvalent antibodies.

Another protein present in a number of exocrine cells is the so-called GRAMP 100 (granule-associated membrane polypeptide of 100 kD) which has been shown in the secretory granule membranes and the apical cell surface in pancreatic acinar cells (Laurie et al., 1992) It, however, does not seem to be related to any of the prostasome antigens. Finally, the possible presence of 5'-nucleotidase in prostasomes (Fabiani and Ronquist, 1993), an ectoenzyme present in a number of cells (see Zimmermann, 1992; LeHir and Kaissling, 1993) should be considered.

5.1.2. Tissue Distribution of the Respective Antigens. The seminosome antibody reacted with a number of elements present in the male genital system, as well as a few extragenital structures. In the seminal pathways, adluminal components such as plasma membranes including kinocilia and stereocilia in epididymal cells, the apical plasma membranes of secretory cells in seminal vesicles and the prostate, as well as the prostatic urethra were strongly immunoreactive. As the acinar precursor cells of the infantile prostate were already immunoreactive, whereas the basal cells were definitely non-reactive, the antibody recognizes one antigen that can be regarded as a marker of prostatic acinar cells. As shown by Western blotting, our antibody is clearly polyvalent, recognizing different antigens.

Relationship to other prostatic antigens: In addition to several secretory proteins, the most significant of those being acid phosphatase, Prostate Specific Antigen (PSA) and ß-microseminoprotein (for review, see: Aumüller and Seitz, 1990), the prostate contains a number of surface or cytoplasmic proteins which may act as viable antigens. There are some reports on prostate cancer cell antigens (e.g. Lowe et al., 1984; Beckett et al., 1991; Lipford and Wright, 1991), which are detected on normal or malignant prostate cells by either monoclonal or polyclonal antibodies. Most of these antibodies, however, cross-react with prostatic secretion and cytoplasm of prostate cells. They rather represent antibodies against soluble antigens. Pastan et al. (1993) have recently published results obtained with a monoclonal antibody (PR1; IgM$_k$ subtype) which reacts uniformly with the surface of several adenocarcinomas of the prostate, hyperplastic and normal prostate cells. Cross-reactivity was seen with apical brush border of the colon, bile ducts, pancreatic acini, pars intermedia of the pituitary, parietal cells of the stomach and principal cells of the renal collecting tubules, i.e. its pattern of immunoreactivity resembles closely that of the prostasome antibody. Regarding the figures of the paper by Pastan et al. (1993), however, immunoreactivity of their antibody is less precisely confined to plasma membranes. Instead mostly all of the cytoplasm, excluding the nuclei is stained. Another antibody that apparently recognizes an antigen closely related to that reacting with our prostasome antibody, has recently been published by Skibinski et al. (1994). Interestingly, this antibody, originally raised against granulophysin (a 40 kD integral membrane glycoprotein present in the dense granules of human platelets) reacted with a broad band of approximately 32 to 37 kD in lysates prepared from prostasomes and stained both acinar cells and luminal contents of the prostate. In addition, it stained epididymal cells. Other cross-reactions are not mentioned in the paper.

5.1.3. Seminosomes as Equivalents of Apocrine Secretion. Apocrine secretion is the release of secretory material from glandular cells through blebs developing from the apical cell portion. We have recently shown (Steinhoff et al., 1994) that apocrine secretion is a hormonally controlled, physiological event in the rat dorsal prostate and coagulating gland. Apocrine secretion was visualized by immunocytochemical staining of transglutaminase, a protein secreted by a non-classical pathway of biosynthesis and release. Contrary to this, in the present paper we show that a particular protein (of around 100 kD) exclusively present in the apical plasma membrane of secretory cells in the human prostate, seminal vesicle and epididymis, is released from the cells in conjunction with secretory proteins (e.g. PSA). In favourably sectioned cells where the cutting plane is tangentially oriented to the apical cell pole, the latter appear as isolated blebs. The question as of yet has been, whether these presumptive blebs are released from the cells in an apocrine fashion or if they represent simple fixation artifacts. As our antibody was prepared from purified prostasomes and is directed (inter alia) against the apical cell pole of prostatic glandular cells, there is circumstantial evidence that the latter are authentic secretory elements. As they are not restricted to the prostate, we prefer to use the term "seminosomes".

5.2. Characteristics of Seminosomal 5′-Nucleotidase

5.2.1. Distribution of 5′-Nucleotidase in Male Genital Organs. 5′-NT is present in the human male genital tract at differing amounts in the various organs. Combining determinations of enzyme activities, Western blotting of organ and tissue extracts, immunohistochemistry for the cellular localization and RT-PCR and RT-PCR for the identification of the sites of biosynthesis, we could unequivocally show the presence of the enzyme at a relatively high concentration and enzymic activity in the epithelium of the human prostate, however, somewhat less in the seminal vesicles and the epididymis. Divergent results were obtained in the testis, where a few spermatids were encountered containing immunoreactive material in the acrosomal cap. Contrary to this no enzyme activity could be measured and the immunoreactive band at 69 kDa was relatively weak in Western blots prepared from testis tissue. RT-PCR of testicular RNA, however, indicated the presence of a signal even in testis. These results may be due to the fact that the number on 5′-NT containing cells is far too low as to generate a signal in Western blots, but is sufficient to be detected both by immunohistochemistry and RT-PCR. The essential result, however, is that no 5′-NT is expressed on testicular spermatozoa. This is in clear-cut contradistinction to the situation in the bull, where the spermatogenic cells, starting from elongate spermatids, constantly contain ecto-5′-NT immunoreactivity (Schiemann et al., 1994).

In epididymis, immunoreactivity for 5′-NT was restricted to the stereocilia of the epididymal principal cells, corresponding to the enzyme activity and the Western blotting results. It was equally present in seminal vesicle epithelium and, at much higher levels in the prostate, indicating that an ecto-form of the enzyme is present throughout all the post-testicular pathways containing secretory elements. In the prostate, the immunolocalization of the enzyme varied somewhat depending on the method used, i.e. in paraffin sections processed with the unlabeled enzyme antibody method, were the secretory cells nearly exclusively labeled. In immunofluorescence studies of semithin cryosections, however, the whole epithelium shows strong immunofluorescence, as do most of the stromal cells. This could indicate that in addition to the ecto-(or soluble?) form, which obviously is relatively fixation-resistant, the cytoplasmic form present in fibroblasts and smooth muscle cells is identified when sufficiently sensitive methods are used.

In previous enzyme histochemical studies at the electron microscope level, we have found 5′-NT activity restricted to the (baso)lateral plasma membrane of prostatic basal cells. Only rarely were labeled granules found in prostatic secretions. This is clearly at variance with the present findings using the more refined immunological and molecular methods, where a high amount of enzyme was present particularly in prostatic secretory cells. The more restricted role of 5'-NT in the adenosine supply of basal cells, as previously suggested, therefore has to be extended into a capable role of the enzyme during the interaction with spermatozoa.

5.2.2. Localization on Spermatozoa. Both at the light and electron microscopic levels, immunoreactivity for 5'-nucleotidase was restricted to the anterior head region of human spermatozoa. This is comparable with the situation in the bull, where an intrinsic (i.e. present already in spermatozoa immediately after spermiation) 5'-nucleotidase immunoreactivity was observed at the tip of the sperm head and the seminal vesicle-derived enzyme simply functioned as a sperm coating protein, covering the whole sperm surface (Schiemann et al., 1994).

5.2.3. Significance of the GPI-Anchor. As has been shown by Fini et al. (1991), 5'-nucleotidase is a glycoprotein which contains a GPI-anchor. This applies not only to the membrane ecto-bound form, which can be removed e.g. from the sperm surface by incubation of bovine spermatozoa with phospholipase C from Bacillus thuringiensis. Removal of the GPI-anchor results in an immediate loss of immunoreactivity (Schiemann et al., in press). The human (seminal) enzyme and its counterpart present on human lymphocytes (CD 73) and placenta, likewise have a PGI-anchoring signal. The mature polypeptide chains are expected to consist of 548 amino acids which seem to exist as dimers (Zimmermann, 1992). Kirchhoff and Hale (1996) have recently pointed to the puzzling fact that the human epididymis synthesizes and secretes a number of GPI-anchored proteins such as CD 59, CD 55, and CD 52 which are incorporated into the plasma membrane on the sperm surface during their passage through the epididymal duct. Using both in situ hybridization and immunofluorescence techniques they have shown that most of these proteins are synthesized in epididymal cells, transported to the fluid secretions, probably bound to membrane vesicles and then incorporated into the sperm membrane (Kirchhoff and Ivell, 1995). Although the function of these GPI-anchored molecules is largely unknown, speculations on their hypothetical role in the protection of spermatozoa from immune attack appears reasonable, as some of these molecules have been found in cells of the immune system.

We have compared the distribution of nucleotidase-immunoreactivity with that of prostasomes/seminosomes, particulate elements present in human semen, which have been related to prostatic secretion. Kirchhoff and Hale (1996) have pointed to the fact that prostasomes contain some immunoreactivity for the CD 55 antigen (Rooney et al., 1993a). Experiments of Rooney et al. (1993b) and Rooney and Atkinson (1994) have shown that prostasomes present in seminal plasma are capable of conveying GPI-anchored molecules located in these particles onto the surface of red cells. This transfer, the mechanism of which is completely unknown as of yet, could also occur during the interaction of seminosomes/prostasomes with spermatozoa. Fabiani and Ronquist (1993) have recently shown that 5'-nucleotidase activity is also associated with prostasomes. It is, therefore, possible that the 5'-nucleotidase immunoreactivity observed on spermatozoa is mainly derived from seminosomes, even if the transfer mechanism is at present a conundrum.

5.3. Functional Considerations

5.3.1. Possible Functional Significance of 5'-Nucleotidase. A number of interesting hypotheses on the functional significance of the enzyme have been proposed in the cases

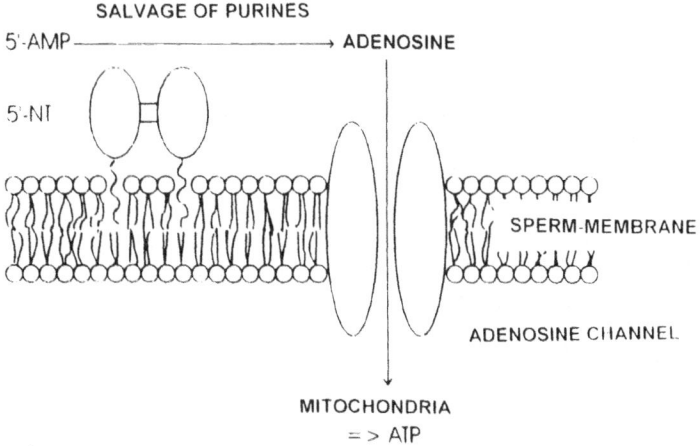

Figure 9. Hypothetical significance of 5'-nucleotidase in the regulation of sperm function.

of different blood cells, such as inhibition of platelet aggregation, suppression of reactive oxygen species in polymorphonuclear leukocytes, activation of macrophages and lymphocytes (for review, see Zimmermann, 1992). A number of these functions may also be likely in other isolated cells diluted in a specific surrounding, namely spermatozoa. During the transit of spermatozoa through the channel system of the male genital tract, they get into close contact with both the stereocilia of the epididymal principal cells as well as the secretions of the accessory sex glands. This could point to an interaction of the sperm surface with the soluble form of the enzyme, as has previously been suggested for bovine spermatozoa (Schiemann et al., 1994).

In this context, the presence of 5'-nucleotidase in the seminosomes is interesting. Prostasome particles have been shown to enhance sperm motility (Fabiani 1994, Fabiani et al., 1994, 1995) and to inhibit the immune response against seminal fluid (Skibinski et al., 1994). The substance(s) responsible for these activities have not yet been identified. Fabiani and Ronquist (1985) have demonstrated that 5'-nucleotidase is present in prostasomes. As this enzyme liberates adenosine which is responsible for a variety of physiological actions, it is one of the most interesting candidates for the interaction between spermatozoa and seminal fluid.

5.3.2. Sperm Motility Regulation by Seminosomal Proteins. Fibronectin is one example of sperm protein which is endogenously present in testicular sperm progenitor cells, but which is also released into the seminal fluid from seminal vesicles, rendering its concentration variable in seminal fluid.

The present immunohistochemical findings show that the distribution of FN on spermatozoa is more diversified than previously described. Glander et al. (1987) in their study only mentioned the EFB, while all other immunohistochemical localization signals, i.e.staining on the neck and mid piece (which can be seen on the pictures in their paper), have not been mentioned.

Immunoelectron microscopy shows that spermatozoa which have completed the acrosomal reaction are devoid of EFB. This may lead to the conclusion that in conjunction to the findings of Hoshi et al. (1994), spermatozoa lacking EFB have lost their adhesive ability. In addition to the suggested role in sperm adhesion of fibronectin, a quite unexpected function of FN was detected in the present study.

Our results concerning the motility parameters, namely, suggest motility modulation by FN. Since FN antibodies increase sperm motility and FN protein decreases sperm motility, an additional effect of FN on sperm functions appears likely. Obviously, the membrane domain responsible for motility regulation is different from that of the EFB. Our findings are difficult to reconcile with the recent observation of Anapliotou et al. (1995), who found a significantly decreased fibronectin-like immunoreactivity in semen from infertile men, even if they were normospermic. These authors did not find any correlation between either fibronectin levels, FSH levels, sperm counts or progressive motility. One cannot exclude the possibility that the quantification method used by the authors (cross immunoelectrophoresis) was responsible for these divergences.

Mitochondria surrounding the axonemal motility system are localized in the midpiece of spermatozoa. Binding of FN to this portion of the spermatozoon may interact with mitochondrial energy production leading to sperm motility changes. A conceivable mechanism could be FN-dependent regulation of calcium influx into the cell. FN and vitronectin, coupled to integrins on leucocyte surface, are able to modify calcium influx into the cell and as a consequence cellular motility (Maxfield, 1993). We presume that such a modification of calcium fluxes may also occur in spermatozoa subsequent to FN binding. Our combined morphological and functional studies revealed a correlation between sperm motility and FN concentration in the sense, that an above-average FN concentration results in impaired sperm motility which may eventually result in diminished fertilization capacity.

As FN is synthesized as an acute phase reactant in rat hepatocytes, pathological effects elicited by elevated FN concentrations may occur in seminal plasma during inflammatory processes of the male genital tract, thereby decreasing sperm motility. The possible relationship between FN levels in semen and sperm motility would open a novel therapeutical avenue.

In addition to the pathophysiological mechanisms discussed, one has to take into consideration that FN plays an important role in sperm-egg adhesion (Hoshi et al., 1994). Manipulation of FN concentrations in semen, therefore, may not only influence sperm motility, but also additional sperm functions, such as sperm-egg interaction.

6. CONCLUSIONS

A 100 kD membrane protein was isolated from human seminosomes, being present in apical membrane differentiations of the secretory cells in epididymis, seminal vesicles and prostate.

Its association with secretory proteins (e.g. 5'- nucleotidase) points to an apocrine release mechanism of secretory and membrane proteins from the human male accessory glands and the subsequent formation of seminosomes.

Using purified seminosomal proteins (5'-nucleotidase, 100 kD membrane protein, fibronectin) and the respective antibodies, the sources of these proteins were identified in the epididymis, seminal vesicle and prostate.

Immunoreactive sites for all three antisera were detected on human spermatozoa. Although no epididymal spermatozoa could be studied to exclude the presence of endogenous antigenic sites on the outer plasma membrane on sperm heads, the distribution of the antigens is in favour of an integration of the seminosomal proteins into the sperm membrane.

Whereas no motility modulation could be achieved when spermatozoa were incubated in vitro with the 100 kD protein or its respective antiserum, fibronectin exerts an inhibitory effect on sperm motility. The concentration of fibronectin in human semen was inversely correlated with sperm motility in fertility patients.

In incubation studies of isolated live human spermatozoa fibronectin at a concentration of 1μg/ml was shown to increase the calcium influx into spermatozoa. This effect could be repeated at least 3 times within 10 min.

5'-Nucleotidase inhibitors and adenosine channel blockers inhibit sperm motility in a dose-dependent manner. A spatial relationship between adenosine channels and 5'-nucleotidase on the plasma membrane of the sperm head is suggested which is effective in the regulation of sperm motility.

7. ACKNOWLEDGMENTS

This work was supported by grant (Az: I/116–825 2653) of the Bundesministerium für Forschung und Technologie (BMFT), by grants of the Kempkes-Stiftung (Medical Faculty, University of Marburg) and by grants of the Deutsche Forschungsgemein-schaft (Au 48/15–1 and 15–2 2, and SFB 286, Projekt B6, Seitz). The authors are grateful to Mr. D. Littauer for his expert help with the biochemical experiments, to Mrs. R. Leineweber for her help in performing CASA measurements and to Dr. S. Groos and Mrs. I. Dammshäuser for their support in immunoelectron microscopy. Special thanks are due to Dr. Christiane Kirchhoff, Hamburg, for discussions on the functional role of 5'-nucleotidase.

8. REFERENCES

Agrawal Y, Vanha-Perttula T 1987 Effect of secretory particles in bovine seminal vesicle secretion on sperm motility and acrosome reaction. Journal of Reproduction and Fertility 79: 409–419.

Anapliotou MLG, Goulandris N & Douvara R 1995 Seminal fibronectin-like antigen and transferrin concentrations in infertile and fertile men. Andrologia 27: 137–142.

Arvidson G, Ronquist G, Wikander G & Ötjeg AC 1989 Human prostasomes membranes exhibit very high cholesterol/phospholipid ratios yielding high molecular order. Biochimica Biophysica Acta 984: 167–173.

Aumüller G & Seitz J 1990 Protein secretion and secretory processes in male accessory sex glands. International Reviews of Cytology 121: 127- 231.

Aumüller G, Seitz J & Riva A 1994 Functional morphology of prostate gland. In: Ultrastructure of Male Urogenital Glands: Prostate, Seminal Vesicles, Urethral, and Bulbourethral Glands (Riva A, Testa Riva F, Motta PM, eds) pp.61–112.

Beckett ML, Lipford GB & Haley CL 1991 Monoclonal antibody PD41 recognizes an antigen restricted to prostate adenocarcinomas. Cancer Research 51: 1326–1333.

Biggers JD, Whitten W & Whittingham DG 1979 The culture of mouse embryos in vitro. In: Methods of Mammalian Reproduction (ed. JC Daniel) pp86–116. WH Freeman, San Francisco.

Brody I, Ronquist G & Gottfried A 1983 Ultrastructural localization of the prostasome - an organelle in human seminal plasma. Upsala Journal of Medical Science 88: 63–80.

Calvete JJ, Solis D, Sanz L, Díaz-Mauriño T, Schäfer W, Mann K & Töpfer-Petersen E 1993 Characterization of two glycosylated boar spermadhesins. European Journal of Biochemistry 218: 719–725.

Fabiani R 1994 Functional and Biochemical Characteristics of Human Prostasomes. Upsala Journal of Medical Science 99: 73–112.

Fabiani R & Ronquist G 1985 Association of some hydrolytic enzymes with the prostasome membrane and their differential responses to detergent and PIPLC treatment. The Prostate 27: 95–101.

Fabiani R & Ronquist G 1993 Characteristics of membrane-bound 5'-nucleotidase on human prostasomes. Clinica Chimica Acta 216: 175–182.

Fabiani R, Johansson L, Lundkvist Ö, Ulmsten U & Ronquist G 1994 Promotive effect by prostasomes on normal human spermatozoa exhibiting no forward motility due to buffer washings. European Journal of Obstetrics & Gynecology and Reproductive Biology 57: 181–188.

Fabiani R, Johannson L, Lundkvist Ö & Ronquist G 1995 Prolongation and improvement of prostasome promotive effect on sperm forward motility. European Journal of Obstetrics & Gynecology and Reproductive Biology 58: 191–198.

Fini C, Ipata PL, Palmerini CA & Floridi A 1983 5'-Nucleotidase from bull seminal plasma. Biochimica Biophysica Acta 748: 405–412.

Fini C & Cannistrado S 1990 5'-Nucleotidase from bull seminal plasma. Biochemical and biophysical aspects. Andrologia 22: 33–43.

Fini C, Coli M, Floridi A 1991 Purification of 5'-nucleotidase from human seminal plasma. Biochimica Biophysica Acta 1075: 20–27.

Fusi FM, Vignali M, Bussaca M & Bronson RA 1992 Evidence for the presence of an integrin cell adhesion receptor on the oolemma of unfertilized human oocytes. Molecular Reproduction and Development 31: 215–222.

Glander H-J, Herrmann K & Haustein U-F 1987 The equatorial fibronectin band (EFB) on human spermatozoa - a diagnostic help for male fertility? Andrologia 19: 456–459.

Hoshi K, Sasaki H, Yanagida K, Sato A & Tsuiki A 1994 Localization of fibronectin on the surface of human spermatozoa and relation to the sperm-egg interaction. Fertility and Sterility 61: 542–547.

Kirchhoff C & Ivell R 1995 Molekulare Aspekte der Spermienreifung im Nebenhoden. Fertilität 11: 167–174.

Kirchhoff C & Hale G 1996 Cell-to-cell transfer of glycosylphosphatidylinositol-anchored membrane proteins during sperm maturation. Molecular Human Reproduction 2: 177–184.

Krause W 1995 Computer-assisted semen analysis systems: Comparison with routine evaluation and prognostic value in male fertility and assisted reproduction. In: Modern Andrology (Ombelet W, Vereecke A, eds.) pp. 60–66 (=Human Reproduction 10) University Press; Oxford.

Laurie SM, Mixon MB & Castle JD 1992 GRAMP 100: A membrane protein concentrated on secretory membranes and the apical cell surface in exocrine acinar cells. Journal of Histochemistry and Cytochemistry 40: 1827–1835.

LeHir M & Kaissling B 1993 Distribution and regulation of renal ecto-5'-nucleotidase: implications for physiological functions of adenosine. American Journal of Physiology 264: (Renal Fluid Electrolyte Physiol 33): F377-F387

Lilja H 1990 Cell biology of semenogelin. Andrologia 22 suppl. 1: 132–141.

Lilja H & Weiber H 1983 Gamma-glutamyltransferase bound to prostatic subcellular organelles and in free form in human seminal plasma. Scandinavian Journal of Clinical Laboratory Investigation 43: 307–312.

Lilja H, Oldbring J, Rannevik G & Laurell C-B 1987 Seminal-vesicle secreted proteins and their reactions during gelation and liquefaction of human semen. Journal of Clinical Investigation 80: 281–285.

Lindahl M, Tagesson C & Ronquist G 1987 Phospholipase A2 activity in prostasomes from human seminal plasma. Urologia Internationalis 42: 385- 389.

Lipford GB & Wright GL jr 1991 Comparative study of monoclonal antibodies TURP-27 and HNK-1: Their relationship to neural cell adhesion molecules and prostate tumor-associated antigens. Cancer Research 51: 2296–2301.

Lowe DH, Handley HH, Schmidt J, Royston I & Glassy MC 1984 A human monoclonal antibody reactive with human prostate. Journal of Urology 132: 780–785.

Matsubara S, Tamada T, Kurahashi K & Saito KT 1987 Ultracytochemical localization of adenosine nucleotidase activities in the human term placenta, with special reference to 5'-nucleotidase. Acta Histochemica Cytochemica 20: 409–419.

Maxfield RF 1993 Regulation of leukocyte locomotion by Ca2+. Trends In Cell Biology 3: 386–391.

Muesch A, Hartmann E, Rohde K, Rubartelli A, Sitia R & Rapoport TA 1990 A novel pathway for secretory proteins? Trends in Biological Sciences 15: 86–88.

Oliw EH & Sprecher H 1989 Metabolism of polyunsaturated fatty acids by an (n-6)-lipoxygenase associated with human ejaculates. Biochimica Biophysica Acta 1002: 283–291.

Olsson I & Ronquist G 1990 Isoenzyme pattern of lactate dehydrogenase associated with human prostasomes. Urologia Internationalis 45: 346–349.

Pastan I, Lovelace E, Rutherford AV, Kunwar S, Willingham MC & Peehl D 1993 PR1 - a monoclonal antibody that reacts with an antigen on the surface of normal and malignant prostate cells. Journal of the National Cancer Institute 85: 1149–1154.

Renneberg H, Konrad L, Dammshäuser I, Seitz J & Aumüller G 1997 Immunohistochemistry of prostasomes from human semen. The Prostate (in press)

Robert M & Gagnon C 1996 Purification and characterization of the active precursor of a human sperm motility inhibitor secreted by the seminal vesicles: Identity with semenogelin. Biology of Reproduction 55: 813–821.

Ronquist G 1987 Effect of modulators on prostasome membrane-bound ATPase in human seminal plasma. European Journal of Clinical Investigation 17: 231–236.

Ronquist G & Brody I 1985 The prostasome: its secretion and function in man. Biochimica Biophysica Acta 822: 203–218.

Ronquist G, Frithz G & Jansson A 1988 Prostasome membrane associated enzyme activities and semen parameters in men attending an infertility clinic. Urologia Internationalis 43: 133–138.

Ronquist G, Nilsson O & Hjerten S 1990 Interaction between prostasomes and spermatozoa from human semen. Archives of Andrology 24: 147–153.

Ronquist G & Stegmayr B 1984 Prostatic origin of fucosyl transferase in human seminal plasma - a study on healthy controls and on men with infertility or with prostate cancer. Urological Research 12: 243–247.

Rooney IA & Atkinson JP 1994 Mechanisms by which cells can acquire complement inhibitors during incubation with seminal plasma. Clinical and Experimental Immunology 97: (suppl. 2) 41 (abstr.)

Rooney IA, Atkinson JP, Krul ES, Schonfeld G, Polakoski K, Saffitz JE & Morgan BP 1993a Physiologic relevance of the membrane attack complex inhibitory protein CD 59 in human seminal plasma: CD 59 is present on extracellular organelles (prostasomes), binds cell membranes, and inhibits complement-mediated lysis. Journal of Experimental Medicine 177: 1409–1420.

Rooney IA, Oglesby TJ, Atkinson JP 1993b Complement in human reproduction: Activation and control. Immunological Research 12: 276–294.

Schiefferdecker P 1917 Die Hautdrüsen des Menschen und der Säugethiere, ihre biologische und rassenanatomische Bedeutung, sowie die Muscularis sexualis. Biologisches Centralblatt 37: 634–662.

Schiemann PJ, Aliante M, Wennemuth G, Fini C & Aumüller G 1994 Distribution of endogenous and exogenous 5'-nucleotidase on bovine spermatozoa. Histochemistry 01: 253–263.

Schiemann PJ, Konrad L, Renneberg H, Fini C & Aumüller G 1997 Distribution of ecto 5'-nucleotidase in the male genital tract as shown by immunohistochemistry and RT-PCR. Biology of Reproduction (submitted)

Seitz J, Keppler C, Rausch U & Aumüller G 1990 Immunohistochemistry of secretory transglutaminase from rodent prostate. Histochemistry 93: 525–530.

Skibinski G, Kelly RW & James K 1994 Expression of a common secretory granule specific protein as a marker for the extracellular organelles (prostasomes) in human semen. Fertility and Sterility 61: 755–759.

Stegmayr B, Brody I, Ronquist G 1982 A biochemical and ultrastructual study on the endogenous protein kinase activity of secretory granule membranes of prostatic origin in human seminal plasma. Journal of Ultrastructural Research 78: 206–214.

Steinhoff M, Eicheler W, Holterhus PM, Rausch U, Seitz J & Aumüller G 1994 Hormonally induced changes in apocrine secretion of transglutaminase in the rat dorsal prostate and coagulating gland. European Journal of Cell Biology 65: 49–49.

Sundermann FW jr. 1990 The clinical biochemistry of 5'-nucleotidase. Annals of Clinical Laboratory Science 20: 123–139.

Vanha-Perttula T 1984 Studies on alanine aminopeptidase, dipeptidyl aminopeptidase 1 und 2 of human seminal fluid and prostasomes. In: Goldberg DM, Werner M (eds.) Selected Topics in Clinical Enzymology, vol 2. Walter de Gruyter Company Berlin-New York pp. 545–565.

Wennemuth G, Schiemann PJ, Krause W, Gressner AM, Aumüller G 1997 Influence of fibronectin on the motility of human spermatozoa. International Journal of Andrology (in press).

Zimmermann H 1992 5'-Nucleotidase: molecular structure and functional aspects. Biochemical Journal 285: 345–365.

40

THE MOLECULAR BIOLOGY OF THE SPERM SURFACE

Post-Testicular Membrane Remodelling

C. Kirchhoff,[1] I. Pera,[1] P. Derr,[1] C.-H. Yeung,[2] and T. Cooper[2]

[1]IHF Institute for Hormone and Fertility Research at the University of
    Hamburg
Grandweg 64, 22529 Hamburg, Germany
[2]Institut für Reproduktionsmedizin der Universität
Domagkstrasse 11, 48149 Münster
Germany

1. SUMMARY

The membrane of testicular spermatozoa undergoes extensive changes in the epididymis, including rearrangement, modification and loss of pre-existing components, addition of new glycoproteins from epididymal secretions, and exchange of lipid constituents. As a result, the membrane of cauda epididymidal spermatozoa has a different composition and different properties, which collectively contribute to male fertility. Special significance has been attributed to sperm surface structures that only appear post-testicularly in the epididymis, the so-called "maturation antigens". Therefore, human post-testicular proteins have been cloned by substractive screening of epididymal cDNA libraries, employing testis as the primary negative control. To date, there is scanty information on their function and mechanism of deposition on the sperm surface. However, the major maturation antigen CD52 seems to bind firmly to the sperm membrane via its GPI anchor. Its synthesis is carefully regulated by the cells of the epididymal epithelium, with temperature and androgens acting synergistically on CD52 mRNA levels.

2. INTRODUCTION

The membrane of mammalian spermatozoa conforms to all accepted notions regarding the structure of membranes in general, consisting of a bilayer of charged and neutral lipids, into which proteins and glycoproteins are inserted. These may span the lipid bilayer or may be restricted to either the external or cytoplasmic faces, or they may be adsorbed by various types of interaction onto the glycocalix. Superimposed upon this general structure are specializations that are unique to the male germ cell, i.e. its differentiation into the well-known domains (for review see Myles, 1993). The sperm membrane domains seem to correlate closely with the underlying organelles implying specialized functions during the fertilization process. Thus, the acrosome membrane may posess special functions involved in the acrosome reaction. The equatorial region of the sperm head may possess fusiogenic properties which enable fusion with the vitellus. The membrane overlaying the outer dense fibres of the principal flagellum piece may be relevant for sperm motility.

It would seem that specialization into domains is established early during spermatogenesis in the testis, but as will be described in the following, membrane composition and surface characteristics of spermatozoa change post-testicularly during epididymal passage, ejaculation and capacitation. This concept of post-testicular "remodelling" (Jones, 1989) is based upon extensive investigation employing various cell surface probes (for review see Eddy and O'Brien, 1994). However, how this heterogeneity is maintained until fertilization, how the characteristics of sperm domains mature post-testicularly in an essentially fluid milieu, and how they respond to external signals is still largely unknown. Maintenance and modification of sperm surface characteristics are energy-consuming processes which require enzymes, sperm-coating proteins, lipids and lipid transfer or exchange molecules. As sperm themselves have very little biosynthetic capability, the surrounding fluids and the male genital tract epithelia must provide the necessary molecules. Many of these molecules seem to be secreted by specialized regions of the epididymal epithelium, and some of them will be described below in greater detail. Others are provided by the seminal plasma during ejaculation and will not be considered here, although they may be equally important during the fertilization process. The examples chosen are mostly limited to the highly abundant post-testicular, mainly epididymis-specific, components that have been cloned or at least been tentatively identified in the human.

3. PROTEOLYTIC PROCESSING AND RELOCATION OF PRE-EXISTING SPERM MEMBRANE PROTEINS

PH-20 and PH-30 (=fertilin) antigens are conserved integral membrane glycoproteins derived from the testis. They have first been identified by monoclonal antibodies in the posterior head region of guinea pig sperm (Primakoff and Myles, 1983). On testicular sperm they are present on the whole cell surface (PH-20) or the whole head surface (PH-30). However, on cauda epididymal sperm, both sperm surface proteins become relocated to the posterior head domain. Simultaneously with this change in location they undergo a proteolytic processing and a change in epitope expression (Blobel et al., 1990). Both proteins are proteolytically cleaved in a region of the epididymis where sperm become fertile, and the maturation process that relocalizes these proteins to the posterior head region can be mimicked *in vitro* by treating testicular sperm with trypsin (Phelps et al., 1990). The hypothesis that specific proteases may be involved in epididymal sperm maturation has long been suggested, and similar proteases may also function in the human male genital tract. The protease(s) acting *in vivo* on PH-20 and PH-30 sperm membrane glycoproteins are not known. For the PH-30 precursor molecule, however, autocatalytical processing has been suggested (Wolfsberg et al., 1993).

4. CHANGES IN GLYCOSYLATION OF SPERM SURFACE MOLECULES

Cell surface labelling techniques and lectin-binding studies suggest that sugar moieties of the sperm membrane are extensively modified as spermatozoa pass through the epididymis (for review see Eddy and O'Brien, 1994; Yanagimachi, 1994). The epididymis seems to be one of the richest sources of extracellular exoglycosidases in the body (Jones, 1989), and there is also an indication for active luminal glycosylation (Tulsiani et al., 1993). Regional differences in glycoprotein-modifying enzymes have been described. Luminal galactosyl-transferase levels are highest within the human distal caput epididymidis (Ross et al., 1993). On the other hand, the luminal enzyme neutral α-glucosidase which has been proposed as a marker for human epididymal potency and function (Cooper et al., 1990; WHO, 1992), appears at higher levels in the human cauda region (Purvis and Edgetveit, 1993).

A convincing example of (a) physiological substrate molecule(s) on the sperm surface has not yet emerged. A $Mr \approx 26$ kDa sperm membrane glycoprotein of rat spermatozoa (SMemG; Hamilton et al., 1986; Moore et al., 1989; Eccleston et al., 1994) has been suggested to represent an important candidate substrate molecule (for review see Myles, 1993, Eddy and O'Brien, 1994). The glycoprotein, containing more than 60% carbohydrate, originally aroused interest because it is the principal component labelled by the galactose oxidase/Na B(^{3}H)$_4$ surface labelling technique on cauda epididymal sperm, but not on testicular or caput sperm (Hamilton et al., 1986). Moreover, rat sperm incorporate galactose from UDP-(^{14}C)galactose into a surface glycoprotein of approximately the same molecular weight, suggesting that it is altered during epididymal maturation by galactosylation of exposed N-acetylglucosamine residues (Hamilton and Gould, 1982).

Despite considerable study, the cell type of origin (spermatogenic or epididymal) remained unknown until recently (see below). Most investigators discussed the possibilty that the protein is already present on the membrane of testicular sperm, but is glycosylated only during epididymal transit. The possibility that it may be synthesized post-testicularly by the epididymal epithelium, shed into the lumen and then bound to sperm in the cauda, was ruled out because it is anchored in the sperm membrane by linkage to glycosylphosphatidyl-inositol (GPI; Moore et al., 1989; Eccleston et al., 1994). Irrespective of its cell type of origin, however, the molecule may well represent an important substrate to carbohydrate modification occuring on the sperm surface.

5. ACQUISITION BY SPERMATOZOA OF SURFACE GLYCOPROTEINS FROM EPIDIDYMAL SECRETIONS

In rat and mouse animal model systems epididymal secretory proteins which interact with the sperm surface have been extensively investigated on the molecular level (for review see Cooper, 1992). A major problem in transferring these results to the human system and cloning homologous gene products lies in their high degree of species specificity, i.e. poor evolutionary conservation, limiting the usefulness of these animal models for the human epididymis. Only recently, an AEG=D/E-related gene product has been cloned from human epididymal tissue employing uni-directional RT-PCR based on a short, conserved sequence stretch of (Hayashi et al., 1996).

To avoid problems caused by lack of sequence information (and lack of tissue) in the human, a strategy was applied which does not depend on sequence homologies. Molecular cloning of the major secretory proteins of the human epididymis was achieved by a substractive screening procedure of cDNA libraries, relying on tissue-specificity and high frequency of the expected products (Kirchhoff et al., 1990). Targetting products of post-testicular origin, human testis mRNA was chosen as the primary negative control tissue. The method has identified several cDNAs derived from highly abundant human epididymal mRNAs (for review see Kirchhoff, 1995). The encoded products were tentatively named HE proteins, for *h*uman *e*pididymis-specific proteins. Most of the cDNAs, namely HE1-HE5, predicted small, major secretory glycoproteins with characteristic N-terminal signal peptides and sites for N-glycosylation. Northern analyses revealed a tissue-specific expression pattern and up-regulation in the epididymis. Employing *in situ* transcript hybridization their mRNAs were localized in the epithelial cells surrounding the lumen of the human epididymal duct (Kirchhoff et al., 1991; 1993; Osterhoff et al., 1994), each gene product showing a distinct spatial pattern of hydridization along the length of this organ (Krull et al., 1993).The regional variation of mRNA levels may lead to regionally varying amounts of the encoded secretory proteins, thus leading to alterations in the luminal microenvironment along the human epididymal duct.

HE1-HE4 and HE6 represented novel human gene products. The HE5 cDNA sequence, however, had already been known from the CD52 lymphocyte surface antigen (Xia et al., 1991; Kirchhoff et al., 1993). The most abundant cDNA, HE1 (Kirchhoff et al., 1996), seemed to be the only highly conserved secretory product identified by the screening procedure. Closely related homologues were found in all mammlian species investigated so far (Uhlenbruck et al., 1993; Ellerbrock et al., 1994). The origin of its peptide sequence in animal evolution, however, seems to be much older. An HE1-related protein has recently been cloned from an insect where it seems to be expressed under steroid hormone control in a highly specialized epithelium (Meszaros and Morton, 1996).

Since conventional protein purification was not readily feasable with human epididymal secretory proteins, verification and localization of the predicted HE proteins in the epididymis and on sperm was achieved by a different strategy (Osterhoff et al., 1994; Uhlenbruck et al., 1994; Kirchhoff et al., 1996). Putatively antigenic sites were predicted from cDNA sequences; these peptide epitopes were created by chemosynthesis and/or recombinant expression; then polyclonal antibodies were raised against the synthetic peptide epitopes. Employing these antibodies in immunohistochemical staining, the predicted HE proteins were localized in the epididymal epithelium, within the lumen of epididymal and deferent ducts, and on the sperm surface. Association with human ejaculated spermatozoa was observed for at least three of the HE proteins, HE2, HE4, and HE5/CD52, and for the human AEG-related protein (see Fig. 1).

HE2, an abundant gene product of the proximal human epididymis (Krull et al., 1993; Osterhoff et al., 1994) seems to be highly specific to the human; no animal homologues have been identified by either cDNA probes (Kirchhoff et al., 1990) or antibodies (Kirchhoff and Osterhoff, unpublished results). Because of the lack of any homologue in other mammalian species, experimental work to elaborate the function of the molecule in an animal model is severely hampered. Antibodies raised against a synthetic HE2 oligopeptide and recombinantly expressed HE2 specifically react with the acrosome and the equatorial region of human ejaculated spermatozoa (Osterhoff et al., 1994), suggesting that HE2 may be involved in sperm-egg fusion.

The HE4 cDNA predicts another small secretory glycopeptide. It is very cysteine-rich and consists of two "whey acidic protein" domains (Kirchhoff et al., 1991). The positions of half-cysteines suggested that it is a two-domain member of the family of "four-disulfide core" proteins which comprises a heterogeneous group of small acid- and

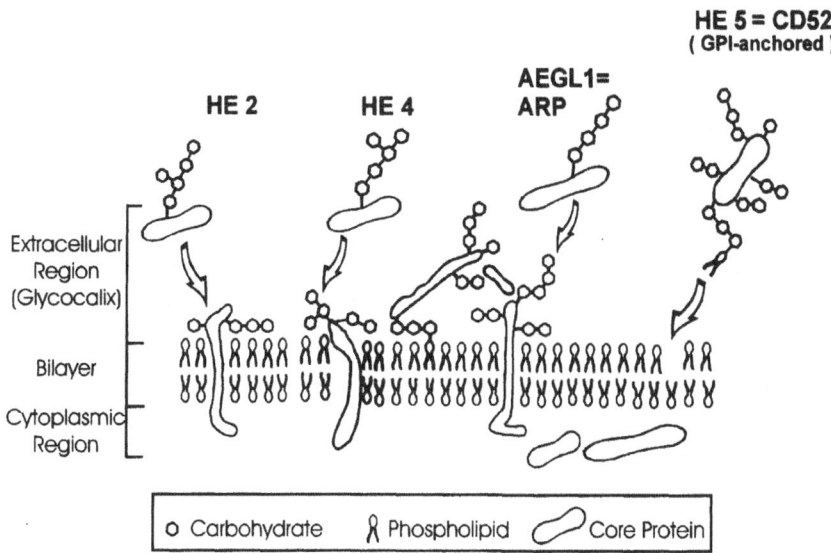

Figure 1. Diagrammatic representation of human epididymal secretory glycoproteins HE2, HE4, HE5/CD52, and AEGL1/ARP which bind to the sperm surface only post-testicularly. Our results suggest that these proteins are novel sperm surface antigens binding to spermatozoa during epididymal transit. To date, there is scanty information as to their function and the mechanism of their deposition on the sperm surface. However, HE5/CD52 which has been known before as "major maturation-associated" antigen, binds to sperm in the distal epididymis by insertion into the membrane via its GPI lipid anchor.

heat-stable proteins of divergent biological functions. The HE4 protein has significant similarity to, but is distinct from the acid-stable proteinase inhibitor of human mucous secretions, antileucoprotease, that has also been identified in human seminal plasma (Stetler et al., 1986; Seemüller et al., 1986). It may protect spermatozoa and/or the epididymal epithelium from proteolytic damage resulting from the premature release of acrosomal proteolytic enzymes or from leucocyte secretory proteases. The HE4 protein seems to coat the entire surface of human spermatozoa (Kirchhoff, unpublished results). Under *in vitro* capacitation conditions, it dissociates from the sperm surface quite easily, suggesting that it might represent a "decapacitation factor" (Osterhoff, unpublished). It is worthy of note that small proteins with proteinase-inhibiting activity (Boettger-Tong et al., 1992) and /or "four disulfide core"-structure, like calthrin II (Coronel et al., 1990) have been identified as decapacitation factors in non-primate species. Different from the HE2 and HE4 antigens which appear to represent "sperm-coating" proteins, showing no direct binding to the sperm membrane, the HE5/CD52 antigen seems to insert into the sperm membrane via its GPI anchor during epididymal transit (Kirchhoff and Hale, 1996). It will be discussed below in greater detail.

6. CHANGES IN LIPID COMPOSITION OF THE SPERM MEMBRANE

Although research in this area has not progressed very far because of insufficient methodology, current evidence suggests that substantial changes of sperm membrane phospholipids, fatty acids and sterols occur during epididymal transit. A phospholipid-binding protein of testicular origin seems to be lost from sperm in the rodent epididymis (Perry et al., 1994), possibly modifying the lipid composition of sperm membranes. However, the human counterpart was found to be expressed only in the brain, not in genital tract tissue. The B/C(=ESP-1) protein of the rodent epididymis has been shown to represent a retinoic acid-binding member of the lipocalin family (Newcomer and Ong, 1992). A related human epididymal product has not yet been observed.

Sperm cholesterol/ phospholipid ratios and lipid flux seem to be crucial for the process of capacitation and for the acrosome reaction. Human sperm membranes can be hyperloaded with cholesterol in the presence of cholesterol-saturated liposomes, and maintained in such a decapacitated state for days without signs of capacitation (Benoff, 1993). Some human spermatozoa that cannot fertilize manifest an unusual failure to loose membrane cholesterol during capacitation (Benoff, 1993). HE1, which is a (the?) major secretory glycoprotein of the primate epididymis (Kirchhoff et al., 1996; Perry et al., 1994; Fröhlich and Young, 1996) seems to be involved in epididymal cholesterol transfer or exchange. A similar protein has been purified from ram epididymal fluid which seems to function as cholesterol transfer protein (CTP, Baker et al., 1993). Its N-terminal amino acid sequence is highly similar to that of HE1 (Kirchhoff et al., 1996). Moreover, all other characteristics (apparent molecular weight, microheterogeneity, glycosylation) seem to be in good agreement. Proteins with an identical peptide backbone have been isolated from non-human primates (Perry et al., 1994; Fröhlich and Young, 1996), representing approximately 20% of the luminal protein content (Fröhlich and Young, 1996). After ejaculation, HE1 seems to be present in the seminal plasma, but only loosely associated with spermatozoa. It may represent the "decapacitation factor" which has long been postulated to maintain the sperm cholesterol content during epididymal transit and storage and to be lost during capacitation in the female tract, accompanied by an efflux of cholesterol from the sperm membrane.

7. ENZYMES PROTECTING THE SPERM MEMBRANE AGAINST LIPID PEROXIDATION

In most tissues the levels of oxygen free radicals are carefully regulated by two enzyme activities: glutathione peroxidase (GPX) and superoxide dismutase (SOD). Although both enzyme activities have a ubiquitous tissue distribution, unique secretory forms are highly expressed in the epididymis of various mammals, including rodents and a non-human primate (Ghyselinck et al., 1991; Perry et al., 1992, 1993). It is noteworthy, that related human products were not found during the differential screening procedure described, suggesting that the enzymes are not among the most abundant human epididymal secretory proteins. On the other hand, a similar screening procedure of a canine epididymal cDNA library resulted in the cloning of the canine secretory GPX which was derived from a highly abundant, epididymis-specific mRNA (Ellerbrock Kirchhoff and Ivell, unpublished). This cDNA cross-hybridized with mRNAs from most mammalian epididymides included in a Northern analysis, but excluding human epididymal RNA. An RT-PCR-cloning strategy was designed, based upon evolutionarily conserved sequence stretches of rat, cynomolgous monkey and dog secretory epididymal GPX, to clone the human homologue. A partial cDNA was obtained, probably representing the human counterpart. Employing this cDNA fragment as a probe, Northern analyses of various human tissue extracts revealed only a faint hybridization signal with poly(A)-enriched epididymal

RNA. Thus, our preliminary results suggest that, unlike other mammalian species, the human epididymis does not produce high amounts of secretory GPX. Additional work will be required to reveal whether this negative result is physiologically significant.

8. POST-TESTICULAR INTEGRATION INTO THE SPERM PLASMA MEMBRANE OF GPI-ANCHORED PROTEINS

Recent evidence shows that some of the GPI-anchored proteins of human spermatozoa, among them HE5/CD52, are not synthesized by sperm themselves, but by the epithelial cells of the genital tract (for review see Kirchhoff, 1995; Kirchhoff and Hale, 1996). Besides lymphocytes, the distal human epididymis is the major site of CD52 expression (Kirchhoff et al., 1993; Hale et al., 1993). Additionally, the GPI-anchored glycopeptide is found on epididymal, but not testicular spermatozoa, and in the ejaculate (Hale et al., 1993; Kirchhoff, 1996). Related GPI-anchored glycopeptides have been observed in other mammals (reviewed in Kirchhoff and Hale, 1996). In each case the pattern of expression in the male reproductive tract is very similar: Transcripts are found only post-testicularly in the epithelial cells of epididymis, deferent duct, and seminal vesicle. The antigen is found in the genital tract epithelium, in the seminal fluid, and on mature spermatozoa. Deviating from the current understanding of sperm membrane biosynthesis this would suggest that CD52 is derived post-testicularly from the epididymal epithelium, integrating via its GPI anchor into the sperm membrane by cell-to-cell transfer.

The homologous gene product of the rat has been cloned and its epididymal origin shown by *in situ* transcript hybridization (Kirchhoff, 1994; 1996). Genetic evidence for its homology to CD52 comes from sequence comparisons of the N-terminal and C-terminal signal peptides and from the non-coding parts of the genes, including a conserved intron location. The peptide backbone sequence predicted from the rat CD52 cDNA includes the sequence of the major rat sperm membrane glycoprotein SMemG (Eccleston et al., 1994; Kirchhoff, 1994; 1996), representing the Mr $\approx$ 26 kDa principal galactose oxidase/NaB(^{3}H)$_4$-labelled surface glycopeptide of rat cauda sperm (Olson and Hamilton, 1978; Hamilton et al., 1986). The appearance of this glycoprotein which can be cleaved from the sperm surface by phosphatidylinositol-specific phospholipase C (Moore et al., 1989; Eccleston et al., 1994) coincides with the sperm acquiring its major physiological properties in the distal epididymis.

9. FACTORS REGULATING THE EXPRESSION OF GLYCOPEPTIDES ON THE SPERM SURFACE

Sperm surface glycopeptides of epididymal origin have long been known to be androgen-dependent (for review see Orgebin-Crist, 1996). Additionally, non-steroidal testicular ("paracrine") factors have been implicated in their regulation. Recently, a striking temperature effect on CD52 mRNA levels was observed (Pera et al., 1996). Body temperature (37°C) as apposed to scrotal temperature (33°C) specifically down-regulates the CD52 mRNA in canine epididymal cells. Epididymal cell cultures, however, differ from the organ *in situ* in many aspects (Cooper et al., 1990; Raczek et al., 1992; for review see Moore and Akhondi, 1996). Therefore, in an attempt to elaborate the physiological significance of our cell culture results, we investigated the regulation of CD52 expression *in vivo* and *in vitro* in another animal model, the rat. CD52 mRNA levels under different cell culture conditions were compared to the *in vivo*-expression pattern in rats subjected to different experimental procedures, e.g. castration, testosterone supplementation, efferent duct ligation/dissection, and exposure of the epididymis to abdominal temperature. In the past, the major identifying criterion for the encoded sperm membrane glycoprotein SMemG has been the labelling of its carbohydrate portion by the galactose oxidase/NaB(^{3}H)$_4$ reaction (Hamilton et al., 1986). We adopted this principle, using a non-radioactive carbohydrate labelling technique (Haselbeck and Hösel, 1990) in order to follow the expression of the CD52 glycoconjugate on spermatozoa and in epididymal tissue by Western blotting.

In primary cell cultures of the rat epididymis, with monolayers of adherent cells spreading from small undisaggregated cell clusters, CD52 mRNA was present during a period of five days. Androgen effects, however, could not be studied *in vitro*, since cells lost their nuclear androgen receptor (AR) soon after the establishment of the explant cultures, independently of the presence of androgens in the culture media. Loss of AR and loss of androgen responsiveness have already been reported for other epididymal cell culture systems. Therefore, to investigate androgen effects on CD52 adult male rats were castrated and Northern analysis of epididymal RNA and proteins performed after 14 days, the most dramatic alterations concerning epididymal weight and RNA expression occuring within this period (Brooks, 1987). Castration significantly reduced the levels of CD52 mRNA especially in the proximal regions of the epididymis, demonstrating androgen dependency of expression. Nevertheless, in the cauda region of castrated animals considerable amounts of the mRNA as well as of the encoded glycopeptide persisted. Testosterone supplementation was sufficient to fully restore both pre-castration levels and spatial expression pattern of epididymal CD52 in castrated animals.

Testosterone supplementation immediately after castration was chosen as rationale to minimize the risk of irreversible damage to the epididymal epithelium, which might cause unspecific effects on epididymal gene expression. Thus, at the time of androgen supplementation the epididymides of the castrated animals still contained testicular factors which have been implicated in paracrine regulation of epididymal function, especially in the caput region (Palladino and Hinton, 1994; Winer and Wolgemuth, 1995). Their possible influence on CD52 expression was investigated by another series of animal experiments involving efferent duct ligation or dissection over a 14-day-period which, however, had no effect on CD52 mRNA levels in either region of the rat epididymis, suggesting that normal circulating levels of serum testosterone are sufficient to support normal CD52 expression *in vivo*. Intraluminal androgens or other testicular factors seemed not to be required.

Body temperature (37°C) as compared to scrotal temperature (33°C) dramatically and specifically down-regulated CD52 mRNA levels in rat epididymal cell cultures. This emphasized the general validity of our recent results on temperature regulation of CD52 mRNA (Pera et al., 1996). In order to assess the physiological significance of this observation, the cryptepididymal condition (Bedford, 1978; Esponda and Bedford, 1986), leaving normal testes in the scrotum, was generated. This allows normal production and transport of spermatozoa which, however, cannot be stored. Comparing sperm protein extracts obtained from unilateral cryptepididymal rats, a pronounced effect on Mr ≈ 26 kDa glycopeptide levels was observed: Spermatozoa isolated from abdominal cauda sections contained significantly lower amounts of the glycopeptide as compared to their scrotal counterparts. As preliminary analyses of flushed cauda epididymal fluid suggested that the glycopeptide was also secreted into the lumen in the cryptepididymal condition, this effect seemed to be caused either by the inability of CD52 to insert into the sperm membrane, or by its selective breakdown at abdominal temperature.

From the negative AR immunostaining results of our rat epididymal cell cultures it may be concluded that the temperature response of CD52 mRNA levels *in vitro* represented the response during androgen deficiency. Synergistic effects on epididymal gene expression of androgen withdrawal and elevated temperature have been described earlier (Esponda and Bedford, 1986; Regalado et al., 1993). Additionally, other non-steroidal factors may be lacking under cell culture conditions. This conclusion was corroborated by results obtained in another series of animal experiments, combining the unilateral cryptepididymal condition with either complete castration or a dissection of the efferent ducts. In castrated animals a complete loss of the persisting CD52 mRNA molecules and of the $Mr \approx 26$ kDa glycopeptide from abdominally located cauda epididymides was observed. In efferent duct-dissected animals, a loss of "long" CD52 mRNA molecules in the cauda region of abdominal organs coincided with a loss of the $Mr \approx 26$ kDa glycopeptide.

A main feature of epididymal CD52 expression *in vivo* in various mammals, including human, is its region-specific pattern with maximum mRNA levels in the distal part of the epididymis (Krull et al., 1993; Pera et al., 1994). The rat epididymis shows an interesting, novel aspect of regionalization, namely a region-specific variation of the CD52 mRNA poly(A) tail length. Besides the "short" mRNA molecules present in all parts of this organ and also in the seminal vesicle, the rat cauda epididymidis contains "long" CD52 mRNA molecules, carrying an extended poly(A) tail. This appears to correlate with the caudal occurance of the principal $Mr \approx 26$ kDa glycopeptide, implying that only the "long" CD52 mRNA molecules are efficiently translated, although regionalization on the level of glycosylation cannot be excluded. Androgens seemed to exert a region-specific effect, depending on the poly(A) tail length of rat CD52 mRNA: Castration of animals predominantly affected molecules with "short" poly(A) tails. Only abdominal translocation of the epididymis, combined with either castration or dissection of efferent ducts, was able to destroy CD52 mRNA molecules with "long" poly(A) tails in the cauda region.

10. CONCLUSION

The functional significance of the regional specialization and the post-testicular modifications of the sperm surface occuring within the epididymis is still unclear. The relationship between sperm maturation and epididymal transit seems to be more flexible in the human than in other mammalian species (Silber et al., 1989). Moreover, the sperm storage capacity of the human epididymis is severely limited (Bedford, 1994). Still, the order with which the sperm surface comes into contact with epididymal secretory proteins may be important for the resultant molecular and physiological effects also in men. Region-specific gene expression in the epididymis implies regionalization of regulatory mechanisms. The tyrosine kinase receptor c-ros has been suggested to be part of a signalling system involved in the regionalization of the mouse caput epididymidis (Sonnenberg-Riethmacher et al., 1996). Its function in the differentiation of the more distal parts of the organ, however, is unknown. Extended studies concerning the mechanisms of gene regulation in the epididymis are impeded in the *in vivo*-system by the inherent complexity, and *in vitro* by a loss of important traits by epididymal cells, e.g. androgen responsiveness. The development of improved *in vitro* sperm co-culture systems combined with the more subtle and specific *in vivo*-methods of targetted gene mutation will enable more profound insights in the future into the mechanisms regulating post-testicular sperm membrane remodelling.

11. ACKNOWLEDGMENTS

The authors are indebted to Professor Dr. J.M. Bedford, New York, for his help with the rat operations, and for stimulating discussions. We thank Drs. C. Osterhoff and R. Ivell, Hamburg, and R. Carballada, Madrid, for helpful discussions, Drs. C. Osterhoff, R. Carballada and P. Saling for making available unpublished results, and Professor Dr. F. Leidenberger for his continuous interest and support. The PG21 polyclonal antibody was a generous gift from Professor G.L. Greene, Chicago; the CAMPATH-1G antibody was kindly provided by Professor G. Hale, Oxford. Surgery was done in compliance with German Animal Welfare Laws under Licence 85/95. The work was supported by a DFG grant given to C.K. (Iv7/4–1).

12. REFERENCES

Baker CS, Magargee SF, Hammerstedt RH 1993 Cholesterol transfer proteins from ram cauda epididymal and seminal plasma. Biology of Reproduction 48: Suppl.1, P-111.

Bedford JM 1994 The status and the state of the human epididymis. Human Reproduction 9: 2187–2199.

Bedford JM 1978 Influence of abdominal temperature on epididymal function in the rat and rabbit. American Journal of Anatomy 152: 509–522.

Benoff S 1993 Preliminaries to fertilization. The role of cholesterol during capaciation of human spermatozoa. Human Reproduction 8: 2001–2008.

Blobel CP, Myles DG, Primakoff P, White JM 1990 Proteolytic processing of a protein involved in sperm-egg fusion correlates with acquisition of fertilization competence. Journal of Cell Biology 111: 69–78.

Boettger-Tong H, Aarons D, Biegler B, Lee T, Poirier GR 1992 Competion between zonae pellucidae and a proteinase inhibitor for sperm binding. Biology of Reproduction 47: 716–722.

Brooks DE 1987 Developmental expression and androgenic regulation of the mRNA for major secretory proteins in the rat epididymis. Mol Cell Endocrinol 53: 59–66.

Cooper TG 1992 Epididymal proteins and sperm maturation. In: Spermatogenesis-Fertilization-Contraception. Molecular, Cellular, and Endocrine Events in Male Reproduction, Schering Foundation Workshop 4 pp 285–318. Eds Nieschlag E et al., Springer Verlag; Berlin.

Cooper TG, Yeung CH, Meyer R, Schulze H 1990 Maintanance of human epididymal epithelial cell function in monolayer culture. Journal of Reproduction and Fertility 90: 81–91.

Eccleston ED, White TW, Howard JB, Hamilton DW 1994 Characterization of a cell surface glycoprotein associated with maturation of rat spermatozoa. Molecular Reproduction and Development 37: 110–119.

Eddy EM, O'Brien DA 1994 The spermatozoon. In: The Physiology of Reproduction, second edition pp 29–77. Eds Knobil E, Neill JD. New York; Raven Press.

Ellerbrock K, Pera I, Hartung S, Ivell R 1994 Gene expression in the dog epididymis: a model for human epididymal function. International Journal of Andrology 17: 314–323.

Esponda P, Bedford JM 1986 The influence of body temperature and castration on the protein composition of fliud in the rat cauda epididymidis. Journal of Reproduction and Fertility 78: 505–514.

Fröhlich O, Young LG 1996 Molecular cloning and characterization of EPI-1, the major protein in chimpanzee (*Pan troglotydes*) cauda epididymal fluid. Biology of Reproduction 54: 857–864.

Ghyselinck NB, Dufaure I, Lareyre JJ, Rigaudiere N, Mattei MG, Dufaure JP 1993 Structural organization and regulation of the gene for the androgen-dependent glutathione peroxidase-like protein specific to the mouse epididymis. Molecular Endocrinology 7: 258–272.

Hale G, Rye PD, Warford A, Lauder I, Brito-Babapulle A 1993 The GPI-anchored lymphocyte antigen CDw52 is associated with the epididymal maturation of human spermatozoa. Journal of Reproductive Immunology 23: 189–205.

Hamilton DW, Gould RP 1982 Preliminary observations on enzymatic galactosylation of glycoproteins on the surface of rat caput epididymal spermatozoa. International Journal of Andrology (Suppl) 5: 73–80.

Hamilton DW, Wenstrom JC, Baker JB 1986 Membrane glycoproteins from spermatozoa: partial characterization of an integral Mr = 24,000 molecule from rat spermatozoa that is glycosylated during epididymal maturation. Biology of Reproduction 34: 925–936.

Haselbeck A, Hösel W 1990 Description and application of an immunological detection system for analyzing glycoproteins on blots. Glycoconjugate 7: 63–74.

Hayashi M, Fujimoto S, Takano H, Ushiki T, Abe K, Ishikura H, Yoshida MC, Kirchhoff C, Ishibashi T, Kasahara M 1996 Characterization of a human glycoprotein with a potential role in sperm-egg fusion: cDNA cloning, immunohistochemical localization, and chromosomal assignment of the gene. Genomics 32: 367–374.

Kirchhoff C 1996 CD52 is the "Major maturation-associated" sperm membrane antigen. Molecular Human Reproduction 2: 9–17.

Kirchhoff C 1994 A major messenger ribonucleic acid of the rodent epididymis encodes a small glycosylphosphatidylinositol-anchored lymphocyte surface antigen. Biology of Reproduction 50: 896–902.

Kirchhoff C, Hale G 1996 Cell-to-cell transfer of glycosylphosphatidylinositol-anchored membrane proteins during sperm maturation. Molecular Human Reproduction 2: 177–184.

Kirchhoff C, Osterhoff C, Young LG 1996 Cloning and characterization of HE1, a (the?) major secretory protein of the human epididymis. Biology of Reproduction 54: 847–856.

Kirchhoff C, Krull N, Pera I, Ivell R 1993 A major mRNA of the human epididymal principal cells, HE5, encodes the leucocyte differentiation CD52 antigen peptide backbone. Molecular Reproduction and Development 34: 8–15.

Kirchhoff C, Habben I, Ivell R, Krull N 1991 A major human epididymis-specific cDNA encodes a protein with sequence homology to extracellular proteinase inhibitors. Biology of Reproduction 45: 350–357

Kirchhoff C, Osterhoff C, Habben I, Ivell R 1990 Cloning and analysis of mRNAs expressed specifically in the human epididymis. International Journal of Andrology 13: 155–167.

Krull N, Ivell R, Osterhoff C, Kirchhoff C 1993 Region-specific variation of gene expression in the human epididymis as revealed by in situ hybridization with tissue-specific cDNAs. Molecular Reproduction and Development 34: 16–24.

Meszaros M, Morton DB 1996 Identification of a developmentally regulated gene, esr 16, in the tracheal epithelium of Manduca sexta, with homology to a protein from human epididymis. Insect Biochemistry and Molecular Biology 26: 7–11.

Moore A, White TW, Ensrud KM, Hamilton DW 1989 The major maturation glycoprotein found on rat cauda epididymal sperm surface is linked to the membrane via phosphatidylinositol. Biochem Biophys Res Commun: 160:460–468.

Moore HDM, Akhondi MA 1996 In vitro maturation of mammalian spermatozoa. Reviews of Reproduction 1: 54–60.

Myles DG 1993 Sperm cell surface proteins of testicular origin: expression and localization in the testis and beyond. In: Cell and Molecular Biology of the Testis pp 452–473 Eds Desjardins C, Ewing LL. University Press, Oxford.

Orgebin-Crist MC 1996. Androgens and epididymal function. In: Pharmacology, Biology, and Clinical Applications of Androgens pp Eds Shalender Bhasin et al. Wiley-Liss, Inc.

Osterhoff C, Kirchhoff C, Krull N, Ivell R 1994 Molecular cloning and characterization of a novel human sperm antigen (HE2) specifically expressed in the proximal epididymis. Biology of Reproduction 50: 516–525.

Palladino MA, Hinton BT 1994 Expression of multiple gamma-glutamyl transpeptidase messenger ribonucleic acid transcripts in adult rat epididymis is differentially regulated by androgens and testicular factors in a region-specific manner. Endocrinology 135: 1146–1156.

Pera I, Ivell R, Kirchhoff C 1994 Regional variation of specific gene expression in the dog epididymis as revealed by in-situ transcript hybridization. International Journal of Andrology 17: 324–330.

Pera I, Ivell R, Kirchhoff C 1996 Body temperature (37C) specifically down-regulates the mRNA for the major sperm surface antigen CD52 in epididymal cell culture. Endocrinology 137: 4451–4459.

Perry ACF, Jones R, Hall L 1995 The monkey ESP14.6 mRNA, a novel transcript expressed at high levels in the epididymis. Gene 153: 291–292.

Perry ACF, Hall L, Bell AE, Jones R 1994 Sequence ananlysis of a mammalian phospholipid-binding protein from testis and epididymis and its distribution between spermatozoa and extracellular secretions. Biochemical Journal 301: 235–242.

Perry ACF Jones R, Hall L 1993 Isolation and characterization of a rat cDNA clone encoding a secreted superoxid dismutase reveals the epididymis to be a major site of its expresssion. Biochemical Journal 293: 21–25.

Perry ACF Jones R, Barker PJ, Hall L 1992 Genetic evidence for an androgen-regulated epididymal secretory glutathione peroxidase whose transcript does not contain a selenocysteine codon. Biochemical Journal 285: 863–870.

Phelps BM, Koppel DE, Primakoff P, Myles DG 1990 Evidence that proteolysis of the surface is an initial step in the mechanism of formation of sperm cell surface domains. Journal of Cell Biology 111: 1839–1847.

Primakoff P, Myles DG 1983 A map of the guinea pig sperm surface constructed with monoclonal antibodies. Developmental Biology 98: 417–428.

Prins GS, Birch L, Greene GL 1991 Androgen receptor localization in different cell types of the adult rat prostate. Endocrinology 129: 3187–3199.

Purvis K, Egdetveit I 1993 Segmental distribution of alpha-glucosidase, ornithine decarboxylase and polyamines in the human epididymis. Journal of Reproduction and Fertility 97: 575–580.

Raczek S, Yeung CH, Wagenfeld A, Hertle L, Schulze H, Cooper TG 1994 Epithelial monolayers from human epididymal and efferent duct tubules; testosterone metabolism and effects of culture conditions on cell height and confluence. Epithelial Cell Biology 3: 126–136.

Regalado F, Esponda P, Nieto A 1993 Temperature and androgens regulate the biosynthesis of secretory proteins from rabbit cauda epididymidis. Molecular Reproduction and Development 36: 448–453.

Ross P, Vigneault N, Provencher S, Potier M, Roberts KD 1993 Partial characterization of galactosyltransferase in human seminal plasma and its distribution in the human epididymis. Journal of Reproduction and Fertility 98: 129–137.

Seemüller U, Arnold M, Fritz H, Wiedenmann K, Machleidt W, Heinzel R, Appelhans H, Gassen HG, Lottspeich F 1986 The acid-stabile protease inhibitor of human mucous secretions. FEBS Letters 199: 43–48.

Silber SJ 1989 Role of epididymis in sperm maturation. Urology 33: 47–51.

Sonnenberg-Riethmacher E, Walter B, Riethmacher D, Gödecke S, Birchmeier C 1996 The c-ros tyrosine kinase receptor controls regionalization and differentiation of epithelial cells in the epididymis. Genes & Development 10: 1184–1193.

Tulsiani DRP, Skudlarek MD, Holland MK, Orgebin-Crist MC 1993 Glycosylation of rat sperm plasma membrane during epididymal maturation. Biology of Reproduction 48: 417–428.

Uhlenbruck F, Sinowatz F, Amselgruber W, Kirchhoff C, Ivell R 1993 Tissue-specific gene expression as an indicator of epididymis-specific functional status in the boar, bull and stallion. International Journal of Andrology 16: 53–61.

WHO laboratory manual for the examination of human semen and sperm-cervical mucus interaction, third edition 1992. Cambridge University Press, Cambridge.

Wolfsberg TG, Bazan JF, Blobel CP, Myles DG, Primakoff P 1993 The precursor region of a protein active in sperm-egg fusion contains a metalloprotease and disintegrin domain: structural, functional, and evolutionary implications. Proceedings of the National Acadademy of Sciences USA 90: 10783–10787.

Winer MA, Wolgemuth DJ 1995 The segment-specific pattern of A-raf expression in the mouse epididymis is regulated by testicular factors. Endocrinology 136: 2561–2572.

Xia MQ, Tone M, Packman L, Hale G, Waldmann H 1991 Characterization of the CAMPATH-1 (CD52) antigen: biochemical analysis and cDNA cloning reveal an unusually small peptide backbone. European Journal of Immunology 21: 1677–1684.

41

PURIFICATION AND STRUCTURAL ANALYSIS OF SPERM CD52, A GPI-ANCHORED MEMBRANE PROTEIN

S. Schröter,[1] C. Kirchhoff,[1] C.-H. Yeung,[2] T. Cooper,[2] and B. Meyer[3]

[1]Institute for Hormone and Fertility Research
University of Hamburg
Grandweg 64, 22529 Hamburg, Germany
[2]Institute of Reproductive Medicine
University of Münster
Domagkstr.11, 48149 Münster, Germany
[3]Institute of Organic Chemistry
University of Hamburg
Martin-Luther-King-Platz 6, 20146 Hamburg
Germany

One of the most prominent changes during epididymal passage is the appearance on human spermatozoa of the so-called "major maturation-associated" antigen HE5 or sperm CD52 (Hale et al., 1993; Kirchhoff, 1996). The mRNA is expressed only post-testicularly by the epididymal epithelial cells. The cDNA sequence turned out to be colinear with that of the lymphocyte surface antigen CD52 (Xia et al., 1991), suggesting that it encodes a glycosylphosphatidylinositol (GPI)-anchored membrane protein with an unusually small peptide core, consisting of only 12 amino acids, which contains a large N-linked carbohydrate moiety. Because the peptide core seemed to be so small and poorly conserved among mammals, we assumed that the function of this molecule may reside in this N-glycan. The structure of the carbohydrate moiety of the corresponding lymphocyte antigen has been studied previously: it consists of a large, partly sialylated tetraantennary polylactosamine core (Treumann et al., 1995). We extracted the antigen from human sperm, seminal plasma, epididymal tissue, and cauda epididymal fluid. It was recognized by the monoclonal antibody CAMPATH-1, raised against the lymphocyte antigen. The epitope consists of the GPI-anchor plus the three last C-terminal amino acids of the peptide core. Partially purified sperm CD52 migrated as multiple bands during SDS-electrophoresis, differences in electrophoretic mobility possibly due to differences in glycosylation and/or association with other proteins/glycolipids. To analyse its glycosylation pattern, sperm CD52 was digested with N-glycanase F in a time-course experiment, and detected by Western blotting employing the CAMPATH-1 antibody. Prior to digestion, the antigen seemed to consist of several glycoforms as visualized by five bands of different electro-phoretic mobility (microheterogeneity between 15 and 23 kDa). In the course of the degly-cosylation reaction, the microheterogeneity disappeared, and a single intensely stained band appeared at approximately 6 kDa, showing that sperm CD52 is N-glycosylated, and that the deglycosylated form was recognized even better by the antibody. The structure of the N-glycan was further studied employing two biotinylated lectins, MAA (from Maackia amurensis) and SNA (from Sambucus nigra), showing a specific affinity for terminal sialic acids in α-2,3- or 2,6-position. SNA detected a pattern of multiple protein bands congruent to that recognized by CAMPATH-1. The MAA detection pattern was different in that it was deplete of the fastest migrating forms. From this pattern it was concluded that all sperm CD52 glycoforms detected on Western blots contained terminal 2,6-linked sialic acid. Additionally, the slower migrating forms contained 2,3-linked sialic acid. To study possible changes in glycosylation during capacitation, Percoll gradient-purified, *in vitro* capacitated human sperm were compared with a non-capacitated control, employing antibody and lectin binding. Comparing equal sperm counts, CD52 was shown to persist on the sperm surface during 6h of *in vitro* capacitation, and was still present after a 24h in-cubation. Moreover, the glycosylation pattern as detected by the two different lectins did not change.

REFERENCES

Hale G, Rye PD, Warford A, Lauder I, & Brito-Babapulle A 1993 The GPI-anchored lymphocyte antigen CDw52 is associated with the epididymal maturation of human spermatozoa. Journal of Reproductive Immunology 23: 189–205.

Kirchhoff C 1996 CD52 is the "Major maturation-associated" sperm membrane antigen. Molecular Human Repro-duction 2: 9–17.

Treumann A, Lifely MR, Schneider P & Ferguson MAJ 1995 Primary structure of CD52. Journal of Biological Chemistry 270: 6088–6099.

Xia MQ, Tone M, Hale G, Waldmann H & Packman L 1991 Characterization of the CAMPATH (CDw52) antigen: biochemical analysis and cDNA cloning reveal an unusually small peptide backbone. European Journal of Immunology 21: 1677–1684.

42

MEASUREMENT OF CALCIUM IN SINGLE HUMAN SPERMATOZOA

G. Wennemuth,[1] S. Eisoldt,[1] H. P. Bode,[2] H. Renneberg,[1] and G. Aumüller[1]

[1]Department of Anatomy and Cell Biology
[2]Department of Pharmacology and Toxicology
Philipps-University Marburg
Robert-Koch-Str. 6, 35037 Marburg
Germany

Calcium which is found ubiquitously in the human body plays a central role in the intracellular signal transduction pathway. In spermatozoa, calcium fluxes are known to mediate a number of divergent effects, such as the acrosomal reaction and induction of hypermotility during capacitation (Yanagimachi et al.,1981, Thomas et al.,1988). A number of compounds have been shown to induce calcium influx into spermatozoa, e.g. progesterone and thapsigargin (Blackmore et al.,1990 and 1993). The channels that are involved in these processes are voltage dependent chloride channels (Roldan et al.,1994).

One major problem which prohibited an adequate analysis of the kinetics was that only a mass effect of several hundred thousand spermatozoa could be recorded and no analysis of single isolated spermatozoa was feasible. There have been a few attempts to fix isolated spermatozoa onto glass slides for microscopic evaluation, but this approach has proven rather inefficient. We, therefore, compared two previously published fixation methods (Ou et al.,1993; Plant et al.,1995) with three newly developed approaches, using laminin and poly-L-lysine as adhesives. The efficiency in fixation of spermatozoa was first monitored. Subsequently, stimulation experiments with progesterone and thapsigargin, a calcium ionophore, were performed.

Sperm Preparation and Loading with Fura - 2\AM. Human semen was obtained by masturbation from healthy donors. A population of >95% motile sperm was prepared by using a discontinuous Percoll gradient as described by Suarez et al. (1986). Spermatozoa were resuspended in an incubation buffer containing Fura - 2\AM at a concentration of 3 μmol. After a loading period of 45 min at 37°C, extracellular Fura - 2\AM was removed by centrifugation through a layer of 40% vol/vol isotonic Percoll. The pellet was washed and resuspended in HHBSS - buffer (Glaum et al.,1990).

Fixation Experiments. We prepared glass slides by coating with laminin, poly-L-lysine, agarose and gelatine (Plant et al.,1995). Another group of slides was prepared ac-

The Fate of the Male Germ Cell, edited by Ivell and Holstein
Plenum Press. New York. 1997

235

cording to Ou et al. (1993) and left untreated. To compare the efficiency of the methods, we counted the number of attached spermatozoa in 100 visual fields per experiment, each time prior to and after the flooding of the measurement chamber.

Stimulation Measurements with Progesterone, Thapsigargin and Fibronectin. Subsequently, we perfomed stimulation experiments on the fixed spermatozoa with different concentrations of progesterone, thapsigargin and fibronectin (range 0,1 - 10,0μg/ml). A Diaphot 300 inverse microscope (Nikon, Düsseldorf) was used for evaluation and documentation.

Comparing the different methods for sperm adhesion with regard to attachment efficiency (number of adherent spermatozoa), we found that immobilization with laminin, poly-L-lysine, agarose and gelatine, respectively, were superior to the method of Ou et al.(1993). Also, when the quality (intensity) of adhesion was checked, the method of Ou et al.(1993) proved less efficient than all other methods used.

The use of laminin and poly-L-lysine was more suitable for sperm fixation rather than agarose and gelatine, in that with the former the heads were attached to the glass slides while the tail was still motile. In addition, this enabled a rapid estimation of the viability of spermatozoa throughout the whole experiment. Following these positive, preliminary results, we performed our subsequent experiments with live spermatozoa adherent to laminin - or poly-L-lysine coated glass slides.

Stimulation experiments with progesterone, thapsigargin and fibronectin: In the experiments which were performed with the respective compounds at concentrations between 0,1 - 10,0 μg/ml, we found that the majority of spermatozoa responds within a few seconds with an intracellular increase in calcium. This effect was reversible and could be repeated two to three times. There were, however, significant differences in the duration and intensity of the reaction between different spermatozoa, as has already been described by Plant et al (1995). Interestingly, fibronectin also elicited an intracellular calcium increase at comparable concentrations.

Our fixation experiments showed that the use of laminin- or poly-L-lysine-coated glass slides yielded a sufficiently strong adhesion of live spermatozoa onto slides at a suitable high number. Also the fixed spermatozoa resisted repeated medium changes. The procedure is, therefore, at least as good as the one described by Plant et al.(1995).

An equally rapid medium exchange was nearly impossible when gelatine or agarose were used, as the spermatozoa were deeply imbedded in the fixation material, thus prohibiting a rapid and direct contact of the spermatozoa with the fluid. Usually, the calcium influx observed after addition of the reactant was always somewhat delayed. An additional disadvantage was that a complete exchange of all of the fluid could not be achieved, due to the poor redistribution of the media from the gelatine. The use of this latter method clearly prohibits the repeated challenge of spermatozoa with a certain compound in a given time.

The novel finding of a repeated influx of calcium into spermatozoa is complicated by the fact that the extent of the reaction in individual spermatozoa is rather different. The reasons why some spermatozoa produce a significant response and others do not, is at present unknown.

A further interesting finding was that fibronectin is capable of inducing calcium influx into spermatozoa. This effect has previously been known only for progesterone, epidermal growth factor and calcium ionophores (Murase et al., 1995; Kotwicka et al., 1996). The visualization of the effect which is also shown for the first time, indicates that there are two major sites on the spermatozoa where the influx occurs: one in the acrosomal region and the other behind the equatorial band. The tail, particularly the principal piece, showed no calcium influx after either stimulation.

Our results point to a significant role of the sperm head membrane in the initiation of sperm motility.

REFERENCES

Blackmore PF, Beebe SJ, Danforth DR, Alexander N 1990 Progesterone and 17α - Hydroxyprogesterone. Journal of Biological Chemistry 265: 1376–1380.

Blackmore PF 1993 Thapsigargin elevates and potentiates the ability of progesterone to increase intracellular free calcium in human sperm: possible role of perinuclear calcium. Cell Calcium 14: 53–60.

Glaum SR, Scholz WK, Miller RJ 1990 Acute and long - term glutamate - mediated regulation of $(Ca^{++})_i$ in rat hippocampal pyramidal neurons in vitro. Journal Pharmacology and Experimental Therapeutics 253: 1293–1302.

Kotwicka M, Warchot P, Filipiak K, Butowska W 1996 Changes in Ca2+ concentration in spermatozoa under effect of steroid hormones. Hum Anat Suppl 178, 91. Verslg. Anat. Gess. Abstract, p130.

Murase T, Roldan ERS 1995 Epidermal growth factor stimulates hydrolysis of phosphatidyl inositol 4'5'-biphosphate, generation of diacylglycerol and exocytosis in mouse spermatozoa. FEBS Letters, 360: 242–246.

Ou MC, Ng HT, Chiang BN, Hong CY, Hsu CT 1993 A motile human sperm head fixation method. Andrologia 25: 67–70.

Plant A, McLaughlin EA, Ford WCL 1995 Intracellular calcium measurements in individual sperm demonstrate that the majority can respond to progesterone. Fertility and Sterility 64: 1213–1215.

Roldan ERS, Murase T, Shi QX 1994 Exocytosis in spermatozoa in response to progesterone and zona pellucida. Science 266: 1578–1581.

Suarez SS, Wolf DP, Meizel S 1986 Induction of the acrosome reaction in human spermatozoa by a fraction of human follicular fluid. Gamete Research 14: 107–121.

Thomas P, Meizel S 1988 An influx of extracellular calcium is required for initiation of the human sperm acrosome reaction induced by human follicular. Gamete Research 20: 397–411.

Yanagimachi R 1981 Mechanism of Fertilization in Mammals. In: Mastroianni, Biggers JD, editors: "Fertilization and Embryonic Development in Vitro" pp 81 - 182: Plenum Press, New York

43

PUTATIVE ROLE OF A SERPIN IN MODULATION OF ACROSOME REACTION

P. Baltes, R. Sánchez, R. Henkel, and W. Miska

Centre of Dermatology and Andrology
Justus-Liebig-University
Gaffkystrasse 14, 35392 Giessen
Germany

INTRODUCTION

The acrosome reaction (AR) is one of the prerequisites for successful mammalian fertilization. This is a modified exocytotic process, terminating in the release of various hydrolytic enzymes, which are thought to have a function for sperm penetration of the outer oocyte investments. Various physiological factors seem to play a role in the induction of the AR. Although zona pellucida, cumulus oophorus and follicular fluid have been shown to be potent inducers of the AR in vitro, the actual inducer in vivo is still unknown. Hence, the aim of these studies was the biochemical and immunological characterisation of the "acrosome reaction - inducing substance" (ARIS) of human follicular fluid and the localisation of its synthesis.

METHODS

Ten μl of the TCA-precipitated protein solution was run on SDS-PAGE gels and either silver-stained or electrotransferred to nitrocellulose. A solution of pure CBG served as positive control. Blotted proteins were labelled with rabbit anti-human CBG and anti-rabbit IgG peroxidase conjugate.

RESULTS AND CONCLUSIONS

Previous experiments performed in our laboratory have yielded several indications for both a protein as well as a steroidal character for ARIS. The results, suggest that the acrosome reaction-inducing activity is a synergistic action of progesterone and a progesterone-binding protein (Miska et al., 1994). Further experiments revealed the immunologi-

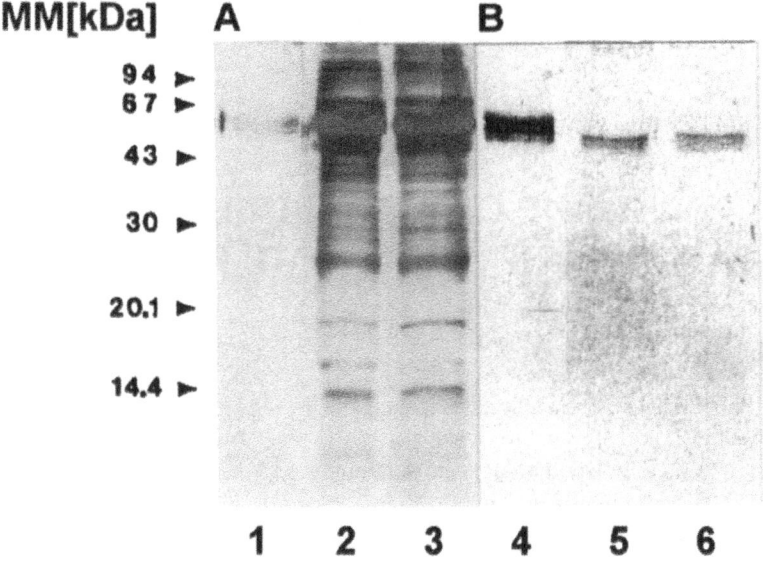

Figure 1. A, SDS-PAGE of hCBG (1,2), hFF (3,4), hCO-CM (5,6). B, Western blot analysis with polyclonal antibody against CBG. hCBG (7,8), hFF (9,10), hCO-CM (11,12).

cal similarity of ARIS with the serine proteinase inhibitor (SERPIN) transcortin (Corticosteroid-binding globulin, CBG) (Fehl et al., 1995). Transcortin has already been described and serves as a transport protein for progesterone and cortisol in the plasma. The AR-inducing effect of follicular cells, such as those of the cumulus oophorus and the granulosa cells has been described earlier (Siiteri et al., 1988), and points to a possible origin for the ARIS of follicular fluid. Additionally to the well-known steroid secretion-activity of cumulus cells, our results demonstrated also protein secretion. An AR-inducing effect could be also shown using the culture medium (CM) of human cumulus oophorus (hCO). This effect could be eliminated by treatment of hCO-CM with monoclonal antibodies against CBG. Western blotting after SDS-PAGE of hCO-CM (Fig. 1) and immuncytochemical experiments strongly indicate that human cumulus cells actively express and secrete a transcortin-like protein.

This study was supported by the BMFT (FKZ 01 KY 9105/6) and Dirección de Investigación y Desarrollo, Universdidad de La Frontera, Temuco, Chile.

REFERENCES

Fehl P, Miska W & Henkel R 1995 Further indications of the multicomponent nature of the acrosome reaction-inducing substance of human follicular fluid. Molecular Reproduction and Development 42: 80–88.
Miska W, Fehl P & Henkel R 1994 Biochemical and immunological characterisation of the acrosome reaction-inducing substance of hFF. Biochemical and Biophysical Research Communications 199: 125–129.
Siiteri JE, Dandekar PV & Meizel S 1988 Human sperm acrosome reaction-initiating activity associated with the human cumulus oophorus and mural granulosa cells. Journal of Experimental Zoology 246: 41–80.

44

IMMUNOELECTRON MICROSCOPIC STUDIES ON OUTER DENSE FIBRES

W. Miska, U. K. Schalles, J. Villegas, and R. Henkel

Centre of Dermatology and Andrology
Justus-Liebig-University
Gaffkystrasse 14, 35392 Giessen
Germany

For a major part of the length of the mammalian sperm tail, the axoneme is surrounded by nine outer dense fibres, thus creating the so-called 9+9+2 cross-sectional pattern. The aim of this study was to evaluate the staining patterns of the Gi-8F8 monoclonal antibody against the 30 kDa outer dense fibre protein (Villegas et al., 1995) in different mammalian spermatozoa using preembedding methods. Previous investigations demonstrated the specificity and cross-reactivity of this monoclonal antibody between human, porcine and bovine spermatozoa at both the light microscopic and ultrastructural level (Schalles et al., 1995). Additionally, parts of the primary structure of the 30kDa ODF protein showed high homologies with the main ODF protein in rat spermatozoa (Stalf et al., 1995; Morales et al., 1995). In order to localise the specific binding sites of the Gi-8F8 moAb at the ultrastructural level, without disturbance of fixation or embedding media, preembedding techniques were applied.

For immunogold electron microscopy, semen samples obtained from normozoospermic men and boar were washed twice with phosphate buffered saline (PBS) and demembranated with Nonidet-P40 or 1% Triton X-100. The sperm pellet was then incubated with Gi-8F8 for lh at room temperature (RT). After washing twice with PBS spermatozoa were incubated using a second antibody conjugated with 5 nm gold particles for lh at RT. Following fixation, samples were dehydrated and embedded in Spurr's resin or LRWhite. Ultrathin sections were examined under a Hitachi HU ST 12 electron microscope at 80 kV.

The target antigen was ultrastructurally localised on the side of the ODF adjacent to the axoneme, which belongs to the medulla and where the cortex is usually absent. In addition, no labelling pattern of the cortex could be detected (Fig. 1). In both longitudinal and transverse sections no labelling pattern of the cortex could be seen. The sperm head and other cytoplasmic structures showed no labelling (data not shown). The presented results clearly demonstrate the immunological distinctness of the medulla and cortex in ODF of human and porcine spermatozoa, indicating that ODF are composed of at least two different proteins.

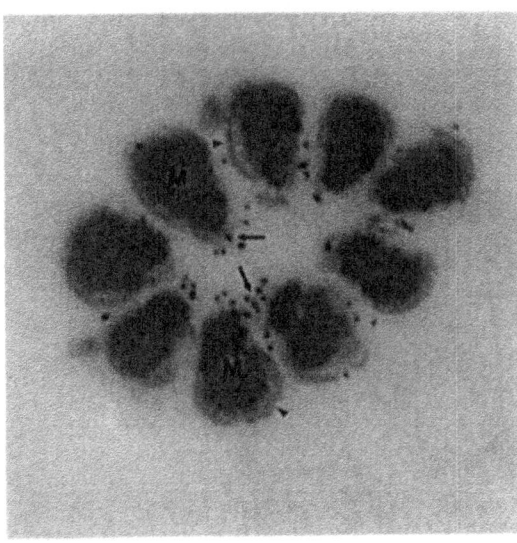

Figure 1. Transverse section through the midpiece of a tail of a human spermatozoon. Medullary substructures are clearly immunoreactive to ODF antibodies (arrows). The cortex (arrowheads) is not labeled and detaches from the medulla (M). Mitochondria and axoneme are solubilized. x 220,000.

ACKNOWLEDGMENTS

This project was supported by the DFG, Project No. Mi 325/7-1 and Mi 325/7–2.

REFERENCES

Villegas J, Miska W, Stalf Th, Schalles UK, Schill WB 1995 Production of monoclonal antibodies against the main protein component of boar outer dense fibers.Reproduction in Domestic Animals, Suppl. 3: 135.

Schalles UK, Villegas J, Henkel R, Miska W 1995 Immunocytochemical localization of the 30 kDa outer dense fiber polypeptide in different mammalian species. Reproduction in Domestic Animals, Suppl. 3: 98.

Stalf Th, Villegas J, Schalles UK, Schill WB, Miska W 1995 Investigations about the primary structure of the major protein of mammalian outer dense fibers. Reproduction in Domestic Animals, Suppl. 3: 96.

Morales CR; Oko R, Clermont Y 1994 Molecular cloning and developmental expression of an mRNA encoding the 27 kDa outer dense fiber protein of rat spermatozoa. Molecular Reproduction and Development 37: 229–240.

45

THE USE OF SPIN-LABELLED PHOSPHOLIPID ANALOGUES TO CHARACTERIZE THE TRANSVERSE DISTRIBUTION OF PHOSPHOLIPIDS AND THE ACTIVITY OF PHOSPHOLIPASE-A2 IN THE CELL MEMBRANE OF BULL SPERMATOZOA

K. Müller,[1] T. Pomorski,[2] P. Müller,[2] and A. Herrmann[2]

[1]Institut für Fortpflanzung landwirtschaftlicher Nutztiere Schönow e. V.
Bernauer Chaussee 10, 16321 Schönow, Germany
[2]Humboldt-Universität zu Berlin
Institut für Biologie
Invalidenstr. 43, 10115 Berlin
Germany

Spin-labelled phospholipids with a NO-group on a short ß-fatty acid chain incorporate readily from the external solution into the outer leaflet of cell membranes. After incorporation of those analogues for phosphatidylserine (PS), -ethanolamine (PE), -choline (PC) and sphingomyelin (SM) in the bull sperm cell membrane, we followed their translocation to the cytoplasmic leaflet by "Back-exchange" of the analogues present in the outer leaflet on BSA. As the spin-labelled phospholipid analogues for PS, PE and PC are very sensitive to hydrolytic processes we also determined their degradation by phospholipase-A_2 activity. The resulting free diffusible spin-labelled fatty acids produce an isotropic signal in the ESR spectra of the supernatant BSA fraction from which one may calculate the relative amount of hydrolysed label (Morrot et al., 1989).

We could establish an asymmetric transverse distribution of phospholipids, which we also found recently in the ram sperm cell membrane (Müller et al., 1994), and which Nolan et al., 1995, found with fluorescent labels in the bull sperm cell membrane: The aminophospholipids PS and PE were rapidly translocated to the cytoplasmic membrane half where they accumulate. The inward movement of PC was slower, presumably reflecting passive diffusion. SM remained completely outside. Such a transverse phospholipid asymmetry is known from several eukaryotic cell membranes (Zachowski, 1993) and seems to be caused also in sperm cells by an ATP-dependent aminophospholipid translocase activity.

Despite the presence of 5 mM diisopropyl fluorophosphate (DFP) and 1 mM EGTA in the medium, we found the label degradation via phospholipase-A_2 activity much higher in fresh bull spermatozoa, than for example, in fresh ram sperm cells. However, when the pure ejaculates were stored for 4 hours before the experiment, the hydrolysis of label was reduced. Further experiments to establish on which membrane half label hydrolysis occurs or to characterize the influence of seminal plasma constituents will provide a better understanding of the unexpectedly high phospholipase-A$_2$ activity in fresh bull spermatozoa.

The large extent of label hydrolysis in bull spermatozoa leads to limitations of the experimental assessment of lipid asymmetry because the correction of spectra for the amount of degraded label is only reasonable in the case of slow and moderate label modification. The fast translocation of the aminophospholipid analogues became already apparent when the label degradation was still at a low level. Data points taken later in the course of the experiments have to be considered with caution because of the large extent of hydrolysis. The effective inhibition of label hydrolysis or the use of phospholipid analogues with ether-linked fatty acids are needed to overcome this problem in future studies.

This work was supported by the Deutsche Forschungsgemeinschaft.

REFERENCES

Morrot G, Hervé P, Zachowski A, Fellmann P & Devaux PF 1989 Aminophospholipid translocase of human erythrocytes: phospholipid substrate specifity and effect of cholesterol Biochemistry 28: 3456–3462.

Müller K, Pomorski T, Müller P, Zachowski A & Herrmann A 1994 Protein dependent translocation of aminophospholipids and asymmetric transbilayer distribution of phospholipids in the plasma membrane of ram sperm cells. Biochemistry 33: 9968–9974.

Nolan JP, Magargee SF, Posner RG & Hammerstedt RH 1995 Flow cytometric analysis of transmembrane phospholipid movement in bull sperm. Biochemistry 34: 3907–3915

Zachowski A 1993 Phospholipids in animal eucaryotc membranes: transverse asymmetry and movement. Biochemical Journal 294: 1–14

46

INTERACTIONS BETWEEN LEUKOCYTES AND THE MALE REPRODUCTIVE SYSTEM

The Unanswered Questions

A. G. Rossi[1] and R. J. Aitken[2]

[1]Department of Medicine
Respiratory Medicine Unit
Rayne Laboratory
University of Edinburgh Medical School
Teviot Place, Edinburgh EH8 9AG
Scotland
[2]MRC Reproductive Biology Unit
Centre for Reproductive Biology
37 Chalmers Street
Edinburgh EH3 9EW
Scotland

It is well established that leukocytes are often present in seminal fluid and consequently there exists an overwhelming potential for interactions between different types of leukocytes and spermatozoa. However, the precise role of leukocytes in the semen and the fate of spermatozoa are not fully understood. For comprehensive descriptions of the biological significance of leukocytes in semen we recommend reviews by Wolff (1995) and Aitken and Baker (1995) and for a discussion of the immunosuppressive mechanisms in semen we suggest a recent review by Kelly (1995). The purpose of this mini-review is to highlight some of the pertinent, largely unanswered, questions regarding the role of interactions between inflammatory cells and the male reproductive system. In addition, we discuss the ill-defined fate of senescent spermatozoa in comparison to the relatively well understood fate of apoptotic somatic cells.

Given the potential immunogenicity of sperm surface antigens, it is remarkable that inflammatory reactions are not more commonly observed in the male reproductive tract. In the testes, male germ cells develop and are sequestered from the immune system by the blood-testis barrier. Thus, the male gamete resides for set periods of time in an immunologically privileged milieu. However, in the rete testes and epididymis no such barrier exists, and yet immunological and inflammatory responses are not mounted against the millions of spermatozoa stored in this organ. The epididymis is clearly an immu-

The Fate of the Male Germ Cell, edited by Ivell and Holstein
Plenum Press, New York, 1997

nologically competent tissue (Pollanen & Cooper 1994) that can elicit inflammatory reactions as in the case of infection generating epididymitis (Purvis & Christiansen 1996). Taken together, this information then begs the question as to why spermatozoa are not recognized as foreign and an immunological response not elicited? Indeed it has been suggested by Tomlinson et al. (1992) that the quality of sperm morphology is directly correlated with the size of the seminal leukocyte population. Such observations have led to the suggestion that phagocytic cells, particularly macrophages, are able to move through spermatozoa stored in the male reproductive tract, identifying morphologically abnormal cells and then removing them. The epididymis would be the logical site for such surveillance to be conducted, although there is little evidence to support this contention.

At some time during the life of most fertile and infertile men significant levels of leukocytes are present in their seminal fluid. The upper limit of normality for leukocyte numbers in semen has been defined as 1×10^6 /ml with values greater than this termed leukocytospermic by the World Health Organization. The precise source of these seminal leukocytes is currently unknown although histological studies have revealed leukocyte populations in virtually all tissues of the male reproductive tract. In order to gain a better understanding of the immunology of the interactions between leukocytes and the male genital tract, the following series of questions should be addressed and ultimately answered. (i) What is the principle site of origin for leukocytes present in semen? (ii) Is there a different route of entry for different leukocyte species? For example, is it possible that the dominant leukocyte species in the ejaculate, the neutrophil (Aitken et al. 1994), enters via the secondary sexual glands whereas the much smaller numbers of macrophages enter via the epididymis? In the absence of secondary sexual gland infection, do they enter the male tract at the level of the rete testes or epididymis or is their entry via the male secondary sexual glands? (iii) If macrophages do enter via the epididymis and are involved in surveying the integrity of the sperm population, how are the abnormal cells located? (iv) What is the activation status of the recruited leukocytes? i.e., are they in a quiescent, primed or activated state? A primed cell which is hyper-responsive for further stimulation by secretagogue agonists or a fully activated leukocyte may lead to liberation of granule enzymes and toxic oxygen metabolites (Rossi et al. 1993, O'Flaherty & Rossi 1993, Kitchen et al. 1996) that are potentially damaging to cells including spermatozoa. (v) Do abnormal or senescent spermatozoa give out chemotactic signals? This latter question has been partly addressed in that seminal fluid and spermatozoa appear to be chemotactic for neutrophils (Maroni et al. 1972) and macrophages (Maroni & Wilkinson 1971) although the precise chemotactic stimuli have not been elucidated. (vi) How do macrophages attracted to the epididymis distinguish the surfaces of normal and abnormal spermatozoa?

During the classical inflammatory response, the body mounts a protective tissue response to injury or damaged tissues which serves to destroy, dilute, partition off or remove the injurious agent and the injured tissues. This type of response, however, is not usually mounted against the allogenic spermatozoa in the male and when millions of spermatozoa are introduced into the female genital tract. In inflammatory situations highly complex, inter-dependent biochemical and physiological mechanisms have evolved to deal with the inflammatory insult. When this crucial and normally beneficial response occurs in an uncontrolled or exaggerated manner the host's own tissues are damaged often resulting in chronic inflammation. Toxic products liberated by recruited leukocytes, such as reactive oxygen species (e.g., O_2^-, H_2O_2, OH^-, NO) and proteases (e.g., elastase and collagenase) can be deleterious to cells and tissues including male germ cells. It is well recognized that seminal plasma contains immunosuppressive and anti-inflammatory agents (e.g., prostaglandins, complement inhibitors, antioxidants, anti-proteases and certain cytokines). Thus the potentially damaging leukocytes in the semen are rendered impotent due to the large repertoire of anti-inflammatory material in the seminal plasma (Kelly 1995). Spermatozoa are extremely susceptible to oxidative stress (Aitken 1996) and the epididymis protects spermatozoa from this kind of damage by secreting unique forms of superoxide dismutase and glutathione peroxidase (Hinton et al. 1996). In light of such factors, it would clearly be extremely counterproductive if macrophages, actively engaged in phagocytosing defective cells, were to liberate products of the oxidative burst. Thus one might expect phagocytosis in the epididymal lumen to be "silent", raising the question of whether there is inhibition of activation of macrophages in the epididymal lumen and how this suppression might be achieved?

Similar considerations might be applied to the interactions of spermatozoa with neutrophils. Spermatozoa coated with antisperm antibodies and complement can interact with neutrophils *in vitro* in such a way that phagocytosis occurs with an inhibited oxidative burst (D'Cruz et al. 1992) and the adhesive events involved in the immune destruction of motile sperm by neutrophils is dependent upon the CD11b/CD18 glycoprotein complex (D'Cruz & Haas 1995). Histochemical studies employing nitroblue tetrazolium to detect superoxide anion generation suggest that, under such circumstances, the oxidative burst is initiated at the point of sperm-neutrophil contact but does not propagate through the cell (D'Cruz et al. 1992).

Questions concerning the ability of spermatozoa to suppress certain aspects of phagocyte function are particularly pertinent in the case of vasectomy. Following this operation spermatogenesis is not impaired and so millions of spermatozoa must be phagocytosed near the ligation site every day. The fact that this process is completely asymptomatic again suggests that the spermatozoa are able to suppress leukocyte activation and prevent the induction of an inflammatory reaction. In an inflammatory context, if recruited leukocytes are not rapidly removed once their primary purpose has been achieved (e.g., destruction of invading bacteria), there is potential for these cells to liberate their toxic intracellular contents resulting in further tissue damage. Indeed, before the seminal work of Wyllie and colleagues (Kerr et al. 1972, Wyllie et al. 1980), who described a physiological mode of cell death termed apoptosis (programmed cell death), it was widely believed that cells simply died by necrosis in an uncontrolled manner (Hurley 1983). If this were the case, most cells, and especially inflammatory cells, would liberate their intracellular contents, many of which can lead to an exaggerated inflammatory response. Virtually all cells, including spermatocytes (Dix et al. 1996) under normal conditions undergo controlled physiological death by apoptosis, a process known to play a fundamental role in almost all aspects of life (Wyllie et al. 1980). Cells that have undergone apoptosis show remarkably similar structural, morphological and biochemical changes. These similarities are perhaps indicative of a common underlying molecular mechanism. Apoptotic cells are often smaller (due to cytoplasmic shrinkage), vacuolated and exhibit major changes on their cell surface. Importantly, apoptotic cells remain intact, retaining their cytoplasmic granules and maintaining plasma membrane integrity such that they exclude vital dyes. It is the ultrastructural changes observed in the nucleus which are strikingly characteristic of an apoptotic cell; there is condensation of nuclear chromatin into dense crescent shaped aggregates with the nucleolus becoming more conspicuous. Biochemically, apoptosis is typically characterized by endogenous endonuclease activation resulting in internucleosomal cleavage of chromatin. When DNA is extracted from apoptotic cells and separated electrophoretically on an agarose gel there is a characteristic "ladder" pattern of DNA fragments, representing multimers of the 180–200 base pairs of DNA associated with nucleosomes. Apoptotic cells have altered cell surfaces; for example, most apoptotic cells express on the external surface of the plasma membrane, phosphatidylserine molecules; a process which appears to be highly regulated, important in phagocyte recognition (see table 1) and useful for assessing apoptosis *per se* (Koopman et al. 1994). Other cell surface changes may be unique to a particular cell system such as shedding of the surface molecule CD16 (FcγRIII) from apoptotic neutrophils (Dransfield et al. 1994, Rossi et al. 1995). Importantly, apoptotic cells are functionally isolated, i.e., they become down regulated to further receptor-dependent stimulation (Whyte et al. 1993)

During *in vitro* culture, apoptotic cells will inevitably undergo disintegration (necrosis) with release of potentially histotoxic contents. This highly undesirable scenario, if occurring *in vivo,* would result in an exaggerated inflammatory response and destruction of host tissue cells leading to an uncontrolled chronic situation where the normally highly effective recognition and clearance mechanisms (see below) are overwhelmed or defective. In contrast, during resolution of inflammation the majority of senescent cells undergo apoptosis leading to rapid recognition and engulfment by local phagocytes. Indeed, it was over one hundred years ago, using recently developed intravital light microscopic techniques on a number of transparent invertebrates, that the Russian biologist Elie Metchnikoff first described phagocytosis of bacteria and senescent leukocytes by macrophages. Although there have been numerous accounts of macrophage phagocytosis of different cells over the last century it is only relatively recently that the major developments regarding recognition of apoptotic cells have been described. Macrophages and other "semi-professional" leukocytes have been demonstrated to phagocytose apoptotic cells but not fresh or aged, non-apoptotic cells, rapidly and efficiently (Savill et al, 1989, Stern et al. 1992, 1996). Isolated apoptotic cells are observed relatively infrequently in inflammatory foci and in order to visualize intact apoptotic cells within a macrophage by electron microscopy the cells have to be fixed within minutes. When macrophages ingest particles such as zymosan (yeast cell walls) *in vitro* their response is to liberate pro-inflammatory mediators (e.g., eicosanoids, granular enzymes and cytokines). However, macrophages, capable

Table 1. Molecular mechanisms of recognition of apoptotic cells by phagocytes

	References
Lectin-like receptors. Specific carbohydrates on the surface of apoptotic cells are recognized by phagocytic cells containing lectin-like receptors such as the asialoglycoprotein, mannose or mannose/fucose receptor.	Duvall et al. 1985, Dini et al. 1992, 1995, Hall et al. 1995
$\alpha_v\beta_3$/CD36/thrombospondin (TSP). It is hypothesized that a TSP binding moiety expressed on the surface of apoptotic cells binds to nearby TSP which in turn acts as a bridging molecule between the apoptotic cell and the ingesting phagocyte. The phagocyte expresses two receptors on its surface; the $\alpha_v\beta_3$ "vitronectin receptor" integrin and an 88 kD monomer termed CD36 which co-operate to bind TSP.	Savill et al. 1989, 1990, 1992, Ren et al. 1995, Ren & Savill 1995
Phosphatidylserine (PS) and PS receptors. Exposure of phosphatidylserine on the surface of apoptotic cells is believed to be recognized by putative phosphatidylserine receptors located on the surface of the ingesting phagocyte. Recently, it has been suggested that members of the scavenger receptor family may act as PS receptors.	Fadok et al. 1992a, 1992b, 1993, Martin et al. 1995, Verhoven et al. 1995, Platt et al. 1996, Ramprasad et al. 1995, Rigotti et al. 1995, Sambrano et al. 1994, Sambrano & Steinberg 1995
61D3 antigen. The mAb 61D3 can specifically attenuate the recognition of apoptotic cells by human monocyte-derived macrophages. The 61D3 antigen has not been fully characterized.	Flora & Gregory 1994

of ingesting multiple apoptotic cells simultaneously, do not liberate pro-inflammatory mediators or potentially injurious products, either derived from the macrophage *per se* or from the apoptotic cell, into the surrounding medium (Meagher et al. 1992). Thus, removal of apoptotic cells by phagocytes, together with limitation of cell function associated with apoptosis (Whyte et al. 1993) effectively neutralizes any existing pro-inflammatory potential. This non-inflammatory clearance was subsequently shown to be due to the molecular mechanisms by which macrophages recognize and ingest apoptotic cells (see Savill & Haslett 1994, Haslett et al. 1994). Since apoptotic cells are avidly ingested by these cells (non-apoptotic cells are not) surface changes on the apoptotic cell that are recognizable by the phagocyte have evolved. There are a number of known molecular mechanisms for recognition of apoptotic cells (see table 1) but for a more detailed description of these mechanisms we recommend reviews by Savill & Haslett 1994 and Hart et al. 1996. It is important to note that these putative mechanisms may only apply to a particular cell system, they may occur in combination with one another and in view of the multiple genes involved in phagocytosis of apoptotic cells in the nematode *Caenorhabditis elegans* it is likely that other mechanisms exist.

Given the information presented in this chapter on the resolution of the inflammatory response, is it possible that spermatozoa do not stimulate release of toxic oxygen radicals and other pro-inflammatory mediators by phagocytic cells because they have undergone some form of programmed cell death? If so, this process would have to be distinguished from the apoptosis seen in somatic cells because spermatozoa are transcriptionally inactive. In spermatozoa, this process would comprise an "outside-in" programmed destruction of the cell rather than the usual "inside-out" apoptotic event. Indeed, Gorczyca et al. 1993 demonstrated the presence of DNA strands and an increased sensitivity of DNA *in situ* to denaturation in abnormal human sperm cells analogous to the effect of endonuclease activity characteristic of programmed cell death of somatic cells. Thus, if spermatozoa undergo a form of apoptosis, it raises questions about the surface changes that occur on senescent spermatozoa that enables them to be recognized by phagocytes. Interestingly, it has recently been demonstrated that apoptotic spermatogenic cells, expressing phosphatidylserine on the outer leaflet of their plasma membranes, are recognised and phagocytosed by Sertoli cells (Shiratsushi et al., 1997). Would the manner in which spermatozoa die influence the subsequent behaviour of phagocytes towards them? It would be interesting to compare the nature of sperm-phagocyte interaction under circumstances where the former had been allowed to undergo a senescent death with spermatozoa that had been killed by other means including heat, freeze-thawing, complement-mediated cytolysis, oxidative stress and incubation with spermicidal compounds including surface-active agents and membrane stabilizers. Such questions become particularly important when we consider the fate of the spermatozoa in the female reproductive tract. At insemination, hundreds of millions of potentially immunogenic cells are deposited in the female reproductive tract. Hundreds of these cells will end up in the pelvic cavity in the search for an ovum while millions will perish lower down the reproductive tract in an abortive attempt to reach the site of fertilization. All of the cells that penetrate the cervix must presumably be phagocytosed with the possible exception of the single spermatozoon that successfully fertilizes the egg. What fate befalls the millions of spermatozoa that enter the female reproductive tract following insemination? The limited studies that have been performed suggest that the response of the female tract is to mount a leukocytic infiltration which is particularly marked in the cervix of vaginal inseminators, such as the rabbit, but is focused on the uterine lumen of species exhibiting an intra-uterine mode of insemination such as the rat. The molecular identity, and origin, of the chemotactic factors, the cellular composition of the leukocytic infiltration and the presence or absence of an oxidative burst are all unknown. If the apparent silence of the phagocytic response depends on senescent spermatozoa undergoing a programmed form of death, what happens in the case of women using synthetic spermicides, when the spermatozoa are killed by detergents such as nonoxynol 9 or benzylkonium chloride?

Answers to such questions are particularly important in relation to the transmission of viral infections such as HIV. Insemination not only deposits free virus or infected T cells in the vagina, it also enhances the risk of transmission by (i) stimulating a local leukocytic infiltration, which may include $CD4^+$ T-cells and (ii) suppressing certain elements of the immunological defense system. Clearly any factors, including spermicides, that influence the nature of the local leukocytic infiltration following insemination and the characteristics of the subsequent immune response mounted by the female tract, will have a bearing on the transmission of HIV.

In conclusion, the interaction between spermatozoa and the immune system is poorly characterized. Leukocytic infiltrations involving large scale phagocytosis of spermatozoa are observed following vasectomy in the male or as a consequence of insemination in the female. However, in neither of these situations is an inflammatory response observed. These results suggest that spermatozoa either suppress, or fail to ignite, key elements of the immune response. Elucidation of the biochemical mechanisms controlling the interaction between spermatozoa and the immune system have important implications for our understanding of immunological infertility and the spread of sexually transmitted disease, including HIV.

ACKNOWLEDGMENTS

We thank Drs Ian Dransfield and Simon P. Hart for helpful discussions and thank the Medical Research Council (UK) for financial support.

REFERENCES

Aitken RJ 1995 Free radicals, lipid peroxidation and sperm function. Reproduction Fertility and Development 7: 659–668.

Aitken RJ & Baker HWG 1995 Seminal leukocytes: passangers, terrorists or good Samaritans. Human Reproduction 10: 1736–1739.

Aitken RJ, West K & Buckingham D 1994 Leucocytic infiltration into the human ejaculate and its association with semen quality, oxidative stress and sperm function. Journal of Andrology 15: 343–352.

D'Cruz OJ, Wang B-L & Haas GG Jr 1992 Phagocytosis of immunoglobulin G and C3-bound human sperm by human polymorphonuclear leukocytes is not associated with the release of oxygen radicals. Biology of Reproduction 46: 721–732.

D'Cruz, OJ & Haas GG Jr 1996 β2-Integrin (CD11/CD18) is the primary adhesive glycoprotein complex involved in neutrophil-mediated immune injury to human sperm. Biology of Reproduction 53: 1118–1130.

Dini L, Autuori F, Lentini A, Oliverio S & Piacentini M 1992 The clearance of apoptotic cells in the liver is mediated by the asialoglycoprotein receptor. Federation of European Biochemical Society letters 296: 174–178.

Dini L, Lentini A, Diez Diez G, Rocha M, Falasca L, Serafino L & Vidal-Vanaclocha F 1995 Phagocytosis of apoptotic bodies by liver endothelial cells. Journal of Cell Science 108: 967–973.

Dix DJ, Allen JW, Collins BW, Mori C, Nakamura N, Poorman-Allen P, Goulding EH & Eddy EM 1996 Targeted gene distruption of *Hsp70–2* results in failed meiosis, germ cell apoptosis, and male infertility. Proceedings of the National Academy of Sciences USA 93: 3164–3268.

Dransfield I, Buckle A-M, Savill JS, McDowall A, Haslett C & Hogg N (1994) Neutrophil apoptosis is associated with a reduction in CD 16 (FcγRIII) expression. Journal of Immunology 153: 1254–1263.

Duvall E, Wyllie AH & Morris RG 1985 Macrophage recognition of cells undergoing programmed cell death (apoptosis). Immunology 56: 351–358.

Fadok VA, Savill JS, Haslett C, Bratton DL, Doherty D, Campbell PA & Henson PM 1992 Different populations of macrophages use either the vitronectin receptor or the phosphatidylserine receptor to recognize and remove apoptotic cells. Journal of Immunology 149: 4029–4035.

Fadok VA, Voelker DR, Campbell PA, Cohen JJ, Bratton DL, & Henson PM 1992 Exposure of phosphatidylserine on the surface of apoptotic lymphocytes triggers specific recognition and removal by macrophages. Journal of Immunology 148: 2207–2216.

Fadok VA, Laszlo DJ, Noble PW, Weinstein L, Riches DWH & Henson PM 1993 Particle digestibility is required for induction of the phosphatidylserine recognition mechanism used by murine macrophages to phagocytose apoptotic cells. Journal of Immunology 151: 4274–4285.

Flora PK & Gregory GD 1994 Recognition of apoptotic cells by human macrophages: inhibition by a monocyte/macrophage-specific monoclonal antibody. European Journal of Immunology 24: 2625–2632.

Gorczyca W, Tragonos F, Jesionowska H & Darzynkiewicz Z 1993 Presence of DNA strand breaks and increased sensitivity of DNA in situ to denaturation in abnormal human sperm cells: analogy to apoptosis of somatic cells. Experimental Cell Research 207: 202–205.

Hall SE, Savill JS, Henson PM & Haslett C. 1994 Apoptotic neutrophils are phagocytosed by fibroblasts with participation of the fibroblast vitronectin receptor and involvement of a mannose/fucose-specific lectin. Journal of Immunology 153: 3218–3227.

Haslett C, Savill JS, Whyte MKB, Stern M, Dransfield I & Meagher LC 1994 Granulocyte apoptosis and the control of inflammation. Philosophical Transactions of the Royal Society B. 345: 327–333.

Hinton BT, Palladino MA, Rudolph D & Labus JC 1996 The epididymis as protector of maturing spermatozoa. Reproduction Fertility and Development 7: 731–745.

Hurley JV 1983 Terminations of acute inflammation I. Resolution, In Acute Inflammation, edn 2, pp 109–117. Ed JV Hurley. Churchill Livingstone, London.

Hart SP, Haslett C & Dransfield I 1996 Recognition of apoptotic cells by phagocytes. Experientia 52: 950–956.

Kelly RW 1995 Immunosuppressive mechanisms in semen: implications for contraception. Human Reproduction 10: 1686–1693.

Kerr JFR, Wyllie AH & Currie AR 1972 Apoptosis: a basic biological phenomenon with wide ranging implications in tissue kinetics. British Journal of Cancer 26: 239–257.

Kitchen E, Rossi AG, Condliffe AM, Haslett C & Chilvers ER 1996 Demonstration of reversible priming of human neutrophils using platelet-activating factor. Blood 88: 4330–4337.

Koopman G, Reutelingsperger CPM, Kuijten GAM, Keehnen RMJ, Pals ST & Van Oers MHJ 1994 AnnexinV for flow cytometric detection of phosphatidyl expression on B cells undergoing apoptosis. Blood 84: 1415–1420.

Maroni ES, Symon DNK & Wilkinson PC 1972 Chemotaxis of neutrophil leukocytes towards spermatozoa and seminal fluid. Journal of Reproduction and Fertility 28: 359–368.

Maroni ES & Wilkinson PC 1971 Selective chemotaxis of macrophages towards human and guinea-pig spermatozoa. Journal of Reproduction and Fertility 27: 149–152.

Martin SJ, Reutelingsperger CPM, McGahon AJ, Rader JA, van Schie RCAA, LaFace DM & Green DR 1995 Early redistribution of plasma membrane phosphatidylserine is a general feature of apoptosis regardless of the initiating stimulus: inhibition by overexpression of bcl-2 and Abl. Journal of Experimental Medicine 182: 1545–1556.

Meagher LC, Savill JS, Baker A, Fuller RW & Haslett C 1992 Phagocytosis of apoptotic neutrophils does not induce macrophage release of thromboxane B_2. Journal of Leukocyte Biology 52: 269–273.

O'Flaherty JT & Rossi AG 1993 5-Hydroxyicosatetraenoate stimulate neutrophils by a stereospecific, G-protein linked mechanism. Journal of Biological Chemistry 268: 14708–14714.

Platt N, Suzuiki H, Kurihara Y, Kodama T & Gordon S 1996 Role for the class A macrophage scavenger receptor in the phagocytosis of apoptotic thymocytes in vitro. Proceedings of the National Academy of Sciences USA 93: 12456–12460.

Pollanen P & Cooper TG 1994 Immunology of the testicular excurent ducts. Journal of Reproductive Immunology 26: 167–1216.

Purvis K & Christiansen E 1996 The impact of infection on sperm quality. Journal of the British Fertility Society 1: 31–41.

Ramprasad MP, Fischer W, Witztum JL, Sambrano GR, Quehenberger O & Steinberg D 1995 The 94- to 97-kDa mouse macrophage membrane protein that recognises oxidised low density lipoprotein and phosphatidylserine-rich liposomes is identical to macrosialin, the mouse homologue of human CD68. Proceedings of the National Academy of Sciences USA 92: 9580–9584.

Ren Y & Savill JS 1995 Proinflammatory cytokines potentiate thrombospondin-mediated phagocytosis of neutrophils undergoing apoptosis. Journal of Immunology 154: 2366–2374.

Ren Y, Silverstein RL, Allen J & Savill JS 1995 CD36 gene transfer confers capacity for phagocytosis of cells undergoing apoptosis. Journal of Experimental Medicine 181: 1857–1862.

Rigotti A, Acton SL & Krieger M 1995 The class B scavenger receptors SR-BI and CD36 are receptors for anionic phospholipids. Journal of Biological Chemistry 270: 16221–16224.

Rossi AG, MacIntyre DE, Jones CJP & McMillan RM 1993 Stimulation of human neutrophil polmorphonuclear leukocytes by leukotriene B$_4$ and platelet-activating factor: an ultrastructural and pharnacological study. Journal of Leukocyte Biology 53: 117–125.

Rossi AG, Cousin JM, Dransfield I, Lawson MF, Chilvers ER & Haslett C 1995 Agents that elevate cAMP inhibit human neutrophil apoptosis. Biochemical and Biophysical Research Communications. 217: 892–899.

Sambrano GR, Parsatharathy S & Steinberg D 1994 Recognition of oxidatively damaged erythrocytes by a macrophage receptor with specificity for oxidized low density lipoprotein. Proceedings of the National Academy of Sciences USA 91: 3265–3269.

Sambrano GR & Steinberg D 1995 Recognition of oxidatively damaged and apoptotic cells by an oxidized low density lipoprotein receptor on mouse peritoneal macrophages: role of membrane phosphatidylserine. Proceedings of the National Academy of Sciences USA 92: 1396–1400.

Savill JS, Henson PM & Haslett C 1989 Phagocytosis of aged human neutrophils by macrophages is mediated by a novel "charge-sensitive" recognition mechanism. Journal of Clinical Investigation 84: 1518–1527.

Savill JS, Wyllie AH, Henson JE, Walport MJ, Henson PM & Haslett C 1989 Macrophage phagocytosis of aging neutrophils in inflammation: programmed cell death in the neutrophil leads to its recognition by macrophages. Journal of Clinical Investigation 83: 865–875.

Savill JS, Dransfield I, Hogg N & Haslett C 1990 Vitronectin receptor-mediated phagocytosis of cells undergoing apoptosis. Nature 342: 170–173.

Savill JS, Hogg N, Ren Y & Haslett C 1992 Thrombospondin cooperates with CD36 and the vitronectin receptor in macrophage recognition of neutrophils undergoing apoptosis. Journal of Clinical Investigation 90: 1513–1522.

Savill JS & Haslett C 1994 Fate of neutrophils, In: Immunopharmacology of Neutrophils, pp 295–314. Eds PG Hellewell & TJ Williams. Academic Press Ltd, London.

Shiratsuchi A, Umeda M, Ohba Y & Nakanishi Y 1997 Recognition of phosphatidylserine on the surface of apoptotic spermatogenic cells and subsequent phagocytosis by Sertoli cells of the rat. Journal of Biological Chemistry 272: 2354–2358.

Stern M, Meagher L, Savill JS & Haslett C 1992 Apoptosis in human eosinophils. Programmed cell death in the eosinophil leads to phagocytosis by macrophages and is modulated by IL-5. Journal of Immunology 148: 3543–3549.

Stern M, Savill JS & Haslett C 1996 Human monocyte-derived macrophage phagocytosis of senescent eosinophils undergoing apoptosis. Mediation by $\alpha_v\beta_y$/CD36/thrombospondin recognition mechanism and lack of phlogistic response. American Journal of Pathology 149: 911–921.

Tomlinson MJ, White A, Barratt CLR, Bolton AE & Cooke ID 1992 The removal of morphologically abnormal sperm forms by phagocytes: A positive role for seminal leukocytes? Human Reproduction 7: 517–522.

Verhoven B, Schlegel RA & Williamson P 1995 Mechanisms of phosphatidylserine exposure, a phagocyte recognition signal, on apoptotic T lymphocytes. Journal of Experimental Medicine 182: 1597–1601.

Whyte MKB, Meagher LC, MacDermot J & Haslett C 1993 Impairment of function in aging neutrophils is associated with apoptosis. Journal of Immunology 150: 5124–5134.

Wolff H 1995 The biological significance of white blood cells in semen. Fertility and Sterility 63: 1143–1157.

Wyllie AH, Kerr JFR & Currie AR 1980 Cell death: the significance of apoptosis. International Review in Cytology 68: 251–306.

OXYTOCIN AND MALE REPRODUCTIVE FUNCTION

R. Ivell,[1] M. Balvers,[1] W. Rust,[1] R. Bathgate,[1] and A. Einspanier[2]

[1]Institute for Hormone and Fertility Research
University of Hamburg
Grandweg 64, 22529 Hamburg, Germany
[2]Deutsches Primatenzentrum
Göttingen, Germany

1. SUMMARY

In the male mammal, the small peptide hormone oxytocin is produced in similar quantities within the hypothalamo-pituitary magnocellular system as in the female, yet for the male little is known about the physiology associated with this hormone. The present review summarizes what is known about the function of oxytocin in the male mammal and tries to take account of both central and systemic effects, and those linked with a local production of oxytocin within the male reproductive organs. In several species a pulse of systemic oxytocin, presumably of hypothalamic origin, appears to be associated with ejaculation. The systemic hormone could act peripherally stimulating smooth muscle cells of the male reproductive tract, but could also reflect central effects in the brain modulating sexual behaviour. In addition to systemic oxytocin, the peptide is also made locally within the testis, and possibly also the epididymis and prostate. In the former tissue it appears to have an autocrine/paracrine role modulating steroid metabolism, but may in addition be involved in contractility of the seminiferous tubules. However, the latter function may involve the mediacy of Sertoli cells which under some circumstances can also exhibit the components of a local oxytocin system. In the prostate of the rat and the dog oxytocin is linked again to steroid metabolism and may also act as a growth regulator. Finally, oxytocin in seminal fluid is discussed and its possible role in respect to the fate of the semen following ejaculation.

2. INTRODUCTION

It is a well established observation that in the hypothalamus of most male mammals there is as much oxytocin expression both at the protein and mRNA level as in females

(Burbach et al., 1987). Yet, whereas in female mammals there is a clearly circumscribed physiology for oxytocin of the hypothalamo-pituitary magnocellular system in regard to birth and lactation, no comparable endocrinology is described for the male. In fact there appears to be a paradox in as much as that it is generally considered that oxytocin expression is principally regulated by ovarian estrogens, albeit not necessarily directly at the level of the oxytocin gene. Whether testosterone in the male can act on the oxytocin gene indirectly via central aromatization is not known, though castration of male rats does not appear to influence peripherally circulating oxytocin (Carter et al., 1988).

There are two possible answers to this apparent paradox. The first is that oxytocin does indeed have a role to play in male reproduction analogous (or ? homologous) to that in female physiology. Alternatively, in both male and female mammals hypothalamic oxytocin may subserve functions quite unrelated to reproductive physiology. For the rat, this may indeed be true, since it has been shown that in this species oxytocin in both sexes is linked with vasopressin gene expression on responding to hyperosmolar stress (Balment et al., 1980; Van Tol et al., 1986), and that oxytocin may be linked to control of sodium homeostasis and ingestion behaviour, also in males (Verbalis et al., 1995). However, neither physiology appears to involve oxytocin in either men or women (Williams et al., 1985). Oxytocin may, however, be involved in a central response to other forms of stress, where at least in rats a sexually dimorphic release of oxytocin under various stress situations is reported (Carter et al., 1986).

Recently, the results of oxytocin gene ablation in the mouse were reported (Nishimori et al., 1996; Luedke et al., 1996; Young et al., 1996) and implied that oxytocin may have no significant role to play in the male. Not only was there a complete absence of phenotype in the homozygous male knock-out animals, but even in females the only phentoypic characteristic of an absence of oxytocin appeared to be a deficiency in milk letdown. There appeared to be no impairment of labour or any other female reproductive function. However, this example illustrates a well-known feature of those functions in which oxytocin is involved; namely that they are informationally highly redundant physiologies. Reproduction, by definition, is that part of physiology which is subjected to highest Darwinian selective pressure, and thus should encourage, through evolution, the acquisition of "fail-safe", fall-back control systems to guarantee the passage of genes to the next generation. In other words, we should expect most reproductive functions to be highly redundant. Thus these knock-out experiments do not support the view that oxytocin has no function in the male, merely that this may be part of a redundant regulatory system.

However, there is a small, dispersed literature which does suggest that oxytocin may be involved in male reproductive physiology. The present review aims to summarize these observations to build up an overall model for oxytocin function in the male.

3. SYSTEMIC OXYTOCIN AND EJACULATION

There are several reports which show that a pulse of oxytocin is measurable in peripheral serum accompanying ejaculation in some male mammals (reviewed by Forsling, 1986, and Wathes, 1989). In the simplest version, exemplified by the study of Murphy et al. (1987) for the human, during penile erection and arousal there appears to be an elevation of plasma vasopressin up to ejaculation, and this is followed, coincident with ejaculation itself, by a single pulse of oxytocin. It seems likely, by analogy to the female, that this pulsatile oxytocin is derived from the hypothalamo-pituitary system, but to date there is little evidence to confirm this, except that in the human this ejaculatory oxytocin pulse can be suppressed by naloxone, which presumably is acting on opiate receptors within the posterior pituitary (Murphy et al., 1990). Oxytocin does not induce ejaculation itself; this process is 100% under nervous control via sympathetic nerves from the lumbar region of the spinal cord, hence the common observation of anejaculation in accident patients with spinal cord damage. This nervous input to the male tract appears to operate largely via a-adrenergic receptors (Marshall et al., 1996).

That the systemic oxytocin pulse is not simply an adjunct of the ejaculatory contraction of the male tract, has been shown in the horse, where also a pulse of systemic oxytocin is measured at ejaculation (Bader and Schams, unpublished; cited in Bettendorf and Breckwoldt, 1989). However, here there is an earlier oxytocin pulse associated with an intromission which was not accompanied by ejaculation. A similar pre-ejaculatory rise in serum oxytocin is reported for the rabbit (Stoneham et al., 1985). Another hint comes from studies on electroejaculation in bulls (Schams et al., 1982). Electrical rectal stimulation of bulls leads to ejaculation accompanied by a systemic pulse of oxytocin. However, the same procedure applied to female cows causes a very similar release of oxytocin, as does copulation in female rabbits (Todd and Lightman, 1986).

One interpretation of these observations, might be that the systemic pulse of oxytocin which is observed in peripheral serum might not itself be physiologically important, but instead an epiphenomenon accompanying a more significant release of oxytocin from the magnocellular nuclei of the hypothalamus within the brain itself. It is well established for female rats, that parallel to the release of oxytocin from the posterior pituitary, there is also release within the hypothalamic nuclei and from axons projecting from these nuclei to other areas of the brain (Neumann et al., 1995; Verbalis et al., 1995). It is also now well established that oxytocin and the oxytocin receptor within the brain are involved in both affiliative, sexual and other behaviour patterns (Insel et al., 1995; Kovacs, 1986), though the majority of such studies have been carried out in female animals. Furthermore, it has been shown in male rats, as well as in rabbits and monkeys, that oxytocin administered cerebroventricularly can induce penile erection (Argiolas, 1992). Thus, it would appear likely that the systemic pulse of oxytocin observed at ejaculation has a correlate within the brain modulating aspects of male sexual behaviour. In those cases where a systemic oxytocin pulse is observed during sexual activity, but not only immediately associated with ejaculation (see above), then these might well correlate with a central effect of oxytocin in relation to sexual behaviour. In a series of studies directly addressing the central role of oxytocin in the male, Lightman and colleagues were able to show in the rat that there is a central release of oxytocin at ejaculation into the cerebrospinal fluid (Hughes et al., 1987), and that infusion of oxytocin into the third ventricle increased the latencies to the first mount and intromission, and lengthened post-ejaculatory refractory periods (Stoneham et al., 1985). And in their study on the effects of naloxone on ejaculation in human male volunteers, the resulting elimination of the serum oxytocin pulse at ejaculation, while not influencing ejaculation per se, led to the subjective consensus of a "decreased arousal and pleasure at orgasm" (Murphy et al., 1990).

Nevertheless, there is some evidence to suggest that systemic oxytocin may have a direct role in the periphery on certain male physiological parameters at ejaculation. For example, Voglmayr (1975) reports that there is an increased fluid flow and an increase in sperm concentration within rete testis fluid directly measured from the canulated ram testis upon systemic injection of oxytocin. In an experimental paradigm where bulls were electroejaculated twice at an interval of 10 minutes, every 2 weeks for an 8 week period, the effect of a bolus injection of oxytocin 10 minutes prior to the first electroejaculation was to shift the higher volume and sperm count from the second of the two ejaculates to the first (Berndtson and Igboeli, 1988). And in the rat systemic infusion of oxytocin reduced the number of intromissions to ejaculation in a dose-dependent fashion (Stoneham et al., 1985). More recently, Frayne and coworkers (Frayne et al., 1996) have shown that by systemically applying an oxytocin antagonist during the first wave of spermatogenesis in late pubertal rats, there is a significant delay in the first appearance of spermatozoa in the epididymis, suggesting again that oxytocin may play a role in spermeation or in regulating the flow of seminal fluid.

An alternative approach to understanding how oxytocin might work in the male reproductive system is to assess the distribution and cellular localization of specific oxytocin receptors. Maggi et al., (1987) using binding studies were able to show the presence of oxytocin receptors in the tunica albuginea of the testis, in the prostatic capsule and in the epididymis of the boar, and Bodansky et al. (1992) could induce contractions of the guinea pig prostate capsule in vitro by addition of oxytocin. In a recent study in the marmoset monkey, Einspanier and Ivell (1997) using two different antibodies specifically recognizing the oxytocin receptor, showed the presence of receptors, as well as in the testis (see below), in the smooth muscle cells of the myoid sheath of the epididymis, and also in myoid cells surrounding the lobules of the bulbourethral gland. Within the prostate of this species, receptors could also be sporadically identified within the stromal tissue. An analysis by RT-PCR for the presence of the specific mRNA for the oxytocin receptor both in the marmoset (Einspanier and Ivell, 1997) and in human tissues (Fig.1), confirmed this distribution with robust signals within the testis, epididymis and prostate. Oxytocin receptor mRNA could also be detected in two cell lines from fetal tract tissues (REP, RVP, Coleman and Harris, 1991; Fig 1) and in two prostatic carcinoma cell lines (LnCAP, DU145; Fig.1).

Summarizing all these observations would suggest that a pulse of systemic oxytocin released at ejaculation or just before could serve to modulate the contraction of ducts and glandular lobules throughout the male tract, thus influencing the fluid volume of different ejaculate components, and possibly shortening the time from intromission to ejaculation. At the same time, oxytocin released centrally into the brain could influence sexual behaviour, subjective appreciation of arousal (orgasm) and latency to subsequent ejaculation.

Until very recently, there was no information as to the causal mechanism which leads to the release of hypothalamic oxytocin at ejaculation. Now, Yanagimoto et al. (1966) have demonstrated for rats that tractile stimulation of the penis, via the dorsal penile nerve elicits a specific activation of oxytocinergic neurones in the paraventricular nucleus of the hypothalamus. This important study thus furnishes the final components of a complete oxytocin-based reflex arc for the male, similar to the milk let-down reflex or the Ferguson reflex in the female.

4. OXYTOCIN AND TESTIS FUNCTION

The testis has been shown to be a site not only of oxytocin receptors but also of local oxytocin production (Pickering et al., 1990). Whole testis extracts from several species have been shown to contain authentic oxytocin peptide (Pickering et al., 1990; Bathgate et al., 1993), as well as oxytocin mRNA (Fig.1; Ang et al., 1991; Foo et al., 1991; Ivell, 1990; Ungefroren et al., 1990). A number of studies have shown that the Leydig cells form an autocrine/paracrine system. Immunohistochemical studies have identified both oxytocin peptide in the Leydig cells of the rat, human, bovine, and marmoset testis (Guldenaar and Pickering, 1985; Weindl et al., 1986; Ungefroren et al., 1994; Kulkarni et al., 1992; Einspanier and Ivell, 1997), as well as the neurophysin molecule which forms a part of the oxytocin precursor

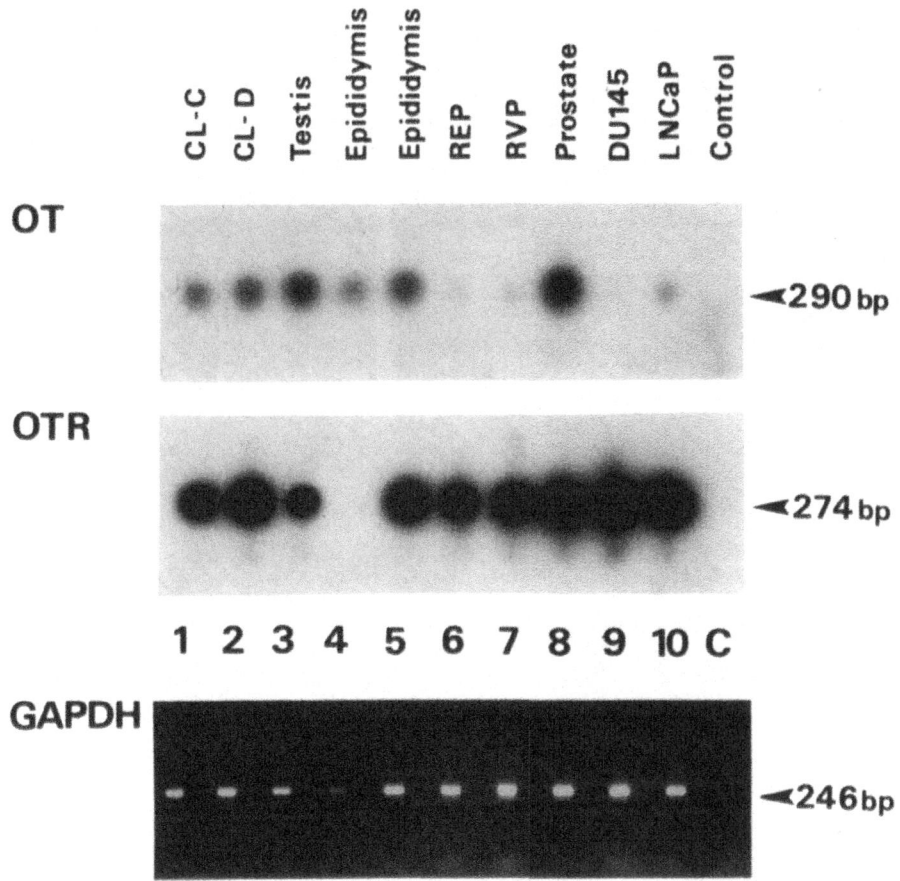

Figure 1. RT-PCR analysis of specific transcripts encoding oxytocin (OT) or oxytocin receptor (OTR) in extracts of human tissues and cell lines. For OT, forward and reverse primers were as in Einspanier & Ivell (1997) and the RT-PCR products visualized as described by hybridizing to an internal oligonucleotide probe, radiolabelled by [γ^{32}P]ATP in the presence of T4 polynucleotidyl kinase. For OTR, forward (5'-GTGTCAGCAGCGTCAAGC) and reverse (5'-TGGCGGAGCAGCACAGG) primers were derived from the human OTR cDNA sequence (Kimura et al., 1992). The resulting RT-PCR fragment, which spanned the third splice junction, was similarly hybrized to a radiolabelled internal oligonucleotide (5'-GCCAACGCGCCCAAGGAAGCCTCGG). CL-C and CL-D: human corpora lutea of the late cycle. REP and RVP are two immortalized cell lines from the fetal tract of the rat (Coleman & Harris, 1991). DU145 and LnCAP are prostatic carcinoma cell lines. The quality of the RNA used as template was checked by evaluating the levels of the mRNA for glyceraldehyde 3-phosphate dehydrogenase (GAPDH), using primers described elsewhere (Einspanier & Ivell, 1997).

polypeptide (Ungefroren et al., 1994; Einspanier and Ivell, 1997). Leydig cells also contain large numbers of oxytocin receptors. This has been demonstrated for the rat (Bathgate and Sernia, 1994) and the marmoset monkey (Fig.2) using direct receptor autoradiography, as well as immunohistochemically in the marmoset using two different specific anti-receptor antibodies (Einspanier and Ivell, 1997), and in the human (Fig.3).

The function of oxytocin in the interstitial compartment of the testis is less clear. Cultured Leydig cells have shown only a variable response to oxytocin in terms of testosterone production (e.g. Sharpe and Cooper, 1986; Frayne and Nicholson, 1995). In con-

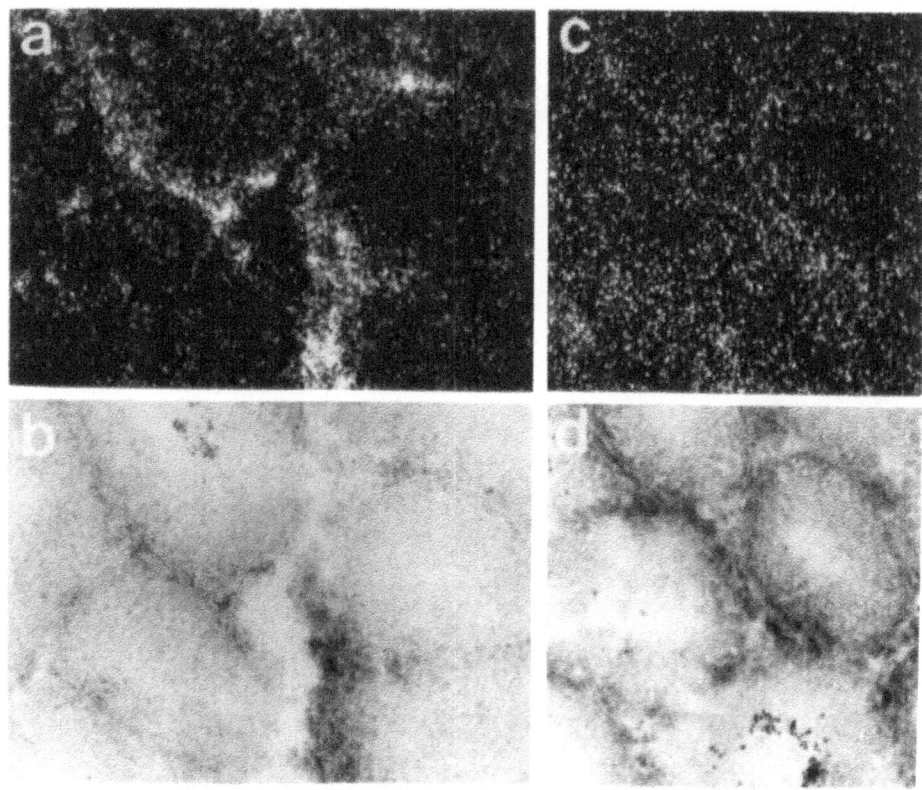

Figure 2. Oxytocin receptor autoradiography of marmoset testis. Autoradiography was performed exactly as described in Bathgate & Sernia (1994) using [^{125}I]OVTA (New England Nuclear, Bad Homburg, Germany) as radioligand. a and b: The interstitial cells are clearly labelled with radioligand; all remaining compartments of the testis being indistinguishable from background. c and d: Control sections where the radioligand was quenched with an excess of unlabelled oxytocin prior to incubation. a and c are darkfield reflectance images of the sections shown in transmitted light in b and d.

trast, a testicular oxytocin implant in the rat led to a shift in androgen metabolism, with a decrease in testosterone production and an increase in the formation of dehydrotestosterone (DHT; Nicolson et al., 1991). In a transgenic mouse, however, which had an over-expression of oxytocin in the tubular compartment of the testis, there appeared to be a specific reduction in both testosterone and DHT within the testis (Ang et al., 1994).

One of the oldest observations on a possible testicular role for oxytocin is its ability to stimulate peristaltic contractions of the seminiferous tubules, both in vivo and in culture (Roosen-Runge, 1951; Niemi and Kormano, 1965; Pickering et al., 1990). This would ap-

⟶

Figure 3. Immunohistochemical staining for specific oxytocin receptors in sections of normal human testis using the monoclonal antibody 2F8 (courtesy of Dr Kimura). Specific immunoreactivity is visualized as dark staining using a modified APAAP method, with NBT/BCIP as chromogen. The sections are not counterstained. A and B: Different regions of the same testis section immunostained for the oxytocin receptor. C: Control section where the mouse primary antibody is replaced by an equivalent amount of mouse IgM.

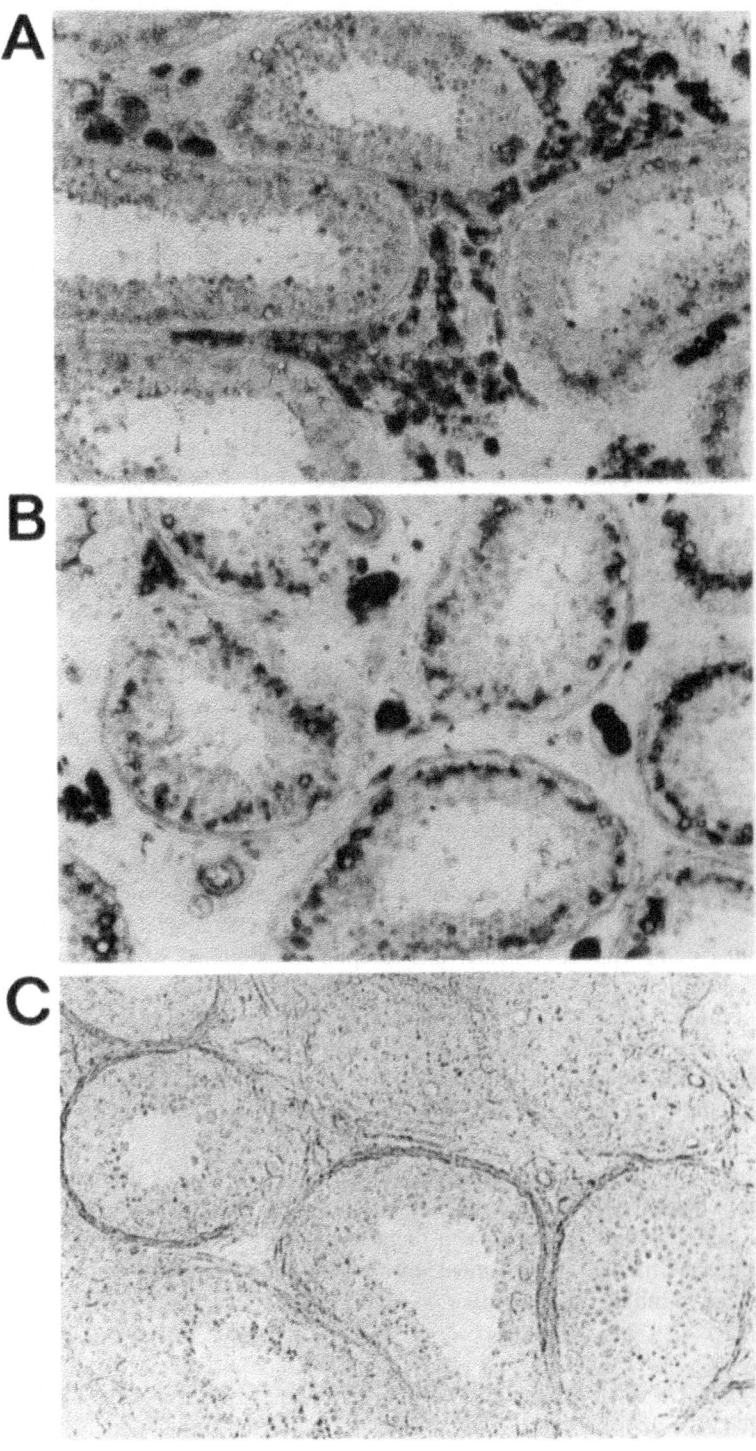

pear logical in view of the ability of this peptide to influence smooth muscle cells in most other parts of the male and female reproductive systems. However, both receptor autoradiography (Bathgate and Sernia, 1994), as well as receptor pharmacology on purified peritubular myoid cells (Howl et al., 1995) and immunohistochemistry using specific anti-receptor antibodies (Einspanier and Ivell, 1997) agree very definitively that the smooth muscle cells of the seminiferous tubules are devoid of oxytocin receptors, but do possess vasopressin V1a (Bathgate and Sernia, 1994; Howl et al., 1995) as well as endothelin receptors (Tripiciano et al., 1996), both of which can mediate tubule contractility. Yet, it appears that even in purified rat tubule preparations in culture only 1 nM oxytocin is effective to induce smooth muscle cell contraction, which is too low a concentration to be acting at any other than specific oxytocin receptors (Pickering et al., 1990).

A solution to this paradox might be found in the recent observations that there may be a second intratesticular compartment which is able to produce oxytocin and has oxytocin receptors, namely the Sertoli cells within the seminiferous tubules. A first hint at this novel system came from studies in the ram where evidence was provided for oxytocin within the testicular tubule fluid from the excurrent ducts (Knickerbocker et al., 1988). This was followed by the study of a transgenic mouse carrying in its genome a bovine oxytocin transgene driven by its own promoter (Ang et al., 1991; 1994). This transgene was expressed at high levels in the Sertoli cells of the mouse testis both at the mRNA and the protein level. An assessment of the bovine testis then showed that oxytocin mRNA was massively expressed also within the seminiferous tubules, yet translation of this mRNA was blocked at the post-transcriptional level (Ungefroren et al., 1994), though a slight oxytocin immunoreactivity could still be detected in Sertoli cells. Subsequently, the Nicholson group showed that by subjecting the rat testis to severe insult by heating or by treatment with methoxyacetic acid caused a substantial increase in oxytocin within the seminiferous tubule compartment, but not in the interstitium (Nicholson et al., 1994). Then using immunohistochemistry in the marmoset testis, weak but specific signals for the oxytocin receptor, as well as in Leydig cells, could also be seen in Sertoli cells (Einspanier and Ivell, 1997). This immunohistochemical staining for the oxytocin receptor is more evident in the human testis (Fig.3), where in regions which are relatively devoid of Leydig cells, there is clear staining for the oxytocin receptor within the Sertoli cells of the tubular compartment. Whether this receptor is functional is not known, but it does mean that seminiferous tubules may be able to respond to oxytocin and by a local paracrine mechanism influence the myoid cells in the lamina propria. The purpose of seminiferous tubule contractility is not clear, but may be associated with spermeation and the transport of sperm-containing fluid toward the rete testis. This would agree with the observations that oxytocin can influence sperm concentrations in rete testis fluid (Voglmayr, 1975), epididymis (Frayne et al., 1996), or in the ejaculate (Berndtson and Igboeli, 1988).

In conclusion, it would appear that within the testis there is a prominent local oxytocin system in the interstitial compartment, which may have a modulatory role on steroidogenesis or 5α-reductase activity. Then there appears to be a lower level local oxytocin system within the tubular compartment, whose function at this time remains obscure, but may be related to tubule contractility.

5. OXYTOCIN EXPRESSION IN OTHER TISSUES OF THE MALE TRACT

There is very little information on oxytocin in other tissues of the male tract. In the marmoset, immunohistochemical analysis failed to show up the presence of the hormone in any part of the tract and accessory organs (Einspanier and Ivell, 1997). For the human, there is definite oxytocin mRNA within both the epididymis at a very low level, and also at a greater level within the prostate (Fig.1). The Nicholson group have recently reported immunohistochemical staining for oxytocin peptide in the epithelial cells of the caput epididymis of the rat, which appears not to be affected by efferent duct ligation (Harris et al., 1996). Additionally, they have provided clear evidence for oxytocin in prostate extracts from the dog and rat (Nicholson and Jenkin, 1995). Interestingly, there is an increase in oxytocin levels in the dog prostate with age paralleling the development of prostatic hyperplasia in this species.

In primary cultures of rat epididymal and prostatic cells, oxytocin appears to stimulate 5α-reductase activity (Nicholson and Jenkin, 1995), much as in the testis. It is interesting to note, that in the dog prostate, the age-dependent increase in oxytocin content accompanies also an increase in endogenous 5α-reductase activity (Nicholson and Jenkin, 1995). In another study, Plecas et al. (1992) have shown that oxytocin can increase cell growth and mitotic activity for rat prostatic cells. Thus a picture is presented for oxytocin within the prostate, and possibly also in other accessory glands, having an autocrine/paracrine modulatory role on cell growth and metabolism. The observation of sporadic receptors within the prostatic stroma in the marmoset (Einspanier and Ivell, 1997) and of receptor mRNA in prostate extracts from both marmoset and human (Fig.1) would support this conclusion.

6. OXYTOCIN IN SEMINAL FLUID

Oxytocin has been measured in the testicular tubular fluid from the ram (Knickerbocker et al., 1988) and from the rat, increasing after physiological insult (Nicholson et al., 1994). Additionally, there is evidence to suggest that in the ram oxytocin may be taken up from seminal fluid by epididymal epithelial cells (Veeramachaneni and Amann, 1990). Ejaculated human semen has also been shown to contain oxytocin (Nicholson et al., 1985). It is not known where this oxytocin is produced, though the prostate seems to be a likely candidate.

There is no function described for seminal fluid oxytocin. However, one possibility is that this oxytocin may react with specific oxytocin receptors on the mucous epithelium of the cervix at ejaculation. In the cow from the cycle, there is good evidence for the presence of such receptors, which appear to vary in number through the estrous cycle peaking at estrus (Fuchs et al., 1996). Under such circumstances one could speculate that as well as prostaglandins, also seminal oxytocin might have a role to play in stimulating the cervical response to ejaculation which initiates the rapid transport of sperm to the oviducts in a cycle dependent manner (see articles by Leyendecker et al., 1997, and Wildt et al., 1997, this volume).

7. ACKNOWLEDGMENTS

We gratefully acknowledge many colleagues for advice and suggestions during the preparation of this review. We thank also Dr Ann Harris (Oxford) for the cell-lines REP and RVP, and Dr Tadashi Kimura (Osaka and Hamburg) for the monoclonal anti-OTR antibody used in Figure 3, Professor Maurice Manning (Toledo, OH) for the generous gifts of oxytocin analogues and Professor Michail Davidoff (Hamburg) for provision of the human testis sample used for immunohistochemistry. This research was funded in parts by grants from the Deutsche Forschungsgemeinschaft (Ei333/2–1 and Iv7/5–1).

8. REFERENCES

Ang HL, Ungefroren H, De Bree F, Foo NC, Carter D, Burbach JP, Ivell R & Murphy 1991 D. Testicular oxytocin gene expression in seminiferous tubules of cattle and transgenic mice. Endocrinology 128: 2110–2117.

Ang HL, Ivell R, Walther N, Nicholson H, Ungefroren H, Millar M, Carter D & Murphy D. 1994 Over-expression of oxytocin in the testes of a transgenic mouse model. Journal of Endocrinology, 140: 53–62.

Argiolas A 1992 Oxytocin-induced penile erection: pharmacology, site and mechanism of action. Annals of the New York Academy of Sciences 652: 194–203.

Balment RJ, Brimble MJ & Forsling ML 1980 Release of oxytocin induced by salt-loading and ist influence on renal excretion in the male rat. Journal of Physiology (London) 308: 439–499.

Bathgate RAD & Sernia C 1994 Characterization and localization of oxytocin receptors in the rat testis. Journal of Endocrinology 141: 343–352.

Bathgate RAD, Sernia C & Gemmell RT 1993 Arginine vasopressin- and oxytocin-like peptides in the testis of two australian marsupials. Peptides 14: 701–705.

Berndtson WE & Igboeli G 1988 Spermatogenesis, sperm output and seminal quality of Holstein bulls electroejaculated after administration of oxytocin. Journal of Reproduction and Fertility 82: 467–475.

Bettendorf G & Breckwoldt M (Eds) 1989 Reproduktionsmedizin. Gustav Fischer Verlag, Stuttgart.

Bodansky M, Sharaf H, Roy JB & Sais SI 1992 Contractile activity of vasotocin, oxytocin and vasopressin on mammalian prostate. European Journal of Pharmacology 216: 311–313.

Burbach JPH, Voorhuis TAM, Van Tol HHM & Ivell R 1987 In situ hybridization of oxytocin messenger RNA: macroscopic distribution and quantitation in rat hypothalamic cell groups. Biochemical Biophysics Research Communications 145: 10–14.

Carter DA, Saridaki E & Lightman SL 1988 Sexual differentiation of oxytocin stress responsiveness: effect of neonatal androgenization, castration and a luteinizing hormone-releasing hormone antagonist. Acta Endocrinologica 117: 525–530.

Carter DA, Williams TDM & Lightman SL 1986 A sex difference in endogenous opioid regulation of the pituitary response to stress in the rat. Journal of Endocrinology 111: 239–244.

Coleman L & Harris A 1991 Immortalization of male genital duct epithelium: an assay system for the cystic fibrosis gene. Journal of cell Science 98: 85–89.

Einspanier A & Ivell R 1997 Oxytocin and oxytocin receptor expression in reproductive tissues of the male marmoset monkey. Biology of Reproduction 56: 416–422.

Foo NC, Carter D, Murphy D & Ivell R 1991 Vasopressin and oxytocin gene expression in rat testis. Endocrinology 128: 2118–2128.

Forsling ML 1986 Regulation of oxytocin release. Current Topics in Neuroendocrinology 19–54.

Frayne J & Nicholson HD 1995 Effect of oxytocin on testosterone production by isolated rat Leydig cells is mediated via a specific oxytocin receptor. Biology of Reproduction 52: 1268–1273.

Frayne J, Townsend D & Nicholson HD 1996 Efects of oxytocin on sperm transport in the pubertal rat. Journal of Reproduction and Fertility 107: 299–306.

Fuchs AR, Ivell R, Fields PA, Chang SMT & Fields MJ 1996 Oxytocin receptors in bovine cervix: distribution and gene expression during the estrous cycle. Biology of Reproduction 54: 700–708.

Guldenaar SEF & Pickering BT 1985 Immunocytochemical evidence for the presence of oxytocin in rat testis. Cell Tissue Research 240: 485–487.

Harris GC, Frayne J & Nicholson HD 1996 Epididymal oxytocin in the rat: its origin and regulation. International Journal of Andrology 19: 278–286.

Howl J, Rudge SA, Lavis RA, Davies AR, Parslow RA, Hughes PJ, Kirk CJ, Michell RH and Wheatley M 1995 Rat testicular myoid cells express vasopressin receptors: receptor structure, signal transduction and developmental regulation. Endocrinology 136: 2206–2213.

Hughes AM, Everitt BJ, Lightman SL & Todd K 1987 Oxytocin in the central nervous system and sexual behaviour in male rats. Brain Research 414: 133–137.

Insel TR, Winslow JT, Wang ZX, Young L & Hulihan TJ 1995 Oxytocin and molecular basis of monogamy. In: Oxytocin. 227–234. Ed. Ivell R & Russell J. Plenum Press, New York.

Ivell R (1991) Vasopressin and oxytocin gene expression in the mammalian ovary and testis. In Vasopressin. 31–38. Ed Jard S & Jamison R. Colloque INSERM / John Libbey Eurotext, Paris.

Kimura T, Tanizawa O, Mori K, Brownstein MJ, Okoyama H 1992 Structure and expression of a human oxytocin receptor. Nature 356: 526–529.

Knickerbocker JJ, Sawyer HR, Amann RP, Tekpetey FR & Niswender GD 1988 Evidence for the presence of oxytocin in the ovine epididymis. Biology of Reproduction 39: 391–397.

Kovacs GL 1986 Oxytocin and behavior. Current Topics in Neuroendocrinology, 91- 128.

Kulkarni SA, Garde SV & Sheth AR 1992 Immunocytochemical localization of bioregulatory peptides in marmoset testes. Archives of Andrology, 29: 87–102.

Luedke C & Muglia L 1996 Targetted inactivation of the murine oxytocin gene. Endocrine Society Abstracts #P2–514, San Francisco, June 1996.

Maggi M, Malozowski S, Kassis S, Guardabasso V & Rodbard D 1987 Identification and characterization of two classes of receptors for oxytocin and vasopressin in porcine tunica albuginea, epididymis, and vas deferens. Endocrinology 120: 986–994.

Marshall I, Burt RP, Green GM, Hussain MB & Chapple CR 1996 Different subtypes of α_{1A}-adrenoceptor mediating contraction of rat epididymal vas deferens, rat hepatic portal vein and human prostate distinguished by the antagonist RS 17053. British Journal of Pharmacology 119: 407–415.

Murphy MR, Seckl JR, Burton S, Checkley SA & Lightman SL 1987 Changes in oxytocin and vasopressin secretion during sexual activity in men. Journal of Clinical Endocrinology and Metabolism 65: 738–741.

Murphy MR, Checkley SA, Seckl JR & Lightman SL 1990 Naloxone inhibits oxytocin release at orgasm in man. Journal of Clinical Endocrinology and Metabolism 71: 1056–1058.

Neumann I, Pittmann QJ & Landgraf R 1995 Release of oxytocin within the supraoptic nucleus: mechanisms, physiological significance and antisense targetting. In: Oxytocin. 173–184. Ed Ivell R & Russell J. Plenum Press, New York.

Nicholson HD, Greenfield HM & Frayne J 1994 The effect of germ cell complement on the presence of oxytocin in the interstitial and seminiferous tubule fluid of the rat. Journal of Endocrinology, 143: 471–478.

Nicholson HD, Guldenaar SEF, Boer GJ & Pickering BT 1991 Testicular oxytocin: effects of intratesticular oxytocin in the rat. Journal of Endocrinology 130: 231–238.

Nicholson HD, Peeling WB & Pickering BT 1985 Oxytocin in the prostate and semen? Journal of Endocrinology 104 (Suppl): 127.

Nicholson HD & Jenkin L 1995 Oxytocin and prostatic function. In: Oxytocin. 529–537. Ed Ivell R & Russell J. Plenum Press, New York.

Niemi M & Kormano M 1965 Contractility of the seminiferous tubule of the postnatal rat testis and its responser to oxytocin. Annales Medicinae Experimentalis et Biologiae Fenniae 43: 40–42.

Nishimori K, Guo Q, Kumar TR & Matzuk MM 1996 A transgenic model to study oxytocin function in vivo. Endocrine Society Abstracts, # P2–513, San Francisco, June 1996.

Pickering BT, Ayad VJ, Birkett SD, Gilbert CL, Guldenaar SEF, Nicholson HD, Worley RTS & Wathes DC 1990 Neurohypophyseal peptides in the gonads: are they real and do they have a function? Reprouction, Fertility and Development, 2: 245–262.

Plecas B, Popovic A, Jovovic D & Hristic M 1992 Mitotic activity and cell deletion in ventral prostate epithelium of intact and castrated oxytocin-treated rats. Journal of Endocrinological Investigation 15: 249–253.

Roosen-Runge EC 1951 Motions of the seminiferous tubules of rat and dog. Anatomical record 153 (Suppl): 109.

Schams D, Baumann G & Leidl W 1982 Oxytocin determination by radioimmunoassay in cattle II Effect of mating and stimulation of the genital tract in bulls, cows and heifers. Acta Endocrinologica 99: 218–223.

Sharpe RM & Cooper I 1987 Comparison of the effects on purified Leydig cells of four hormones (oxytocin, vasopressin, opiates and LHRH) with suggested paracrine roles in the testis. Journal of Endocrinology 113: 89–96.

Stoneham MD, Everitt BJ, Hansen S, Lightman SL & Todd K 1985 Oxytocin and sexual behaviour in the male rat and rabbit. Journal of Endocrinology 107: 97–106.

Todd K & Lightman SL 1986 Oxytocin release during coitus in male and female rabbits: effect of opiate receptor blockade with naloxone. Psychoneuroendocrinology 11: 367–371.

Tripiciano A, Palombi F, Stefanini M, Ziparo E & Filippini A 1996 Both endothelin receptors subtypes in myoid cells mediate seminiferous tubule contractility. European Testis Workshop Miniposter # C15, Geilo, Norway, May 1996.

Ungefroren H, Davidoff M & Ivell R 1994 Post-transcriptional block in oxytocin gene expression within the seminiferous tubules of the bovine testis. Journal of Endocrinology, 140: 63–72.

Van Tol HHM, Voorhuis TAM & Burbach JPH 1986 Oxytocin gene expression in discrete hypothalamic magnocellular cell groups is stimulated by prolonged salt loading. Endocrinology 120: 71–76.

Veeramachaneni DNR & Amann RP 1990 Oxytocin in the ovine ductuli efferentes and caput epididymidis: immunolocalization and endocytosis from the luminal fluid. Endocrinology 126: 1156–1164.

Verbalis JG, Blackburn RE, Hoffman GE & Stricker EM 1995 Establishing behavioral and physiological functions of central oxytocin: insights from studies of oxytocin and ingestive behaviors. In: Oxytocin. 209–226. Ed. Ivell R & Russell J. Plenum Press, New York.

Voglmayr JK 1975 Output of spermatozoa and fluid by the testis of the ram and its response to oxytocin. Journal of Reproduction and Fertility 43:119–122.

Wathes DC 1989 Oxytocin and vasopressin in the gonads. Oxford Reviews in Reproductive Science 11: 225–283.

Weindl A, Braun J & Rust M 1986 Immunohistochemical demostration of oxytocin in the testis. Acta Endocrinologica S274: 105–106.

Williams TDM, Abel DC, King CMP, Jelley RY & Lightman SL 1986 Vasopressin and oxytocin responses to acute and chronic osmotic stimuli in man. Journal of Endocrinology 108: 163–168.

Yanagimoto M, Honda K, Goto Y, Negoro H 1996. Afferents orginating from the dorsal penile nerve excite oxytocin cells in the hypothalamic paraventricular nucleus of the rat. Brain Research 733: 292–296

Young WS, Shepard E, Amico J, Hennighausen L, Wagner KU, LaMarca ME, McKinney C & Ginns EI 1996 Deficiency in mouse oxytocin prevents milk ejection, but not fertility or parturition. Journal of Neuroendocrinology 8: 847–854.

48

SENSORY INNERVATION OF THE HUMAN PENIS

Z. Halata and A. Spaethe

Department of Functional Anatomy
University of Hamburg (UKE)
Martinistrasse 52, 20246 Hamburg
Germany

INTRODUCTION

Male reproductive behavior has a complex neural organisation. The penile erection and ejaculation require the integration and proper sequencing of somatic, sympathetic, and parasympathetic innervation (Johnson and Halata, 1991). In the present poster we try to answer two questions: 1. which type of sensory nerve endings occur and 2. how they are distributed. We have concentrated on the sensory innervation of the glans penis.

MATERIAL AND METHODS

In total 4 specimens of human penis were studied. One penis was fixed in 10% neutral-buffered formaldehyde, after embedding in paraffin, 8 μm thick sections were obtained and stained with a silver impregnation technique (Spaethe, 1984). Small cubes (side length ca. 3 mm) from 3 cases of sex reversal, aged between 22 and 28 years, were processed for electron microscopy.

RESULTS AND SUMMARY

The glans penis skin is covered by a thin keratinized stratified squamous epithelium (Fig.1). The epithelium has rete ridges varying in height depending on the region of glans and the age. The epithelium over the ventral surface surrounding the frenulum is thinner than that of the dorsal aspect. The rete ridges are higher and narrower in younger than in older individuals.

The dermis underlying the epithelium consists of a dense connective tissue. The papillary layer forms the overlying rete ridges. The dense reticular layer of the dermis blends with the trabeculae separating the blood sinuses of the corpus spongiosum. Sensory nerve endings

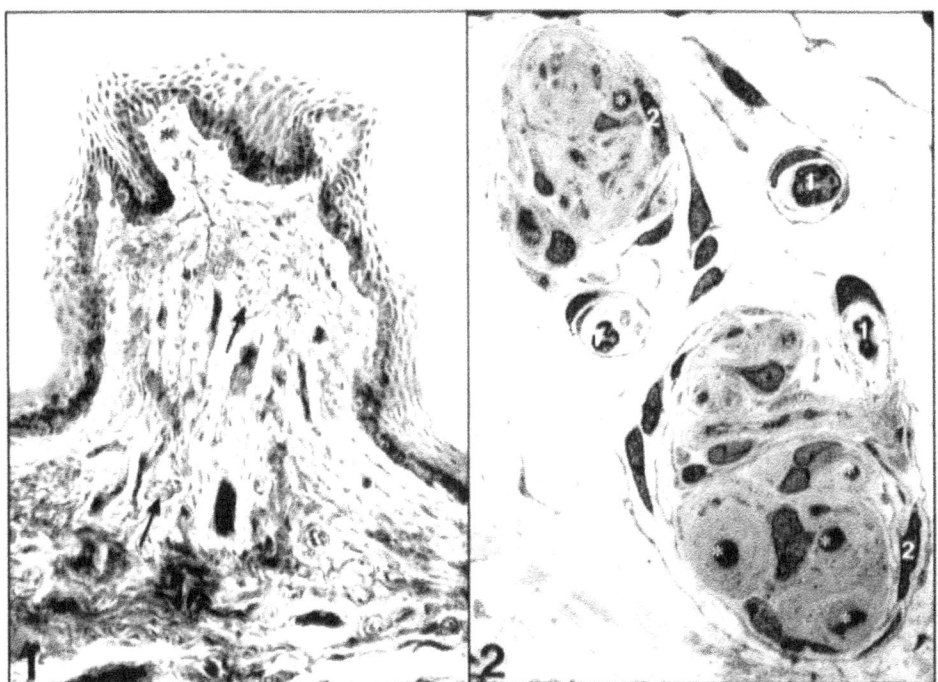

Figures 1 and 2. (See text for details.)

in the skin of glans penis are located in all tissue layers. In the epidermis and dermis free nerve endings are to be found (Fig. 1 asterisks); in the dermis genital end bulbs (Fig 1 arrows; Fig. 2 asterisks), Pacinian corpuscles and rarely Ruffini corpuscles can be identified. Free nerve endings are located throughout the glans penis and represent the most common type of sensory nerve endings. They are derived from thin myelinated axons (diameter 1–3 μm) or from unmyelinated C fibers. Two different types of genital end bulbs are to be recognized. Corpuscles tightly abutting the overlaying epithelium are small and lack a perineural capsule. Large genital end bulbs are situated deeper in the dermis and are completely surrounded by a multilayered perineural capsule (Fig. 2-2). Occasionally small and large Pacinian corpuscles are to be found deep in the dermis of the dorsal aspect and corona glandis region. Ruffini corpuscles are only occasionally to be observed in the dense connective tissue of the dermis by means of electron microscopy. The afferent myelinated axon has a diameter of 2–4 μm (Fig. 2-1). Most of the corpuscles have an incomplete perineural capsule.

In contrast to typical glabrous skin, the so-called mucocutaneous tissue of the glans penis contains a predominance of free nerve endings, numerous genital end bulbs and rarely Pacinian und Ruffinian corpuscles. Merkel nerve endings and Meissner corpuscles are not present.

REFERENCES

Johnson RD & Halata Z 1991 Topography and ultrastructure of sensory nerve endings in the glans penis of the rat. Journal of Comparative Neurology 312: 299–310.
Spaethe A 1984 Eine Modifikation der Silbermethode nach Richardson für die Axonfärbung in Paraffinschnitten. Verhandlungen der Anatomischen Gesellschaft 78: 101–102.

49

THE UTERINE PERISTALTIC PUMP

Normal and Impeded Sperm Transport within the Female Genital Tract

G. Kunz,[1] D. Beil,[2] H. Deiniger,[2] A. Einspanier,[4] G. Mall,[3] and G. Leyendecker[1]

[1]Department of Obstetrics and Gynecology
[2]Department of Radiology I
[3]Department of Pathology
Klinikum Darmstadt
Academic Teaching Hospital to the University of Frankfurt
Grafenstr. 9, 64283 Darmstadt, Germany
[4]Deutsches Primatenzentrum GmbH
Kellnerweg 4, 37077 Göttingen, Germany

1. SUMMARY

Rapid as well as sustained sperm transport from the cervical canal to the isthmical part of the fallopian tube is provided by cervico-fundal uterine peristaltic contractions that can be visualized by vaginal sonography. The peristaltic contractions increase in frequency and presumably also in intensity as the proliferative phase progresses. As shown by placement of labeled albumin macrospheres of sperm size at the external cervical os and serial hysterosalpingoscintigraphy (HSSG) sperm reach, following their vaginal deposition, the uterine cavity within minutes. In the early follicular phase a large proportion of the macrospheres remains at the site of application, while a smaller proportion enters the uterine cavity with even a smaller one reaching the isthmical part of the tubes. In the mid-follicular phase of the cycle with increased frequency and intensity of the uterine contractions the proportion of macrospheres entering the uterine cavity as well as the tubes has significantly increased. In the late follicular phase with maximum frequency and intensity of uterine peristalsis the proportion of macrospheres entering the tube increases further at the expense of those at the site of application as well as within the uterine cavity. The transport of the macrospheres into the tube is preferentially directed into the tube ipsilateral to the dominant follicle, which becomes apparent in the mid-follicular phase as soon as a dominant follicle can be identified by ultrasound. Since the macrosphere are inert particles the directed sperm transport into the tube ipsilateral to the dominant follicle is not functionally related to a mechanism such as chemotaxis but is rather provided by uterine contraction of which the direction may be controlled by a specific myometrial architecture in combination with an asymmetric distribution of myometrial oestradiol receptors.

Women with infertility and mostly mild endometriosis display on VSUP a uterine hyperperistalsis with nearly double the frequency of contractions during the early and mid- as well as midluteal phase in comparison to the fertile and healthy controls. During midcycle these women display a considerable uterine dysperistalsis in that the normally long and regular cervico-fundal contractions during this phase of the cycle have become more or less undirected and convulsive in character. Hyperperistalsis results in the transport of inert particles from the cervix into the tubes within minutes already during the early follicular phase, and may therefore constitute the mechanical cause for the development of endometriosis in that it transports detached endometrial cells and tissue fragments via the tubes into the peritoneal cavity. Moreover, dysperistalsis may contribute to the infertility in these patients since it results in a break down of sperm transport within the female genital tract.

2. INTRODUCTION

There is no doubt that the ultimate fate of the successful male germ cell is to impregnate the female oocyte, where its genetic material will fuse with that of the oocyte to result in fertilization and embryo formation. This particular sperm is usually deposited five to zero days prior to ovulation (Wilcox et al., 1995) in the posterior vaginal fornix in close vicinity to the external os of the cervical canal from where it reaches its final site of destination, the place of sperm-oocyte encounter within the tube ipsilateral to the dominant follicle.

Usually, the uterus is considered to be specialized, first, for the reception of the blastocyst by the endometrium and the continuous nourishment of the developing fetus and, second, for the eventual expulsion of the fetus (Romanini, 1994). Considering the fact that the sperm has to cover a long distance within the female genital tract from the external os of the cervix to the site of fertilization within the tube it is surprising that the facilitation and the guidance of this journey has only recently been considered a genuine and active function of the uterus (Kunz et al., 1996a). Previously, the ascension of the sperm within the female genital tract to the site of fertilization was regarded more or less a functional capacity of the sperm itself with the uterus serving merely as a passive canal, although a functional importance had been ascribed to uterine contractions (Moghissi, 1977; Harper, 1994). With respect to sperm ascension, attention was mostly directed towards the cervical canal with its glands and cyclically changing secretion. These are assumed to provide optimal conditions for the penetration of the sperm into the female genital tract around ovulation and serve, with its crypts, as a sperm reservoir, from where constant release for sustained sperm transport could occur. In this communication, the mechanism of uterine peristalsis will be discussed, and its role in sperm transport within the female genital tract will be outlined. It will be demonstrated that this function of the uterus is of fundamental importance in the process of reproduction and that disturbances of the uterine mechanism of sperm transport may result in infertility.

3. UTERINE PERISTALSIS

Rapid sperm transport from the vagina to the Fallopian tubes within minutes has been described in many species including man (Hartman, 1962; Mortimer, 1983; Hunter,

1987; Drobnis and Overstreet, 1992; Harper, 1994). Since the velocity of sperm movement does not itself account for covering such a long distance through the female genital tract within a few minutes, rapid sperm transport is considered a passive phenomenon and has been ascribed to uterine contractions (Moghissi, 1977; Harper, 1994; Kunz et al., 1996a; Leyendecker et al., 1996).

3.1. Vaginal Sonography of Uterine Peristalsis (VSUP)

Contractile activity of the non-pregnant uterus has been known for many decades (Hendricks, 1966; Cibils, 1967; Martinez-Gaudio et al., 1973). High resolution sonography has made it possible to demonstrate these contractions without invasive techniques. These contractions involve mostly the subendometrial layer of the myometrium and may be detected only by endometrial movements (Birnholz, 1984). Following their first description they have been further characterized (Oike et al., 1988; De Vries et al., 1990; Lyons et al., 1991; Fukuda and Fukuda, 1994). The contractions increase in frequency and in intensity as the follicular phase progresses with an inverse pattern during the luteal phase. The peristaltic waves of the endometrium and the subendometrial layer of the myometrium are directed from the cervical canal to the fundal part of the uterus, while only during menstruation do they exhibit a fundo-cervical direction (Lyons et al., 1991).

In our own study (Kunz et al., 1996a; Leyendecker et al.,1996), with measurements of the uterine peristalsis during the menstrual period, the early, mid- and late follicular as well as mid-luteal phases of the cycle, respectively (Fig. 1) it was demonstrated that there was a steady increase in peristaltic activity ranging from roughly 1.2 contraction per minute during the menstrual period and early follicular phase to 2.8 contractions per minute in the late follicular phase. During the mid-follicular and mid-luteal phases, respectively, the frequency averaged 1.5 contractions per min. Over the same time period, the proportion of fundo-cervical contraction waves decreased significantly from 43% during the menstrual period to less than 1% in the periovulatory phase. Thus, almost all peristaltic

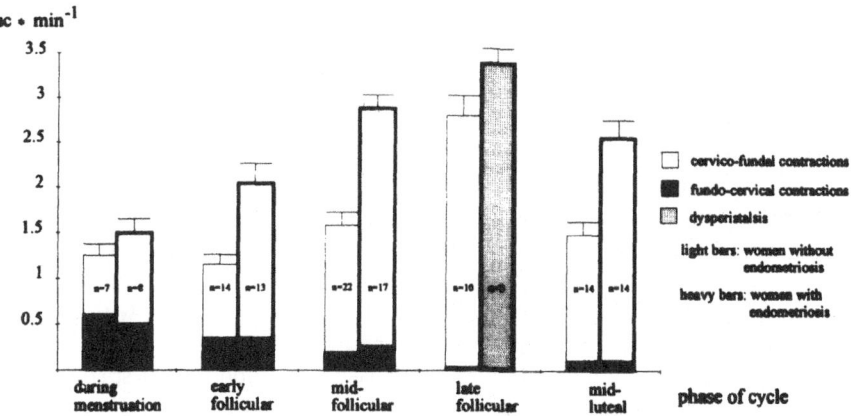

Figure 1. A graphical demonstration of the frequency of the subendometrial uterine peristaltic waves during menstruation, the early, mid- and late follicular and mid-luteal phases of the cycle, respectively, as determined by vaginal ultrasonography (contractions/min ± SEM). The graph shows also the relative distribution of fundo-cervical contractions versus cervico-fundal contractions during these different phases of the cycle (from Leyendecker et al., 1996).

waves of the uterus during the late follicular phase had a cervico-fundal direction. They also appeared to be more intense than during the other parts of the follicular phase, which might, however, be related to the thickness of the endometrium rendering the movements more pronounced. In comparison to the early follicular phase, however, a thicker proportion of the myometrium appeared also to be involved in the contractions. Because the frequency, intensity and direction of the uterine peristalsis depend upon the phase of the cycle, an endocrine control of this phenomenon by the ovary may be assumed. In this regard oxytocin and prostaglandins may function as mediators (Eliasson and Posse, 1960; Hein et al., 1973; Karim et al., 1973; Fuchs et al., 1985; Lefebvre et al., 1994a, Lefebvre et al., 1994b). This view is supported by the finding that administration of oestradiol valeriate to hypogonadal women yielding a pattern of serum oestradiol values similar to that of the normal cycle could completely mimic the cyclic changes of uterine peristalsis and that the frequency of the uterine peristaltic contractions could be significantly increased during the follicular phase of the cycle by the administration of an i.v. bolus of oxytocin. The increase in the frequency of the peristaltic contractions could be totally attributed to the peristaltic waves with cervico-fundal direction, which may be related to the high density of oxytocin receptors in the cervical tissue (unpublished).

Peristaltic contractions with the same frequency as described above were also observed with transvesical scanning (Birnholz, 1984). Thus, it is very unlikely that the uterine peristalsis was induced by the vaginal ultrasound examination. In contrast, it can be assumed that uterine peristalsis during the menstrual cycle is a continuous phenomenon with varying frequency, intensity and direction of the contraction waves depending on the phase of the cycle and does not require a specific stimulus for initiation. These studies, however, do not exclude a coital enhancement of uterine contractions.

3.2. Hysterosalpingoscintigraphy (HSSG)

It is reasonable to assume that the uterine peristaltic activity provides the forces that are required for the transport of spermatozoa from the external os of the cervix into the tubes within minutes. Using hysterosalpingoscintigraphy (Itturalde and Venter, 1981; Becker et al., 1988), rapid sperm transport was studied by placing technitium-labelled albumin macrospheres of sperm size at the external os of the uterine cervix and following their path through the female genital tract (Kunz et al., 1996a; Leyendecker et al., 1996).The albumin macrospheres used in our study resemble spermatozoa in their size. Thus, the demonstration of their ascension through the genital was considered to represent passive sperm transport.

According to these data (Fig. 2–4) the following concept of the dynamics of rapid sperm ascension within the female genital tract was proposed. Rapid sperm ascension occurs immediately following deposition of the ejaculate at the external os of the cervix. As early as one minute thereafter spermatozoa have reached the intramural and isthmical part of the tube. Quantitatively, however, the extent of ascension increases with the progression of the follicular phase. While only a few spermatozoa enter the uterine cavity and even fewer the tubes during the early follicular phase, the proportion of spermatozoa that enters the uterine cavity increases dramatically during the mid-follicular phase with still a limited entry into the tube. During the late follicular phase there is a considerable ascension of spermatozoa into the tubes.

Furthermore the HSSG revealed the preferential direction of rapid sperm transport into the tube ipsilateral to the dominant follicle. This corresponds with recent findings during surgery that the number of sperm around ovulation was higher in the tube ipsilat-

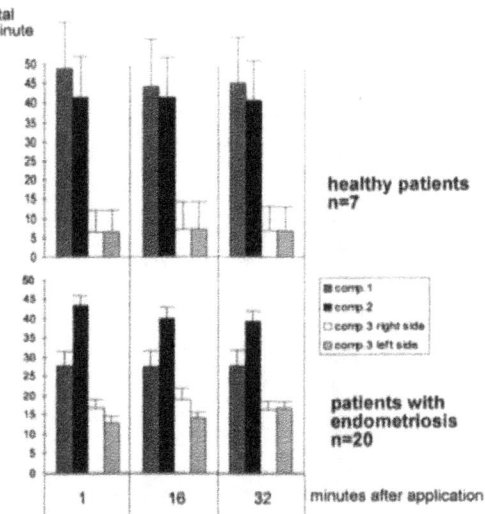

Figure 2. The distribution of the percentage of total counts, representing the labeled albumin macrospheres, within the female genital tract (compartment 1, 2 and 3 being the upper vagina, the uterine cavity and the isthmical part of the tubes respectively) following 1, 16 and 32 minutes after vaginal application during the early follicular phase. With respect to compartment 3, the right and left tubes were differentiated. The amount of radioactivity transported into the tubes was significantly higher in patients with endometriosis in comparison with healthy controls (P< 0.01) (from Leyendecker et al., 1996).

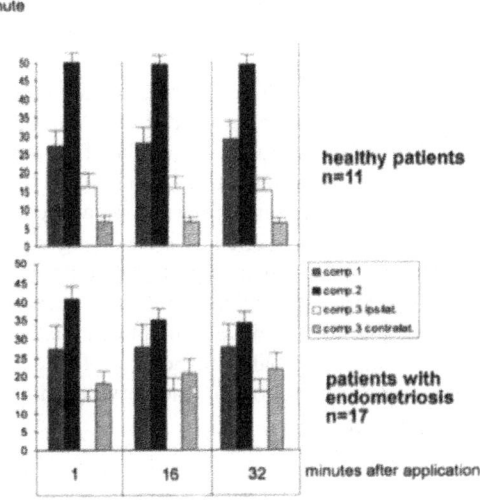

Figure 3. The respective distribution pattern of radioactivity, as in Fig. 2, obtained during the mid-follicular phase of the cycle. While in patients without endometriosis the labeled macrospheres preferentially entered the tube ipsilateral to the dominant follicle, in patients with endometriosis the macrospheres preferentially entered the tube contralateral to the dominant follicle. The difference in ascension into the contralateral tube between the two groups of patients was significant (P<0.01) (from Leyendecker et al., 1996).

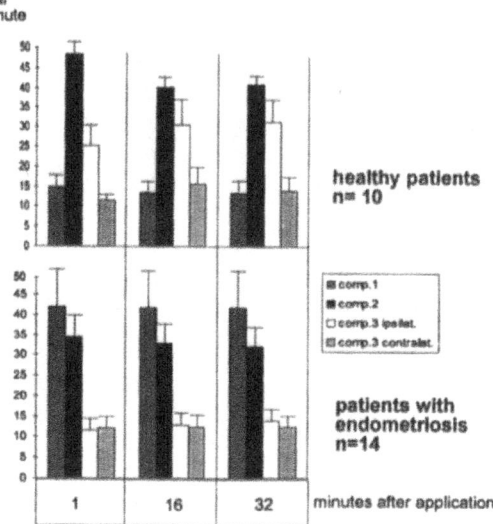

Percent of total
counts per minute

healthy patients
n= 10

□ comp.1
■ comp.2
□ comp.3 ipsilat.
▨ comp.3 contralat.

patients with
endometriosis
n=14

minutes after application

Figure 4. The distribution pattern of macrospheres, as in Fig. 2, during the late follicular phase. While the macrospheres preferentially entered the dominant ipsilateral tube in healthy women, the dysperistalsis observed in patients with endometriosis during this stage of the follicular phase resulted in a break down of transport capacity, leading to a distribution pattern of the macrospheres that resembled the pattern for normal women during the early follicular phase (Fig. 2; upper panel) (from Leyendecker et al., 1996).

eral to the dominant follicle than on the other side (Williams et al., 1993). This directed passive ascension of sperm (macrospheres) into the "dominant" tube begins already during the mid-follicular phase as soon as a dominant follicle can be demonstrated (Fig. 3; upper panel).

In these normal females with proven tubal patency there were no indications that the albumin macrospheres entered the peritoneal cavity to a large extent. In contrast, they were rather retained in the isthmical part of the tube. This arrest of passive sperm transport in the isthmical part of the tube appears to be of significance in the physiological process of reproduction (Harper, 1994). Electronmicroscopical studies have demonstrated significant morphological and, by inference, functional changes of the tubal epithelium during the menstrual cycle (Jansen, 1980; Amso et al., 1994a; 1994b; Crow et al., 1994). Jansen (1980) first described the preovulatory appearance of secretory cells in the isthmical part of the tube being responsible for the development of an isthmical mucous plug during this phase of the cycle. This plug is probably responsible for the isthmo-tubal arrest of passive sperm ascension and may serve, following the cervical mucus and crypts, as a secondary tubal sperm reservoir (Harper, 1994).

4. DIRECTED SPERM ASCENSION AND SIDE-SPECIFIC ASYMMETRY OF MYOMETRIAL HORMONE RECEPTORS

Our studies with inert particles suggested that the directed ascension of sperm into the "dominant" tube is not a property of the sperm and is thus not provided by mechanisms such as chemotaxis but rather constitutes a specific utero-tubal function controlled by the dominant follicle in that the uterine myometrium with its specific architecture (Goerttler, 1930) is activated and contracts in a manner providing this directed transport. In order to elucidate the mechanisms that govern directed sperm ascension, the estrogen receptor distribution within the myometrium was studied. It could be demonstrated in removed uteri that the percentage of estrogen receptor positive nuclei of smooth muscle cells in the myometrium was significantly higher on the side of the dominant ovarian structure in comparison with the contralateral side (Kunz et al., 1996b).

The strong correlation between receptor distribution and the site of the dominant ovarian structure suggested that the side-specific asymmetric estrogen receptor distribution is induced by the dominant follicle. This preferential induction of estrogen receptors in the myometrium on the side of the dominant ovarian structure is presumably not effected by the oestradiol concentration in the systemic circulation. Rather, it may involve a more direct endocrine influence mediated, for example, by the utero-ovarian vascular counter-current system (Einer-Jensen, 1988) that may provide higher oestradiol concentrations in the uterus on the side of the dominant follicle. This finding may also indicate that other hormone receptors in the uterus, pertinent to its contractile function, might be expressed asymmetrically with respect to the localization of the dominant ovarian structure. Preliminary data show that oxytocin receptors, exhibiting high concentrations in cervical tissue, like those for oestradiol, are asymmetrically distributed within the fundal part of the myometrium relative to the side of the dominant follicle (unpublished).

The asymmetric estrogen receptor distribution being presumably responsible for the directed sperm transport appears to be superimposed on a basal, more evenly distributed level of estrogen receptors under the influence of systemic oestradiol. This may be derived from the observation that the systemic administration of exogenous estrogen to hypogonadal women results in a uterine peristaltic activity which is, according to VSUP, indistinguishable from normal

5. UTERINE HYPERPERISTALSIS AND DYSPERISTALSIS

In studying infertile women suffering from mostly mild endometriosis, which is considered not to be a cause of infertility (Hull et al., 1986; Adamson and Pasta, 1994), a fundamental disturbance of uterine peristaltic activity was observed (Leyendecker et al., 1996) (Fig 1). In VSUP, these women displayed a considerable degree of hyperperistalsis in that, during the early and mid follicular as well as mid-luteal phase of the cycle, respectively, the frequency of peristaltic contraction was nearly doubled in comparison to normal. During the late follicular phase the frequency was increased further but less pronounced as compared to the early and mid-follicular phase of the cycle. The character of the peristaltic activity, however, had completely changed. While in the fertile controls long and regular cervico-fundal peristaltic waves prevailed, the contractions displayed a convulsive character in the infertile women. Some of the contraction waves started in the middle portion of the uterine cavity, while in other patients the contractions started at different sites at the same time, and some vanished before reaching the fundal part of the uterine cavity. Thus, in comparison with the regular and frequent cervico-fundal contractions of healthy women, the impression of a dysperistalsis prevailed in patients with infertility and endometriosis (Leyendecker and Kunz, 1996).

These abnormalities of the uterine peristaltic activity in women with endometriosis had a profound impact on the uterine transport function as demonstrated by HSSG, which, at least in part, may account for the infertility of these women. Already during the early follicular phase of the cycle, there was a rapid transport of inert particles through the uterine cavity into the tubes. This was further increased during the mid follicular phase. However, there was no directed transport into the tube ipsilateral to the dominant follicle. During the late follicular phase, when the uterine contractions had become dysperistaltic, a breakdown of the uterine transport function occurred in that most of the particles remained at the site of application and only a few entered the tubes without a preference for the "dominant" one (Fig. 2–4).

6. IMPLICATIONS REGARDING THE FUNCTION OF THE CERVICAL MUCUS

These data are also pertinent in reconsidering some of the functions of the cervical mucus with regard to sperm ascension.. It is generally assumed that the sperm actively penetrate the cervical mucus and that the scant and viscous cervical mucus of the early follicular phase acts as a barrier in this respect (Moghissi, 1977). The HSSG demonstrates that already in the early follicular phase, in the presence of scant cervical mucus with little spinnbarkeit, a rapid transport of inert particles through the cervical canal occurs (Fig 2). Moreover, the distribution pattern of the labeled macrospheres within the genital tract of women with endometriosis and hyperperistalsis in the early follicular phase (Fig. 2; lower panel) resembles that of the healthy controls with normoperistalsis in the mid follicular phase of the cycle (Fig. 3; upper panel). Thus, it is not so much the quality of the cervical mucus but rather the power of the uterine peristalsis that determines the amount of sperm ascension through the cervical canal.

According to in vitro studies sperm penetrate the cervical mucus at a speed of 0.1 to 3 mm/min depending upon the phase of the cycle. There is no doubt on the basis of our studies that the sperm's own velocity is of little importance with respect to the ascension through the cervical canal. Irrespective of whether there is a function at all to the sperm's active movement at this stage of reproduction, the interaction of the sperm with the physico-chemical properties of the mucus enable viable sperm to enter the cervical crypts as a primary reservoir for later release. Of course, our model using inert albumin macrospheres cannot account for effects that have to be attributed to the functional capacity of healthy sperm.

7. AN INTEGRAL VIEW ON SPERM ASCENSION WITHIN THE FEMALE GENITAL TRACT AND ITS POSSIBLE DISTURBANCES

The clinician has been familiar for a long time with rhythmical contractions of the non-pregnant uterus. Upon cervical inspection during the preovulatory phase rhythmical protrusions of the abundant cervical mucus can be observed. Only recently, due to high resolution ultrasound examination, it was possible to relate these cervical activities to uterine peristaltic waves that originate in the cervix and are propagated towards the cornual section of the uterus. With the placement of labeled inert particles of sperm size at the external cervical os and following their path through the female genital tract by HSSG it was possible to demonstrate the enormous power and transport capacity of this uterine peristaltic pump. Furthermore, the directed transport of the particles preferentially into the tube ipsilateral to the dominant follicle demonstrated the surprising sophistication of this uterine system of sperm transport (Kunz et al., 1996a).

On the basis of the data obtained in our studies and available from literature, we hypothesize that rapid as well as sustained sperm is controlled by uterine peristaltic activity. Uterine contractions aspirate sperm into the cervical mucus and the uterine cavity, and provide further transport into the isthmical part of the tubes. In the mid- and late follicular phases of the cycle this transport is directed preferentially into the tube ipsilateral to the dominant follicle. This indicates that the mechanism of rapid and passive sperm transport is under the endocrine control of the dominant follicle. Some sperm, probably the most motile ones, follow, by their own movement, the filamentous structures of the cervical mucus and enter the cervical crypts as a primary reservoir. This results in a partial sequestration of the sperm increasing the proportion of less motile and immobile sperm that reach the tubes rapidly. This observation has probably lead to the notion that rapid sperm ascension might not be essential for fertilization (Mortimer, 1983; Hunter, 1987). With the progression of the follicular phase there is an increasing release of sperm from the primary reservoir as they are flushed and squeezed out of the crypts due to the cervical secretion which becomes more profuse and the rhythmical contractions that originate within the cervix, respectively. Entering the "main stream" of cervical secretion they are caught by the "uterine peristaltic system" and rapidly transported in an aliquot of mucus (Fukuda and Fukuda, 1994), which protects the sperm from leukocyte degradation within the uterine cavity (Harper, 1994), to the tube with its isthmic mucus as the secondary reservoir. Dilatation of the external cervical os, maximum cervical secretion and rhythmical protrusion of the mucus around ovulation enlarge the zone of contact between a fresh ejaculate and the uterine peristaltic pump. At the same time preovulatory mucorrhea, together with the rhythmical contractions of the cervix, prevents by large motile sperm from entering the cervical crypts and thus ensures, in combination with maximally increased uterine peristalsis, that no or only minor sequestration of sperm can occur and that motile sperm are directly transported into the isthmical mucus (Jansen, 1980) of the tube ipsilateral to the dominant follicle where they are available for fertilization.

There is, in our opinion, no principle difference between the mechanisms of rapid and sustained sperm transport, respectively. Both aim at the availability of viable sperm at the site of fertilization around ovulation and both rely, in this respect, on the continuous peristaltic activity of the uterus. Sustained sperm transport utilizes the cervical crypts as a primary reservoir, from where later release occurs. The reduced power of the peristaltic pump several days prior to ovulation in comparison to the preovulatory phase might, together with a more viscous cervical mucus at this time, facilitate the migration of motile sperm into the cervical crypts. No data are available, which of the two reservoirs, the cervical or the tubal mucus, have a preponderance in the function of sperm preservation. If there is any preponderance at all, one may assume that it may shift from the cervix to the tube with the progression of the preovulatory phase. In any event, the fundamental importance of sperm preservation within the genital tract and sustained sperm transport for the overall process of reproduction is documented by the observation that intercourse several days prior to ovulation may result in a pregnancy with a considerable probability ranging from about 10% with intercourse five to more than 30% with intercourse two days prior to ovulation, respectively (Wilcox et al., 1995).

These data and considerations show that the availability of sperm at the site of fertilization at the appropriate time depends to a large extent on coordinated uterine peristaltic contractions that cyclically change in quality and frequency. At a low frequency and power, they may favor sperm preservation within the cervical mucus, at a higher preovulatory frequency and power of contractions, the uterine peristaltic pump provides rapid and directed transport of sperm either from the reservoir or from a freshly deposited ejaculate into the tube ipsilateral to the dominant follicle.

Recently, it could be shown that this fine-tuned system is fundamentally disturbed in women with infertility and, mostly, mild endometriosis. Both, the hyper- and dysperistalsis of the uterine peristaltic pump observed in these women may contribute to their reduced fertility. Hyperperistalsis may prevent the development of an adequate pool of preserved sperm within the reservoir of the cervical crypts, and preovulatory dysperistalsis impedes, by a breakdown of sperm transport, the formation of an adequate sperm reservoir in the mucus of the isthmical part of the tube from where sperm migrate to the final site of fertilization.

Independent of its effects on sperm transport and fertility uterine hyperperistalsis may promote the detachment and exfoliation of endometrial cells and tissue fragments and their transtubal transport into the peritoneal cavity and may, therefore, propagate the development of endometriosis (Leyendecker et al., 1996).

8. CONCLUSIONS

Uterine peristalsis is of fundamental importance in the process of reproduction in that it serves sperm transport from the external os of the cervix to the mucus of the isthmical part of the tube ipsilateral to the dominant follicle. This mechanism is controlled by the dominant follicle. This newly disclosed and described uterine function is of clinical importance in that a dysfunction of this functional system may result in infertility and may propagate the development of endometriosis.

9. REFERENCES

Abramovicz JS & Archer DF 1990 Uterine endometrial peristalsis - a transvaginal ultrasound study. Fertility and Sterility 54:451–454

Adamson GD & Pasta DJ 1994 Surgical treatment of endometriosis-associated infertility: Meta analysis compared with survival analysis. American Journal of Obstetrics and Gynecology 171: 1488–1505

Amso NN, Crow J & Shaw RW 1994a Comparative immunohistochemical study of oestrogen and progesterone receptors in the fallopian tube and uterus at different stages of the menstrual cycle and the menopause. Human Reproduction 9: 1027 - 1037

Amso NN, Crow J, Lewin J & Shaw RW 1994b A comparative morphological and ultrastructural study of endometrial gland and fallopian tube epithelia at different stages of the menstrual cycle and the menopause. Human Reproduction 9: 2234 - 2241

Becker W, Steck T, Alber P & Borne W 1988 Hystero-salpingo-scinitraphy: A simple and accurate method of evulating fallopian tube patency. Nuklearmedizin 27: 252–257

Birnholz J. (1984) Ultrasonoc visualization of endometrial movements. Fertility and Sterility 41:157–158

Cibils L.A. (1967) Contractility of the non-pregnant human uterus. Obstetrics and Gynecology. 30: 441 - 461

Crow J, Amso NN, Lewin J & Shaw RW 1994 Morphology and ultrastructure of fallopian tube epithelium at different stages of the menstrual cycle and the menopause. Human Reproduction 9: 2224 - 2233

De Vries K, Lyons EA, Ballard G, Levi CS & Lindsay DJ 1990 Contractions of the inner third of the myometrium. Amercan Journal of Obstetrics and Gynecology 162:679–682

Drobnis EZ & Overstreet JW 1992 Natural history of mammalian spermatozoa in the female reproductive tract. Oxford Review of Reproductive Biology 14: 1–46

Einer-Jensen N 1988. Countercurrent transfer in the ovarian pedicle and it's physiological implications. Oxford Reviews of Reproductive Biology 10:348–381

Eliasson R & Posse N 1960 The effect of prostaglandins on the nonpregnant uterus in vivo. Acta Obstetrica et Gynecologica Scandinavica 39: 112–117

Fuchs A-R, Fuchs F & Soloff M.S 1985 Oxytocin receptors in nonpregnant human uterus. Journal of clinical Endocrinology and Metabolism 60: 37–41

Fukuda M & Fukuda K 1994 Uterine endometrial cavity movement and cervical mucus. Human Reproduction 9:1013–1016

Goerttler K 1930 Die Architektur der Muskelwand des menschlichen Uterus und ihre funktionelle Bedeutung. Gegenbaurs Morphologisches Jahrbuch 65:45–52

Harper MJK 1994 Gamete and zygote transport. p123–187 In The physiology of reproduction. Eds Knobil E. & Neill, JD Raven Press, New York,

Hartman CG 1962 Science and the safe period. A compendium of human reproduction. Williams and Wilkins, Baltimore

Hein PR, Eskes TKAB, Stolte LAM, Braaksma JF, Jansens J & Hoek JM 1973 The influence of steroids on uterine motility in the nonpregnant human uterus. p107–140 In Uterine contractions - side effects of steroidal contraceptives. Ed Josimovich, J.B. New York: Wiley and Sons

Hendricks CH 1966 Inherent motility patterns and response characteristics in the nonpregnant uterus. Amercan Journal of Obstetrics and Gynecology 96: 824–841

Hull ME, Moghissi KS, Magyar DF, & Hayes MF 1986 Comparison of different treatment modalities of endometriosis in infertile women. Fertility and Sterility 47: 40–44

Hunter RHF 1987 Human fertilization in vivo with special reference to progression, storage and release of competent spermatozoa. Human Reproduction 2: 329–332

Iturralde M & Venter PP 1981 Hysterosalpingo-radionuclide scintigraphy. Seminars in Nuclear Medicine 11: 301 - 314

Jansen RPS 1980 Cyclic changes in the human fallopian tube isthmus and their functional importance. American Journal of Obstetrics and Gynecology 136: 292–308

Karim SMM & Hillier K 1973 The role of prostaglandins in myometrial contraction. p141–169 In Uterine contraction - side effects of steroidal contraceptives. Ed Josimovich JB New York: Wiley and Sons

Kunz G & Leyendecker G 1996 Uterine peristalsis throughout the menstrual cycle. Physiological and pathophysiological aspects. Human Reproduction update (in press)

Kunz G, Beil D, Deininger H, Wildt L & Leyendecker G 1996a The dynamics of rapid sperm transport through the female genital tract. Evidence from vaginal sonography of uterine peristalsis and hysterosalpingoscintigraphy Human Reproduction 11: 627–632

Kunz G. Mall G & Leyendecker G 1996b Asymmetry of oestrogen receptor expression in the myometrium relative to the site of the dominant follicle and corpus luteum during the menstrual cycle in healthy women. Human Reproduction 11: Abstract book 1, p 158

Leyendecker G, Kunz G, Wildt L, Beil D & Deiniger H 1996 Uterine hyperperistalsis and dysperistalsis as dysfunctions of the mechanism of rapid sperm transport in patients with endometriosis and infertility. Human Reproduction, 11: 1542–1551

Lefebvre DL, Farookhi R, Larcher A, Neculcea J & Zingg HH 1994a Uterine oxytocin gene expression. I. Induction during pseudopregnancy and the estrous cycle. Endocrinology 134: 2556–2561

Lefebvre DL, Farookhi R, Giaid A, Neculcea J & Zingg HH 1994b Uterine oxytocin gene expression. II. Induction by exogenous steroid administration. Endocrinology 134: 2562–2566

Lyons EA, Taylor PJ, Zheng, XH, Ballard G, Levi CS & Kredentser JV 1991 Characterization of subendometrial myometrial contractions throughout the menstrual cycle in normal fertile women. Fertility and Sterility 55:771–775

Martinez-Gaudio M, Yoshida T & Bentsson LP 1973 Propagated and non propgated myometrial contractions in normal menstrual cycles. American Journal of Obstetrics and Gynecology 115: 107–111

Moghissi KS 1977 Sperm migration through the human cervix. In The uterine cervix in reproduction p146–165. Eds Insler V & Bettendorf G. Georg Thieme Publishers Stuttgart

Mortimer D 1983 Sperm transport in the human female reproductive tract. Oxford Reviews of Reproductive Biology 5: 40–61

Oike K, Obata S, Tagaki K, Matsuo K, Ishihara K & Kikuchi S 1988 Observation of endometrial movement with transvaginal sonography. Journal of Ultrasound in Medicine 7: 99

Romanini C 1994 Measurement of uterine contractions. P337–355 In The uterus. Eds Chard J & Grudzinskas JG Cambridge Reviews in Reproduction, Cambridge University Press

Williams M, Hill C.J, Scudamore I, Dunphy B, Cooke ID & Barratt CLR 1993 Sperm numbers and distribution within the human fallopian tube around ovulation. Human Reproduction 8: 2019–2026

Wilcox AJ, Weinberg CR & Baird DD 1995 Timing of sexual intercourse in relation to ovulation - effects on the probability of conception, survival of the pregnancy, and sex of the baby. New England Journal of Medicine 333: 1517–1521

50

EGG-CUMULUS-OVIDUCT INTERACTIONS AND FERTILIZATION

R. Einspanier, B. Lauer, C. Gabler, M. Kamhuber, and D. Schams

Institute of Physiology
FML Weihenstephan - TU Munich
85350 Freising, Germany

1. ABSTRACT

In this communication we approach the events leading to fertilization in mammals by examining the triangle of egg, sperm and oviductal cell taking account of the local physiology and focussing on auto/paracrine interactions. The expression of growth factors and extra-cellular matrix (ECM) - components in bovine ovarian granulosa- and theca-cells, the oocyte-cumulus complex (OOC) and oviductal epithelium, as well as some of the corresponding secreted proteins can be detected through the estrous cycle. Components of the insulin-like (IGF), fibroblast (FGF) and transforming (TGF) growth factor systems, and also metalloproteinase 1 (MMP1) and urokinase (uPA) are found to be modulated in these cells prior to fertilization. Different expression levels between the cell types are found, each representative of a specific reaction window within that particular stage of the cycle. Our findings support the concept that most of the observed tissue in the reproductive tract is dependent upon on the effects of gonadotropins or steroids, but that the fine-regulation is conveyed by, for example, growth factors and ECM-components. We suggest a sophisticated, auto/paracrine and species-specific crosstalk of growth factors and ECM components between the different cell types involved, enabling fertilization and development of the embryo at the right time and in the right location.

2. INTRODUCTION

The event of fertilization in mammals has to be preceded by different maturation processes of the gametes and of the contributing tissues, such as ovarian and oviductal cells. Here interactions between the oocyte and the cumulus / oviductal cells will be highlighted and in vitro data concerning sperm-oviduct interaction are presented. The study focusses on details elaborated in the bovine system: starting at the ovary, the site of oocyte maturation, and ending in the oviduct, where fertilization takes place. The actors of this

The Fate of the Male Germ Cell, edited by Ivell and Holstein
Plenum Press, New York, 1997

play are: the ovarian follicle, where the oocyte matures; the oocyte, surrounded by specialised granulosa cells, the so called cumulus; the oviduct, where fertilization occurs, and the
ejaculated sperm. Important interactions are suggested between these compartments under
the control of the gonadotropins and steroids, which finally lead to a new individual, the
embryo. Our main interest will be the triangle between the oocyte-cumulus-complex
(OCC), oviductal cells, and sperm just before fertilization. Maturation of the oocyte in the
ovarian follicles and the capacitation of ejaculated sperm have to precede the process of
fertilization.

3. KNOWN ASPECTS OF THE OOCYTE, CUMULUS, FOLLICLE, OVIDUCT AND THEIR INTERACTIONS

In recent years the fertilizing process in mammals has attracted much scientific interest, helped by the rapid development of new powerful molecular techniques. In the following, data are described on the supporting components which contribute to successful
fertilization. It should be emphasized that many of the interactions mentioned here are
thought to be species-specific.

3.1. The Oviduct

The oviduct as the place where the gametes meet is described as an organ of limited
storage, and final capacitation of sperm in the cow (Lefebvre et al.; 1995). Tissue-specific
proteins have been found in the oviductal fluid: the structures of estrogen-dependent glycoproteins have been structurally analysed for the human, baboon, sheep and cattle (Arias
et al.; 1994, Donnelly et al.; 1991, DeSouza & Murray; 1995, Sendai et al.; 1994). Secreted oviductal proteins, like oviductin, are known to associate with the hamster oocyte
(Robitaille et al.; 1988, Boatman et al.; 1994) and sustain the viability of bovine sperm
(Satoh et al.; 1995). Similarly, other distinct proteins adhere to bovine sperm during their
in vitro capacitation (McNutt et al.; 1992). Furthermore, important embryotrophic factors
secreted by the mammalian oviduct have been described by some authors (Gandolfi et al.;
1989, Noda et al.; 1993, Morishige et al.; 1993). Mitogens, such as the growth factors, are
found in variable amounts in an estrous cycle-dependent manner (Schmidt et al.; 1994). A
detailed review on the bovine oviduct as an important reproductive organ, is provided by
Ellington (1991).

3.2. The Oocyte

The maturation of the oocyte is a complex hormone-dependent process which takes
place largely within the ovary. The selection of follicles to become a preovulatory or
Graafian follicle, containing a mature oocyte, is still not clearly understood. The development of the growing follicle and the surrounding cumulus cells is suggested to be trig-
gered by endocrine stimuli in addition to the presence of the oocyte (Tirone et al., 1993);
and a cumulus-expansion enabling factor has been isolated from mouse oocytes (Eppig et
al.; 1993). The oocytes of most species represent a large reservoir of mRNA, e.g. encoding ZP3 in the mouse, but its concentration declines rapidly during the course of final
maturation up to fertilization (Roller et al.; 1989). During this period the egg is always
surrounded by the cumulus cell layer, which comprises specialised granulosa cells, and
which is retained by the oocyte after ovulation. Many of the detailed events involved here

remain unknown, especially mechanisms leading to the selection of the dominant oocyte and the removal of its meiotic block.

3.3. The Cumulus

The cumulus cell is derived from ovarian granulosa cells, but its main histological distinction is the direct contact with the oocyte. It is presumed that these cells are essential for the maturation of the oocyte, since it is known that a denuded oocyte will not be fertilized (Lorenzo et al.; 1995) and that the presence of cumulus or even normal granulosa cells will enhance in vitro fertilization (Fukui; 1990, Sirard; 1990). Gonadotropins have a marked influence on the development and priming of the cumulus: cumulus cell proliferation or expansion will occur in vivo after administration of LH, but in vitro only after FSH (Chen et al.; 1994). Downs and coworker (1988) could show for the mouse that this FSH effect is exclusively mediated by the cumulus cell, presumably due to distinct receptors for LH and FSH on the cumulus, granulosa or theca cells.

Likewise, different growth factors have been found to be expressed in the cumulus: TGFα in the mouse (Brucker et al.; 1991), EGF and TGFß in the porcine (Singh et al.; 1993) and bFGF in the human (Watson et al.; 1992). It is still unclear if and how long the cumulus surrounds the bovine oocyte after ovulation (Lorton & First; 1979, Brackett et al.; 1980), and whether its mitotic activity is blocked in the oviduct (Schuetz et al.; 1989). A cumulus cell layer persisting after ovulation could represent a barrier for the spermatozoon, or alternatively form a terminal sperm selection device as has been discussed elsewhere (Bedford & Kim; 1993).

3.4. The Ovarian Follicle Cells

For most of the time, the mammalian oocyte remains in a quiescent stage within the ovarian follicle and during this time, the surrounding tissue provides all the necessary nutrition. Oocyte maturation is completed within the follicle. The activation of the oocyte is modulated through the influence of endocrine and auto/paracrine effectors (hormones, growth factors) (for a review see Eppig; 1991). For most of its time in the follicle the oocyte is subject to meiotic arrest. Two main cell types are suggested to control this process: the granulosa and theca cell. Important interactions between the granulosa and theca layer are known, and recently Kotsuji et al. (1994) demonstrated that the meiotic block of the oocyte in the follicle is augmented by the granulosa cell under control of the theca. Different methods of in vitro maturation (IVM) have been introduced but these appear to be accompanied by asynchrous development of the oocyte, at least in the cow (Sundstrom et al.; 1988). Nevertheless, these techniques are of major interest to facilitate assisted reproduction in animal breeding and medicine.The following details some of our own findings in this context using the bovine system.

4. EXPERIMENTAL EVIDENCE

4.1. Oocyte Maturation

The maturation of the oocyte starts in the selected follicle under the endocrine control of FSH/LH. We have tried to correlate the appearance of selected growth factors with the stage of development of the ovarian follicle. The separated theca and the granulosa

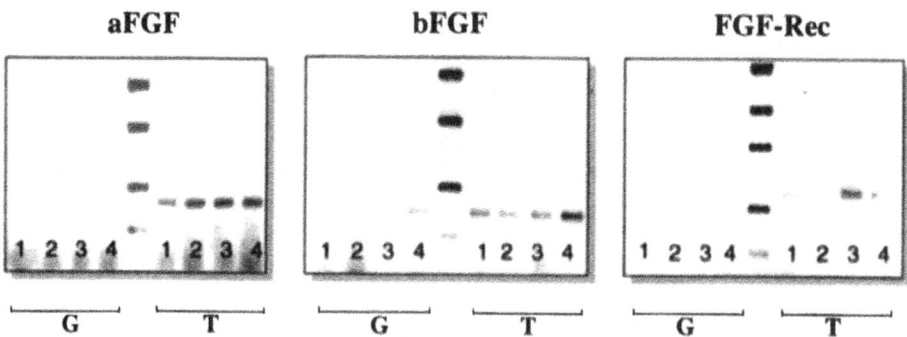

Figure 1. Detection of mRNAs of subdivided granulosa (G) and theca (T) cells in bovine ovarian follicles. Follicle classes are: 1= <0.5 , 2= 0.5–5, 3= 6–20, 4= >20 ng estradiol / ml. Specific PCR products (aFGF 28x cycles, bFGF 30x cycles, FGF-receptor 30x cycles) are separated on 1% agarose gels.

cell layers were of major interest and after determining the developmental stage of each individual follicle based on the estradiol- concentration in the follicular fluid, total RNA was isolated from each cell type. Detection of the distinct component gene products for the FGF - system in these cells is shown in Figure 1.

A remarkable difference between granulosa (G; low expression) and theca cells (T; higher expression) can be detected during follicle growth: levels for a/bFGF and their receptor increase only slightly in the granulosa just before ovulation. The presence of the native proteins in the different compartments was confirmed by immunohistochemistry (data not shown), supporting the PCR-data. From these and other unpublished data we know that granulosa cells are different but dependent upon theca cells. Furthermore, for bFGF differences between the cumulus and granulosa cells could be detected for bFGF: the cumulus expresses high amounts of this protein. During in vitro maturation (IVM) of bovine OCC a time-dependent increase of bFGF is evident. The expression of bFGF during cumulus expansion is one of the earliest responses to the FSH-surge and is then followed by elevated levels of matrixmetalloproteinase 1 (MMP1) and its inhibitor TIMP1. A typical result during a standard 24 h IVM is depicted in Figure 2.

Further experiments indicate that bFGF is a potent initiator of cumulus expansion in vitro. Immunohistochemistry of maturated OCC shows that bFGF-specific staining is clearly located in the cumulus cells surrounding the oocyte (data not shown). From these experiments we deduce that bFGF may be an important auto/ paracrine factor for the maturation of bovine oocytes. Although complete cumulus expansion is known to be essential to ensure high fertility rates in vitro, confusion exists whether the cumulus remains stable after ovulation in vivo.Initial results using an oviductal co-culture system, suggests that the oocyte is not necessarily denuded directly after ovulation and thus interactions between the cumulus and epithelial cells lining the oviduct can be predicted. As yet we know nothing about the fate of the cumulus after being sheared away from the oocyte within the oviduct.

4.2. The Oviductal Environment

The oviductal epithelial cell layer forms the next microenvironment for the ovulated oocyte. This environment must of necessity be very conducive for fertilization and early

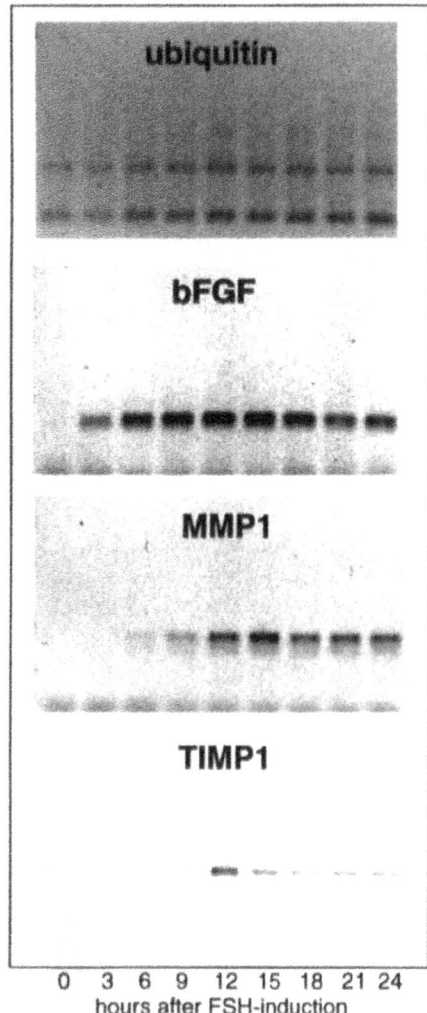

Figure 2. Gene expression during in vitro maturation of bovine oocyte cumulus complexes. PCR products for ubiquitin (quality and quantity control), bFGF (28x cycles), MMP1 (38x cycles) and TIMP1 (30x cycles) are separated on 1% agarose gels.

embryonic development. From recent research we know that a cycle-dependant regulation of secretory products appear to be quite prominent in primates, though less evident in the cow (Ellington; 1991, Schmidt et al.; 1994, Gabler & Einspanier; 1995, Einspanier et al.; 1995). As shown by RNA-quantification of the FGF-system using RNAse Protection Assays (RPA) there is a pronounced regulation of aFGF against which bFGF and the receptor appear to be unmodulated in the oviductal epithelium (Fig. 3). Comparing the specific RNA concentrations in the cells with the levels of the secreted proteins, there appears to be a discrepancy (Fig. 4): a timeshift between the RNA and the soluble growth factor is evident. How might this be explained? Examination of some extracellular matrix proteinases, which are known to liberate the heparin-binding FGFs, indicated the following fine-regulation (Fig. 5): urokinase (uPA) and MMP1 as proteinases are cycle-dependently upregulated, but their inhibitors PAI1 and TIMP1 show only low steady state expression. Thus the proteinases could facilitate the liberation of glycoprotein-bound FGFs. In sum-

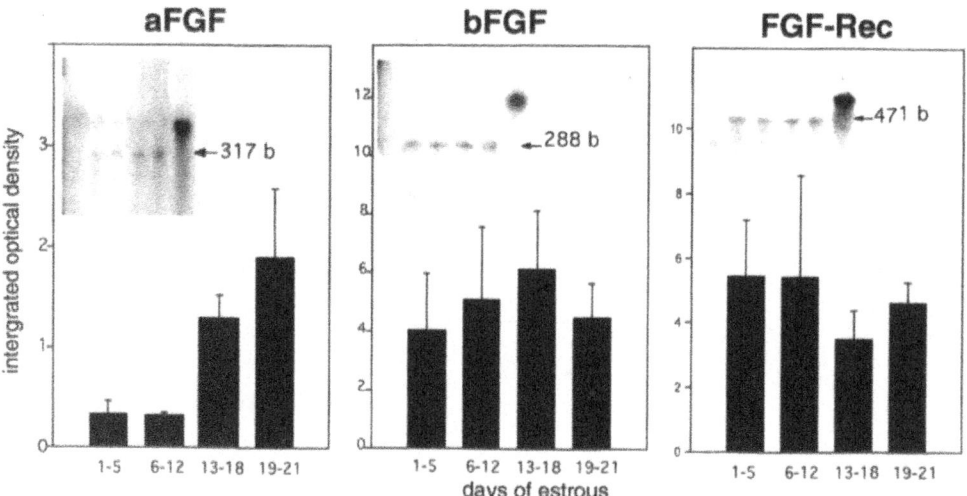

Figure 3. Expression of components of the FGF system in bovine oviducts obtained using RNAse protections assays. Four different cycle stages were analysed each representing five oviducts.

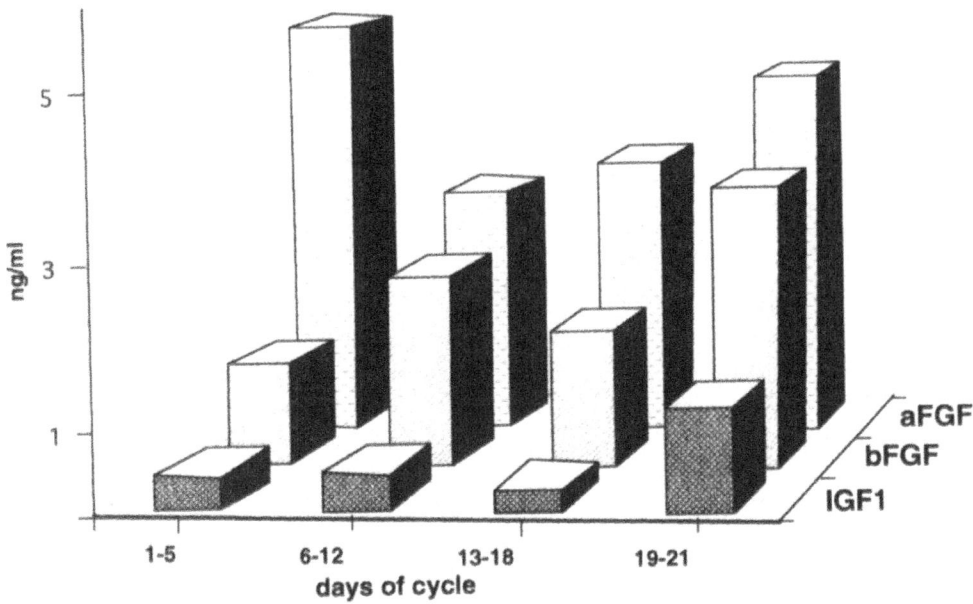

Figure 4. Growth factor protein concentrations in bovine oviductal fluid during the estrous cycle. Depicted are mean data from a minimum of 5 flushings each.

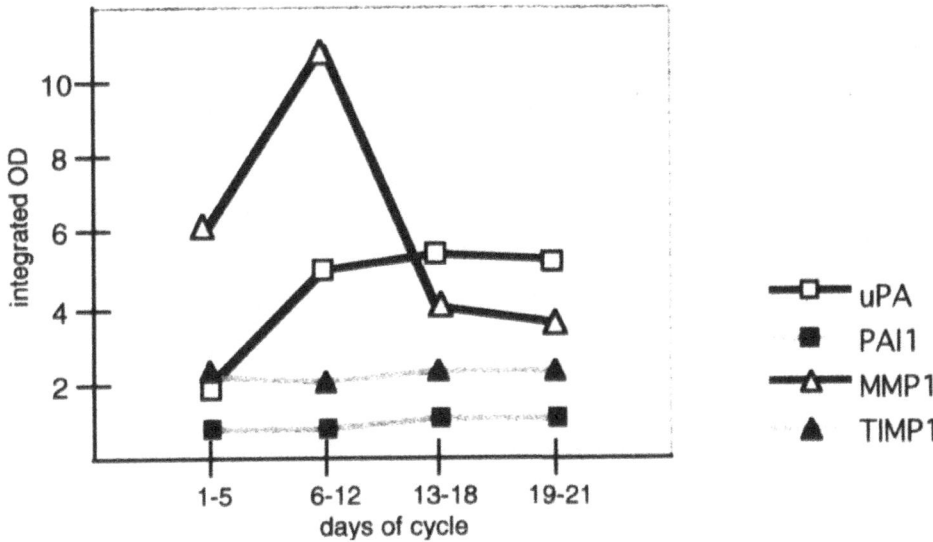

Figure 5. Expression of selected ECM-components in bovine oviducts during estrous cycle. PCR-product intensity was analysed from 5 individual samples (uPA=30x cycles, MMP1. TIMP1 and PAI1= 35x cycles).

mary, therefore, a functional network of ECM components regulating growth factor actions can be suggested within the oviduct. This could then influence the interactions of the gametes and the embryo in a cycle-dependant manner.

4.3. Oviduct-Sperm Interactions

Data will now be presented concerning the interaction between the oviduct and sperm. The binding of oviductal proteins to the sperm surface has been reported. To identify specific sperm- surface proteins which are liberated under oviductal influence, we have developed an oviduct-perfusion system simulating in vivo conditions. After application of biotin-labelled sperm a specific liberation of three surface proteins could be detected within the oviduct. These proteins were isolated and identified by N-terminal sequencing as the known bovine seminal proteins BSP1, 2 and 3 from oviductal flushings (Lauer et al.; 1995). Further experiments indicate that heparin-like structures in the bovine oviducts promote the removal of such proteins, possibly leading towards terminal capacitation (Lauer & Einspanier; 1996).

5. COMPARISION OF GROWTH FACTOR AND ECM EXPRESSION

Finally, when all available RNA-expression data from ovarian granulosa cells, the OCC and oviductal epithelium are surveyed, a complex network of differentially regulated growth factors and ECM components appears (Table 1). Many potent factors are found in all reproductive compartments investigated so far. Striking differences of expression levels were detected in the cumulus for a/bFGF, TGFß, TIMP1 and PAI1 and in the oviductal

Table 1. Relative expression strenght of different growth factors and ECM-components in cows reproductive tissue. Classes of expression levels are: (−) below detection limit, (±) low, (+) medium, (++) high. Superscript (m) indicates modulation of mRNA concentration; data are based on PCR - experiments

	Ovarian granulosa cells	Oocyte+cumulus	Oviductal cells
aFGF	−	−	$+^m$
bFGF	±	$++^m$	+
FGF-Res	$+^m$	$+^m$	++
IGF-1	$++^m$	+	$+^m$
IGF-2	±	$±^m$	±
IGF-Rec	$++^m$	+	$++^m$
TGFa	$±^m$	$±^m$	$+^m$
TGFb	$+^m$	$++^m$	$+^m$
MMP1	−	$±^m$	$±^m$
TIMP1	$±^m$	$++^m$	$+^m$
PAI	±	$++^m$	±

epithelium for aFGF and uPA. Most of these proteins are expressed in a time and cycle-dependent manner in bovine tissues. For bFGF pronounced differences were found between ovarian granulosa and cumulus cells, possibly mediated by FSH and suggesting in the bovine a different expression capacity of the cumulus. Auto/paracrine crosstalk between the mature OCC and the oviduct based on the growth factor bFGF and the protease-inhibitors PAI1 and TIMP1 can also be postulated.

6. FUTURE EXPERIMENTAL PROSPECTS

Modern molecular techiques offer interesting approaches for detecting new expression patterns initiated by acute cell-cell contact. Preliminary results using the RAP-PCR method point to the identity of new oviductal proteins which are expressed after exposure to the gametes. Following a random amplification of mRNAs from treated oviduct cells some differentially expressed molecules are present after sperm or OCC induction, respectively (Fig. 6). Some induced new oviductal mRNAs are provoked by both gametes, suggesting a rapid reaction time by the oviduct within hours. The structure and function of these new products will be of major interest in order to understand the changing metabolism of the oviduct during the cycle. Comparable experiments should also help to elucidate the fate of the cumulus cells within the oviduct after ovulation.

7. CONCLUSION

A complex network of hormones and cellular interactions leads to successful fertilization in mammals. Most of this still needs to be eludicated. Here we show that growth factors and ECM-components are centrally involved in the auto/paracrine control of oocyte maturation, oviductal metabolism and fertilization. Following the detection of a cycle- and tissue-specific pattern of expression for both the IGF and FGF growth factor systems, additional developmental relevance can be attached to a/b FGF for bovine follicles, the OCC and oviducts. A marked timeshift between the expression of the specific

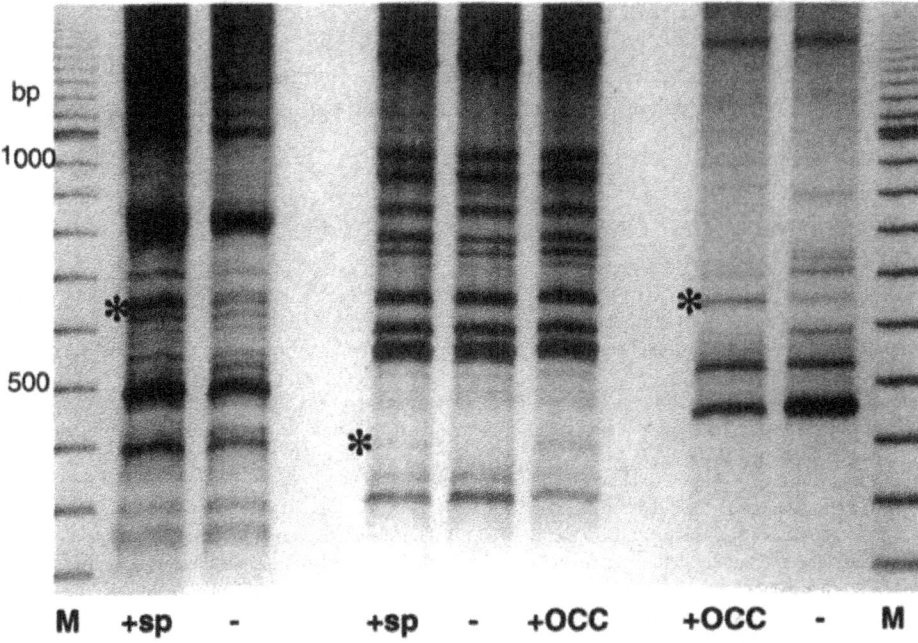

Figure 6. Randomly expressed mRNA detected by RAP-PCR in bovine oviductal epithelial cells (-) and these cells cocultured with oocyte-cumulus (OCC) or sperm (+sp). RAP-PCR products were separated on a 10 % PAGE-gel and silverstained. Differentially expressed RNAs are indicated by asterisks (*).

mRNA and free FGF protein is observed in bovine oviducts, and points to a cycle-dependent parcipitation of ECM-proteases. The cumulus cells show a distinctly elevated bFGF and TGFß expression pattern when compared with ovarian granulosa cells, indicating functional differences between these histologically similar cell types. Further observations lead to the suggestion that growth factors and ECM-components are auto/paracrine effectors mediating the cycle-dependent gonadotropin surge which results in oocyte maturation and ovulation. Elevated bFGF, PAI1 and TIMP in the bovine OCC may additionally interact with the components of the micro environment secreted into the oviduct by the oviductal epithelium. A conditioning of the oviduct by growth factors acting in concert with the ECM, could be involved in creating the optimal milieu for the reception of the gametes. This could also be reflected, as shown here, by the specific modification of the sperm surface after oviductal passage. The perfect crosstalk between the cell-types involved will result in fertilization in vivo and development of the early embryo within the oviduct. To our knowledge, in vitro techniques like IVM or IVF lack most of the extracellular and nutritional components present in vivo under natural condition. Once the gametes have contacted the oviductal epithelium, new mRNAs appear to be expressed within only a few hours. New molecular techniques should enable a rapid progress in discovering the complete structure and functions of these still unknown gene products. It will be of great interest for the future to investigate the cell-to-cell communication pathways in greater details, hopefully leading to a better understanding of the fine-regulation of this fascinating aspect of mammalian reproduction.

8. ACKNOWLEDGMENTS

We would like to thank Dr. Karin Wollenhaupt (Dummerstorf), Andrea Zuber and Sandra Rode (Weihenstephan) for their motivated cooperation. This work was supported by the DFG (Ei 296/4–1).

9. REFERENCES

Arias EB, Verhage HG & Jaffe RC 1994 Complementary deoxyribonucleic acid cloning and molecular characterization of an estrogen-dependant human oviductal glycoprotein. Biology of Reproduction 51: 685–94.

Bedford JM & Kim HH 1993 Cumulus oophorus as a sperm sequestering device in vivo. Journal of Experimental Zoology 265: 321–8.

Boatman DE, Magnoni GE & Robbins RS 1994 Modulation of spermatozoa and zona pellucida properties by the soluble acrosome reaction-inducing factor of the ovulated egg-cumulus complex. Molecular Reproduction and Development 38: 410–20.

Brackett BG, Yon KO, Evans JF & Donawick WJ 1980 Fertilizqtion and early development of cow ova. Biology of Reproduction 23: 189–205.

Brucker C; Alexander NJ; Hodgen GD; Sandow BA 1991 Transforming growth factor-alpha augments meiotic maturation of cumulus cell-enclosed mouse oocytes. Molecular Reproduction and Development 28: 94–8.

Chen L; Russell PT; Larsen WJ 1994 Sequential effects of follicle-stimulating hormone and luteinizing hormone on mouse cumulus expansion in vitro. Biology of Reproduction 51: 290–5.

DeSouza MM & Murray MK 1995 An estrogen-dependant secretory protein, which shares identity with chitinases, is expressed in a temporally and regionally specific manner in the sheep oviduct at the time of fertilization and embryo development. Endocrinology 136: 2485–96

Donnelly KM, Fazleabas AT & Verhage HG 1991 Cloning of a recombinant complementary DNA to a baboon (Papio anubis) estrodiol-dependant oviduct-specific glycoprotein. Molecular Endocrinology 5: 356–64.

Downs SM; Daniel SA; Eppig JJ 1988 Induction of maturation in cumulus cell-enclosed mouse oocytes by follicle-stimulating hormone and epidermal growth factor: evidence for a positive stimulus of somatic cell origin. Journal of Experimental Zoology 245: 86–9.

Einspanier R, Gabler C & Einspanier A 1995 Fibroblast growth factor (a/bFGF), insulin-like growth factor-1 (IGF-1) and transforming growth factor-a (TGFa) mRNAs are differentially expressed in the oviduct of the marmoset monkey. Biology of Reproduction 52 (Supplement): 242

Ellington JE 1991 The bovine oviduct and ist role in reproduction: a review of the literature. Cornell Veterinarian 81: 313–328.

Eppig JJ 1991 Intercommunication between mammalian oocytes and companion somatic cells. Bioassays 13: 569–74.

Eppig JJ, Wigglesworth K & Chesnel F 1993 Secretion of cumulus expansion enabling factor by mouse oocytes: relationship to oocyte growth and competence to resume meiosis. Developmental Biology 158: 400–9.

Fukui Y 1990 Effect of follicle cells on the acrosome reaction, fertilization, and developmental competence of bovine oocytes matured in vitro. Molecular Reproduction and Development 26: 40–6.

Gabler C & Einspanier R 1995 Acidic and basic fibroblast growth factor (aFGF / bFGF) in the bovine oviduct: indication for different expression pattern. Experimental and Clinical Endocrinology (Supplement) 103: 64.

Gandolfi F, Brevini TAL & Moor RM 1989 Effect of oviduct environment on embryonic development. Journal Reproduction and Fertility 38 (Supplement): 107–115.

Kotsuji F; Kubo M; Tominaga T 1994 Effect of interactions between granulosa and thecal cellson meiotic arrest in bovine oocytes. Journal of Reproduction and Fertility 100: 151–6.

Lauer B, Wollenhaupt K & Einspanier R 1995 Detection of a new estradiol-mediated bovine oviduct protein and sperm-oviduct interactions using a tissue-perfusion system. Journal of Reproduction and Fertility (Supplement) 15: 94.

Lauer B & Einspanier R 1996 Are distinct levels of soluble heparin-analogues responsibel for different alterations of sperm in the bovine oviduct ? Experimental Clinical Endocrinology and Diabetes (Supplement) 104: 131.

Lefebvre R, Chenoweth PJ, Drost M, LeClear CT, MacCubbin M, Dutton JT & Suarez SS 1995 Characterization of the oviductal sperm reservoir in cattle. Biology of Reproduction 53: 1066–1074.

Lorenzo PL, Illera MJ, Illera JC & Illera M 1995 Influence of growth factors on the time-dependent meiotic progression of the bovine oocytes during their in vitro maturation. Reviosiones Espania Fisiologica 51: 77–83.

Lorton SP & First NL 1979 Hyaluronidase daoes not disperse the cumulus oophorus surrounding bovine ova. Biology of Reproduction 21: 301–308.

McNutt T, Rogowski L, Vasilatos-Younken R & Killian G 1992 Adsorption of oviductal fluid proteins by the bovine sperm membrane during in vitro capacitation. Molecular Reproduction and Development 33: 313–323

Morishige K, Kurachi H, Amemiya K, Adachi H, Adachi K, Sakoyama Y, Miyake A & Tanizawa O 1993 Menstrual stage-specific expression of epidermal growth factor and transforming growth factor alpha in human oviduct epithelium and their role in early embryogenesis. Endocrinology 133: 199–207

Noda Y, Narimoto K, Umaoka Y, Natsuyama S & Mori T 1993 Analysis of oviduct-derived embryonic growth stimulator activity. Interational Journal of Memopausal Study 38: 57–64

Robitaille G, St Jacques S, Potier M & Bleau G 1988 Characterization of an oviductal glycoprotein associated with the ovulated hamster oocyte. Biology of Reproduction 38: 687–94.

Roller RJ, Kinloch RA, Hiraoka BY, Li SS & Wassarman PM 1989 Gene expression during mammalian oogenesis and early embryogenesis: quantification of three messenger RNAs abundant in fully grown mouse oocytes. Development 106: 251–61.

Satoh T, Abe H, Sendai Y, Iwata H & Hoshi H 1995 Biochemical characterization of a bovine oviduct-specific sialoglycoprotein that sustains sperm viability in vitro. Biochemica Biophysica Acta 1266: 117–23.

Schmidt A, Einspanier R, Amselgruber W, Sinowatz F & Schams D 1994 Expression of IGF1 in the bovine oviduct during the oestrous cycle. Experimental and Clinical Endocrinology 102: 364–69.

Schuetz AW; Whittingham DG; Legg RF 1989 Alterations in the cell cycle characteristics of granulosa cells during the periovulatory period: evidence of ovarian and oviductal influences. Journal of Experimental Zoology 249: 105–10.

Sendai Y, Abe H, Kikuchi M, Satoh T & Hoshi H 1994 Purification and molecular cloning of bovine oviduct-specific glycoprotein. Biology of Reproduction 50: 927–34.

Singh B; Barbe GJ; Armstrong DT 1993 Factors influencing resumption of meiotic maturation and cumulus expansion of porcine oocyte-cumulus cell complexes in vitro. Molecular Reproduction and Development 36: 113–9.

Sirard MA 1990 Effects of granulosa cell co-culture on in-vitro meioticresumption of bovine oocytes. Journal of Reproduction and Fertility 89: 459–65.

Sundstrom P; Nilsson BO 1988 Meiotic and cytoplasmic maturation of oocytes collected in stimulated cycles is asynchronous. Human Reproduction 3: 613–9.

Tirone E, Siracusa G, Hascall VC, Frajese G & Salustri A 1993 Oocytes preserve the ability of mouse cumulus cells in culture to synthesize hyaluronic acid and dermatan sulfate. Developmental Biology 160: 405–12.

Watson R; Anthony F; Pickett M; Lambden P; Masson GM; Thomas EJ 1992 Reverse transcription with nested polymerase chain reaction shows expression of basic fibroblast growth factor transcripts in human granulosa and cumulus cells from in vitro fertilisation patients. Biochemical and Biophysical Research Communications 187: 1227–31.

51

THE CELL BIOLOGY OF FERTILIZATION

R. J. Aitken

MRC Reproductive Biology Unit
Centre for Reproductive Biology
37 Chalmers Street
Edinburgh EH3 9EW, Scotland

1. SUMMARY

Research into the cell biology of mammalian fertilization has been stimulated by the desire to provide a theoretical framework for the development of novel approaches to contraception and the need to understand the cellular basis of human infertility. The results of such studies have revealed a complex cascade of interactions initiated by the contact between capacitated spermatozoa on the oocyte-cumulus complex and culminating in sperm-oocyte fusion. In this review we shall examine our current understanding of the fertilization process, highlighting the strategic importance of recent findings and key areas where information is lacking.

2. CAPACITATION

Immature spermatozoa entering the caput epididymis are incapable of fertilization. It is only after these cells have completed their maturation and are stored in the cauda epididymis that they possess the potential to fertilize the oocyte. Even caudal epididymal spermatozoa only possess a latent potential to participate in the process of fertilization, the realization of this potential requires a period of incubation in the female reproductive tract before full functional competence is attained. This final period of maturation, during which spermatozoa acquire the competence to respond to signals originating from the cumulus-oocyte complex by undergoing the acrosome reaction, is referred to as capacitation.

2.1. Capacitation and cAMP

Although the need for spermatozoa to undergo capacitation has been recognized in biological terms for some time, the biochemical mechanisms controlling these events have not yet been elucidated. To date, two major metabolic changes have been detected in

The Fate of the Male Germ Cell, edited by Ivell and Holstein
Plenum Press, New York, 1997

mammalian spermatozoa as a consequence of epididymal maturation, that appear to be critical to the induction of capacitation. The first is a change in the competence of spermatozoa to generate cAMP following their release from the epididymal environment. Thus, when mature caudal epididymal spermatozoa are released into a simple Krebs Ringer bicarbonate buffer, there is a sudden rise in the cAMP content of these cells that plateaus after around two hours incubation (Figure 1). This rise in cAMP parallels the capacitation of the spermatozoa as reflected by the development of a hyperactivated form of motility (White & Aitken, 1989). If mature spermatozoa are released into a medium containing a phosphodiesterase inhibitor such as isobutylmethylxanthine, the intracellular cAMP rise is accelerated and hyperactivated motility appears earlier. Significantly, immature spermatozoa from the caput epididymis do not exhibit this change in cAMP levels, do not capacitate and do not express hyperactivated motion (White & Aitken, 1989).

The cellular mechanisms underlying this increase in cAMP levels in mature spermatozoa are not understood. One component of the activation process is probably an influx of extracellular calcium during capacitation since cAMP levels do not change in mature spermatozoa released into calcium depleted media (White & Aitken, 1989). This calcium requirement is presumably a reflection of the calcium dependency of sperm adenylate cyclase (Fraser & Monks, 1990). The changes in sperm biochemistry that enable mature epididymal spermatozoa to express this calcium dependent rise in intracellular cAMP, but do not permit such changes in immature cells from the caput epididymis, have not yet been elucidated. However it has been established that differences in the cAMP generating capacity of caput and caudal epididymal spermatoza are not related to differences in ATP content or intracellular pH (White & Aitken, 1989).

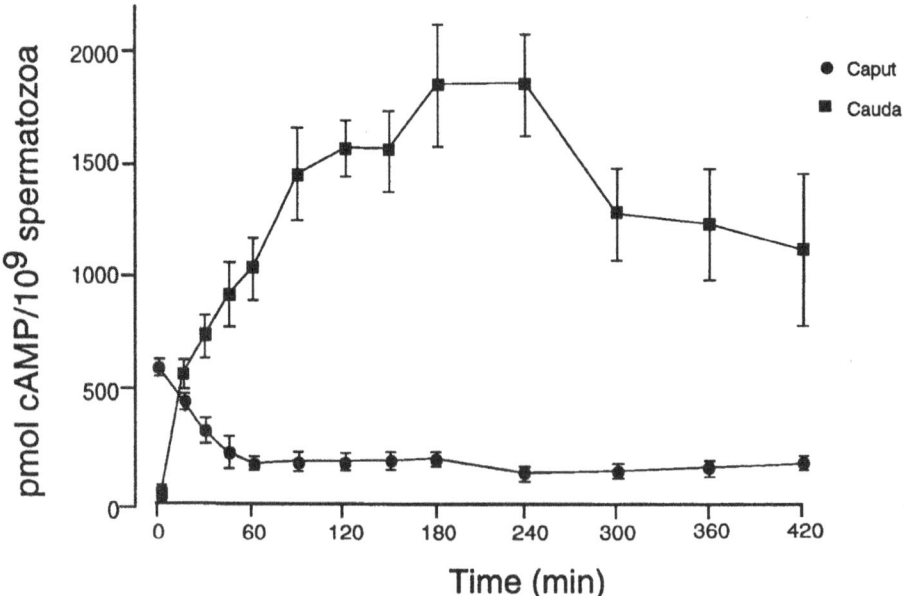

Figure 1. cAMP levels in hamster spermatozoa isolated from the caput and cauda epididymis and incubated in a modified Krebs Ringer bicarbonate buffer (White and Aitken, 1989).

2.2. Capacitation and Hydrogen Peroxide

Another major change known to occur to spermatozoa as they undergo epididymal maturation is the development of a capacity to generate hydrogen peroxide (Fisher & Aitken, 1997). Hamster, rat and mouse spermatozoa have all been shown to generate this oxidant following their release from the cauda epididymis. Immature spermatozoa recovered from the caput and corpus epididymis do not exhibit this capacity. The fact that catalase can disrupt the capacitation of both hamster (Bize et al., 1991) and human spermatozoa (Griveau et al., 1994; Aitken et al., 1995) suggests that hydrogen peroxide is directly involved in the mechanisms regulating capacitation. This conclusion is also supported by the ability of hydrogen peroxide, or peroxide-generating systems employing glucose oxidase, to promote sperm capacitation (Aitken et al., 1995). Furthermore, the stimulation of endogenous reactive oxygen species generation by increasing the availability of NADPH, has been shown to promote the capacitation of human spermatozoa, stimulating both the basal level of sperm activation and the responsiveness of these cells to physiological agonists such as ZP3 and progesterone (Aitken et al., 1995, 1996). Such results suggest that capacitation is a redox regulated event involving the generation of oxidizing conditions within the spermatozoa by virtue of an intrinsic capacity to generate hydrogen peroxide (MacLeod, 1943; Aitken et al., 1995, 1996). Conversely, reducing conditions can suppress sperm capacitation. Thus, membrane permeant thiols such as dithiothreitol (DTT) and 2-mercaptoethanol have the capacity to inhibit sperm capacitation, blocking the responses of human spermatozoa to biological agonists such as progesterone (Aitken et al., 1996). DTT has also been shown to suppress the capacitation of guinea pig spermatozoa, even when this process is promoted by the membrane motility agent A2C (Fleming et al., 1982; Yanagimachi et al., 1983). Such results clearly suggest that the oxidation of sperm SH groups by peroxide may play an important role in the capacitation of mammalian spermatozoa.

2.3. Capacitation and Tyrosine Phosphorylation

While the data cited above indicate that sperm capacitation is a redox regulated event, they do not shed any light on the nature of the processes being controlled in this manner. One possibility would be that oxidizing conditions favour the activation of phospholipase A2 (Goldman et al., 1992) Thus when membrane lipids undergo peroxidation, phospholipase A2 moves in to cleave the oxidized fatty acid from the parent phospholipid. This is a cellular defence mechanism that enables the oxidized fatty acid to be processed subsequently by glutathione peroxidase. The product of phospholipase A2 action is the creation of a lysophopholipid, the presence of which will tend to destabilize the sperm plasma membrane thereby facilitating membrane fusion events such as the acrosome reaction.

A second aspect of sperm biochemistry that is under redox control is tyrosine phosphorylation (Aitken et al., 1995). The spermatozoa of all mammalian species examined to date appear to show a spontaneous increase in tyrosine phosphorylation during capacitation. In mouse spermatozoa, this process appears to involve a dominant protein species of 95 kDa (Leyton et al., 1992) that may be a unique form of hexokinase (Kalab et al., 1994). In human spermatozoa, capacitation results in the spontaneous phosphorylation of a cohort of bands the most dominant of which exhibits a relative molecular mass of approximately 100 kDa (Aitken et al., 1995, 1996; Leclerc et al., 1996). Tyrosine phosphorylation is stimulated by exposing spermatozoa to the ionophore, A23187, suggesting that a rise in intracellular calcium is involved in the induction of phosphotyrosine expression during capacitation (Aitken et al., 1995). Involvement of hydrogen peroxide, but not superoxide, in the mechanisms by which spermatozoa stimulate tyrosine phosphorylation during capacitation is suggested by the suppression of phosphotyrosine expression achieved with catalase. This effect is dose dependent and renders the spermatozoa non-responsive to physiological agonists such as ZP3 and progesterone, while sperm viability and motility are undisturbed (Aitken et al., 1995, 1996). Since superoxide dismutase has no such inhibitory effects, these results suggest that there are causative links between hydrogen peroxide generation by human spermatozoa, tyrosine phosphorylation and sperm capacitation. This conclusion is supported by the fact that exposure of human spermatozoa to exogenous hydrogen peroxide results in the stimulation of tyrosine phosphorylation and the attainment of a capacitated state (Aitken et al., 1995). Stimulation of endogenous reactive oxygen species generation with NADPH also enhances both phosphotyrosine expression and capacitation status (Aitken et al., 1995). Conversely if the oxidizing conditions generated by hydrogen peroxide during sperm capacitation are counteracted by the addition of catalase or membrane permeant reducing agents (2-mercaptoethanol and dithiothreitol) then both capacitation and tyrosine phosphorylation are inhibited (Aitken et al., 1996).

Tyrosine phosphorylation and sperm capacitation can be induced by elevating intracellular cAMP levels as well as stimulating hydrogen peroxide generation (Leclerc et al., 1996). Thus addition of either phosphodiesterase inhibitors or membrane permeant analogues of cAMP induces both tyrosine phosphorylation and capacitation of human spermatozoa (Leclerc et al., 1996). Moreover a positive role for cAMP in the induction of sperm capacitation has been suggested for other species including the hamster (Visconti et al., 1990) mouse (Visconti et al., 1995a,b) and bull (Parrish et al., 1994). In the mouse clear linkages have again been established between cAMP generation, tyrosine phosphorylation and capacitation. Treatments that suppress capacitation including removal of BSA, Ca^{2+} or bicarbonate are associated with a loss of tyrosine phosphorylation (Visconti et al., 1995a). However both tyrosine phosphorylation and and capacitation can be restored under any of these conditions by the addition of membrane permeant cAMP analogues (Visconti et al., 1995a). The involvement of protein kinase A (PKA) in the mechanisms by which cAMP regulates capacitation is suggested by the ability of PKA inhibitors to block sperm capacitation and the capacitation-dependent increase in tyrosine phosphorylation. Cross talk between the cAMP- and tyrosine phosphorylation- dependent signal transduction pathways has also been described for the stimulation of progesterone secretion by ovarian granulosa cells (Aharoni et al., 1993). In this case the synergism between these pathways appears to involve changes in the cytoskeleton.

Similarly in spermatozoa, the elevations of intracellular cAMP and tyrosine phosphorylation observed during sperm capacitation are associated with changes in the sperm cytoskeleton involving the formation of actin filaments on the plasma and outer acrosomal membrane. PIP2-specific phospholipase C becomes attached to this membrane bound F-actin as capacitation proceeds (Spungin and Bretibart, 1996) and subsequently plays an important role in the induction of the acrosome reaction, as discussed below.

In summary, a key component of capacitation is the induction of tyrosine phosphorylation. This phosphorylation event is an essential component of capacitation since if this process is blocked by tyrosine kinase inhibitors, such as genistein, then the spermatozoa lose their capacity to respond to physiological triggers such as progesterone (Aitken et al., 1996). As soon as the spermatozoa are released from the cauda epididymis they initiate tyrosine phosphorylation through the generation of two second messengers, cAMP and reactive oxygen species, particularly hydrogen peroxide.

Redox regulation of tyrosine phosphorylation is widely recognized and appears to involve both the stimulation of tyrosine kinase activity and inhibition of the corresponding phosphatases (Montiero and Stern, 1996). Specific examples of tyrosine kinase receptors that are known to be redox regulated include the insulin receptor (Koshio et al., 1988), which may be present on human spermatozoa (Silvestroni et al., 1992), Ltk (Bauskin et al., 1991) IRS-1 (Wilden and Broadway, 1995), p72 (syk) (Schieven et al., 1993), Lck (Hardwick anbd Sefton, 1995) and EGF (Gamou and Shimizu, 1995).

In addition to the involvement of redox and cAMP regulated mechanisms for the stimulation of tyrosine phosphorylation during capacitation, it is also known that simply adding protein in the form of BSA or foetal bovine serum to media containing mammalian spermatozoa will enhance tyrosine phosphorylation, and thence, capacitation (Visconti et al., 1995a,b). The mechanisms by which exogenous protein exerts this effect probably involves the removal of cholesterol from the membrane and a concomitant increase in membrane fluidity. Cholesterol has long been known to inhibit sperm capacitation and appears to be the major decapacitation factor in seminal plasma (Cross, 1996). It is possible that the removal of cholesterol from the sperm plasma membrane during capacitation allows putative tyrosine kinase type receptors to move freely within the lipid bilayer and oligomerize. Aggregation of this type of receptor is known to induce activation and might contribute to the increased tyrosine phosphorylation seen during the capacitation of mammalian spermatozoa in the presence high concentrations of protein.

The fact that at least three different pathways exist to stimulate tyrosine phosphorylation during capacitation emphasises the importance of phosphotyrosine expression in programming the spermatozoa for fertilization. Analyses of the proteins phosphorylated on tyrosine during capacitation have suggested that multiple components are involved, particularly in the human (Aitken et al., 1995; Leclerc et al., 1996). The focus of current research in this area is to elucidate the nature of these phosphorylated proteins and determine their role in the attainment of a capacitated state.

2.4. Tyrosine Phosphorylation and the Acrosome Reaction

On the basis of experiments conducted using progesterone as a physiological agonist we have concluded that tyrosine phosphorylation must occur prior to the ionic fluxes induced by this reagent for the acrosome reaction to occur (Aitken et al., 1996). If tyrosine phosphorylation has been promoted during capacitation by stimulating the generation of reactive oxygen species, then only 5 minutes exposure to progesterone is needed to induce a maximal biological response (Aitken et al., 1996). Using a cell free system for studying the fusion of plasma and outer acrosomal membranes, Spungin et al. (1995) also concluded that key phosphorylation events must occur during capacitation before the acrosome reaction can be induced by the changes in intracellular calcium and pH induced by physiological agonists such as the zona pellucida or progesterone. A clear understanding of the cellular mechanisms regulating sperm capacitation must now involve identifying those proteins that are tyrosine phosphorylated and determining their role in priming spermatozoa for the acrosome reaction.

One of the few defined proteins that is known to be tyrosine phosphorylated during capacitation is phospholipase C (PLC). Upon agonist induced cell stimulation this enzyme induces the hydrolysis of phosphatidyl inositol 4,5-bisphosphate (PIP2) to release two biologically active metabolites: inositol 1,4,5-trisphosphate (IP$_3$), which induces the release of calcium from intracellular stores, and diacylglycerol, which activates protein kinase C (PKC). Studies with bovine and murine spermatozoa indicate that this enzyme becomes restricted to the sperm head during capacitation as a consequence of binding to F-actin filaments that form on the plasma and outer acrosomal membrane during sperm capacitation (Spungin et al., 1995). The presence of this F-actin framework ensures that the plasma and outer acrosomal membranes cannot fuse during capacitation; the acrosome reaction can only occur when the F-actin has been dispersed. Inhibition of this dispersal with phalloidin disrupts membrane fusion and the acrosome reaction in intact cells (Spungin et al., 1995). It is in effecting the dispersal of F-actin from the plasma and acrosomal membranes that PLC is thought to be of central importance to the acrosome reaction.

During capacitation PLC becomes tyrosine phosphorylated, presumably as a consequence of the cAMP and redox regulated pathways referred to above. As a consequence of tyrosine phosphorylation, PLC becomes bound to the F-actin cytoskeleton associated with the plasma and acrosomal membranes. Capacitation of mouse spermatozoa is associated with a shift in the electrophoretic mobility of PLCγ, which would be consistent with this phosphorylation step (Tomes et al., 1996). Once anchored to the cytoskeleton of the sperm head, this enzyme is in a position to respond to physiological inducers of the acrosome reaction, such as ZP3, with a sudden increase in enzymatic activity. The capacity of sperm PLC to respond to ZP3 is entirely dependent on tyrosine phophorylation since it can be blocked by tyrosine kinase inhibitors such as tyrphostin A1 (Tomes et al., 1996). Stimulation of the anchored PLC is probably brought about by the elevation of intracellular calcium, induced by the binding of progesterone or ZP3 to sperm surface receptors. In a cell free system it has been shown that at pH 7.4 around 200 μM calcium is sufficient to stimulate F-actin release from the plasma membrane and initiate membrane fusion (Spungin et al., 1995). The ways in which activation of PLC, and the consequential generation of DAG and IP_3, might lead to lead to membrane fusion include:

i. a direct effect of DAG on the fluidity and fusogenicity of the plasma and outer acrosomal membranes,

ii. the activation of protein kinase C (PKC) by DAG. Isolated sperm plasma membranes contain a calcium channel that can be activated by PKC (Spungin and Breitbart, 1996) and could be responsible for secondary increases in intracellular calcium in the acrosomal domain, that are thought to important for triggering the acrosome reaction (Tesarik et al., 1996).

iii. the increase in IP3 could also lead to the generation of secondary calcium transients from internal, thapsigargin sensitive calcium stores.

iv. the possible activation of phospholipase A_2 by DAG with the generation of highly fusogenic lysophospholipids (Roldan & Fragio, 1994)

v. F-actin dissolution through the release of actin-severing proteins. Several such proteins (eg. gelsolin, profilin cofilin etc) bind PIP and PIP2 and in this state are unable to bind to F-actin and depolymerize this molecule. PLC that has been tyrosine phosphorylated during capacitation and activated by a calcium transient would be capable of hydrolzing PIP and PIP_2, thereby promoting the activation of these actin severing proteins (Spungin et al., 1995).

3. CONCLUSIONS

A possible sequence of events involved in the capacitation of mammalian spermatozoa and the induction of the acrosome reaction is summarized in Figures 2 and 3. Following ejaculation, mammalian spermatozoa engage in a sensitization process known as capacitation. The kernel of this process is the induction of tyrosine phosphorylation as a consequence of 3 different biochemical processes:

i. the loss of cholesterol from the plasma membrane generating an increase in membrane fluidity and increased spontaneous aggregation of tyrosine kinase type receptors that then spontaneously phosphorylate.

ii. an increase in cAMP levels and

iii. the generation of hydrogen peroxide. The generation of hydrogen peroxide and cAMP may both be dependent on a moderate increase in intracellular calcium during capacitation (White & Aitken, 1989). Under the influence of cAMP, and hence tyrosine phosphorylation, calcium is pumped from the cytoplasm into an intracellular calcium calcium store (Spungin & Breitbart, 1996). Concomitant with the stimulation of tyrosine phosphorylation a network of F-actin is deposited on the plasma and outer acrosomal membrane. This stabilizes these membranes so that premature fusion cannot occur and acts as a scaffolding for binding tyrosine phosphorylated PLC.

Thus, once capacitation has been achieved, the spermatozoon should have a replete intracellular calcium store as well as plasma and acrosomal membranes that are stabilized by a cytoskeleton of F-actin, onto which phosphorylated PLC has become bound. The integrity of the cytoskeleton is ensured by the inactivation of actin severing proteins by PIP2 (Figure 2).

The induction of an acrosome reaction by physiological agonists such as progesterone and ZP3 primarily involves the creation of a calcium transient (Osman et al., 1989; Blackmore et al., 1990; Florman, 1994). According to the model proposed in Figure 3, the sudden rise in intracellular calcium results in the activation of PLC which then catalyses the conversion of PIP2 to DAG and IP3. The hydrolysis of PIP2 is held to activate actin severing proteins, resulting in the dissolution of the cytoskeletal network and increased membrane fluidity. DAG further enhances membrane fusogenicity through the stimulation

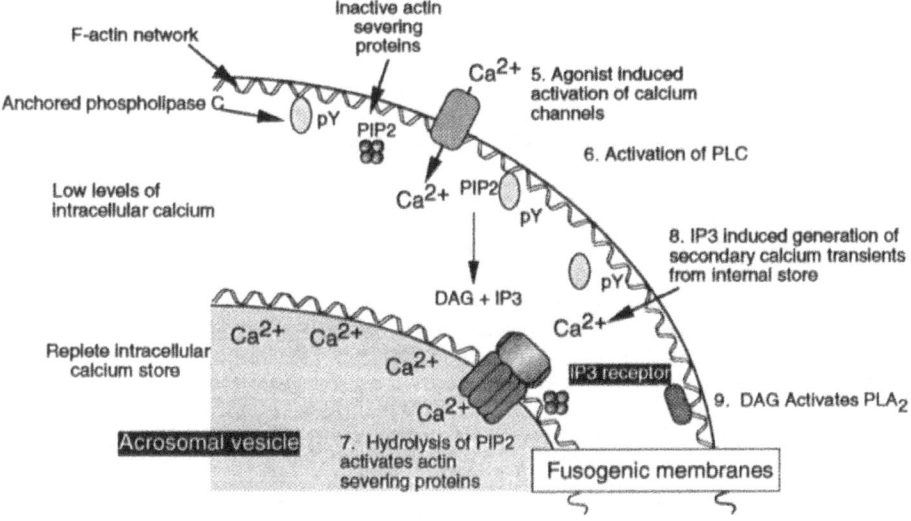

Figure 2. A proposed sequence of events involved in sperm capacitation.

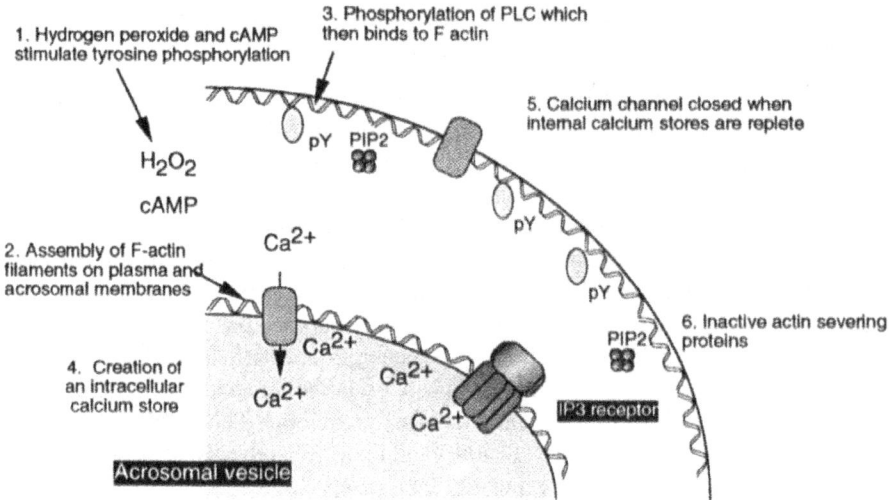

Figure 3. A possible sequence of events during the acrosome reaction.

of PLA2 and the subsequent creation of an unstable lipid bilayer. The IP3 releases calcium from the intracellular store, and it is this secondary calcium release that rapidly precipitates the fusion process that completes the acrosome reaction (Tesarik et al., 1996). The mechanisms by which this secondary transient induces membrane fusion, the nature of tyrosine phosphorylated proteins that enable such oscillations to occur and the points at which these mechanisms break down in infertile patients are all, as yet, unknown.

4. REFERENCES

Aharoni D, Dantes A & Amsterdam A 1993 Cross-talk between adenylate cyclase activation and tyrosine phosphorylation leads to modulation of the actin cytoskeleton and to acute progesterone secretion in ovarian granulosa cells. Endocrinology 133: 1426–1436.

Aitken RJ, Buckingham DW, Harkiss D, Fisher H, Paterson M & Irvine DS 1996 The extragenomic action of progesterone on human spermatozoa is influenced by redox regulated changes in tyrosine phosphorylation during capacitation. Molecular and Cellular Endocrinology 117: 83–93.

Aitken RJ, Paterson M, Fisher H, Buckingham DW & van Duin M 1995 Redox regulation of tyrosine phosphorylation in human spermatozoa and its role in the control of human sperm function. Journal of Cell Science 108: 2017–2025.

Bauskin AR, Alkalay I & Ben-Neriah Y 1991 Redox regulation of a tyrosine kinase in the endoplasmic reticulum. Cell 56: 685–696.

Bize I, Santander G, Cabello P, Driscoll D & C Sharpe 1991 Hydrogen peroxide is involved in hamster sperm capacitation in vitro. Biology of Reproduction 44: 389–403.

Blackmore PF, Beebe SJ, Danforth DR & Alexander N 1990 Progesterone and 17α-hydroxy-progesterone. Novel stimulators of calcium influx in human spermatozoa. Journal of Biological Chemistry 265: 1376–1380.

Cross NL 1996 Human seminal plasma prevents sperm from becoming acrosomally responsive to the agonist, progesterone: cholesterol is the major inhibitor. Biology of Reproduction 54: 138–145.

Fisher HM & Aitken RJ 1996 Comparative analysis of the ability of precursor germ cells and epididymal spermatozoa to generate reactive oxygen metabolites. Journal of Experimental Zoology (in press).

Fleming AD, Kosower NS & Yanagimachi R 1982 Promotion of capacitation of guinea pig spermatozoa by the membrane mobility agent, A2C, and inhibition by the disulphide reducing agent, DTT. Gamete Research 5: 19–33.

Florman HM 1994 Sequential focal and global elevations of sperm intracellular Ca^{2+} are initiated by the zona pellucida during acrosomal exocytosis. Developmental Biology 165: 152–164.

Fraser LR & Monks NJ 1990 Cyclic nucleotides and mammalian sperm capacitation. J Reprod Fert Suppl 42: 9–21.

Gamou S & Shimizu N 1995 Hydrogen peroxide preferentially enhances the tyrosine phosphorylation of epidermal growth factor receptor. FEBS Letters 357: 161–165.

Goldman R, Ferber E. & Zort U 1992 Reactive oxygen species are involved in the activation of cellular phospholipase A_2. FEBS Letters. 309: 190–192.

Griveau JF, Renard P & Le Lannou D 1994 An in vitro promoting role for hydrogen peroxide in human sperm capacitation. International Journal of Andrology 17: 300–307.

Hardwick JS & Sefton BM 1995 Activation of the Lck tyrosine protein kinase by hydrogen peroxide requires the phosphorylation of Tyr-394. Proceedings of the National Academy of Science USA 92: 4527–4531.

Kalab P, Visconti P, Leclerc P & Kopf G 1994 p95, the major phosphotyrosine containing protein in mouse spermatozoa is a hexokinase with unique properties. Journal of Biological Chemistry 269: 3810–3817.

Koshio O, Akanuma Y & M Kasuga. 1988 Hydrogen peroxide stimulates tyrosine phosphorylation of the insulin receptor and its tyrosine kinase activity in intact cells. Biochemical Journal 250: 95–101.

Leclerc P, de Lamirande E, Gagnon C 1996 Cyclic adenosine3',5' monophosphate-dependent regulation of protein tyrosine phosphorylation in relation to sperm capacitation and motility. Biology of Reproduction 55: 684–692.

Leyton L, LeGuen P, Bunch D & Saling PM 1992 Regulation of mouse gamete interaction by a sperm tyrosine kinase. Proceedings of the National Academy of Science USA 89: 11692–11695.

MacLeod J 1943 The role of oxygen in the metabolism and motility of human spermatozoa. Amercican Journal of Physiology 138: 512–518.

Montiero HP & Stern A 1996 Redox regulation of tyrosine phosphorylation-dependent signal transduction pathways. Free Radical Biology &Medicine 21: 323–333.

Osman RA, Andria ML, Jones AD & Meizel S 1989 Steroid induced exocytosis: the human sperm acrosome reaction. Biochemical and Biophysical Research Communications 160: 828–834.

Parrish JJ, Susko-Parrish JL, Uguz C & First NL 1994 Differences in the role of cAMP during capacitation of bovine sperm by heparin or oviduct fluid. Biology of Reproduction 51: 1099–1108.

Roldan ERS & Fragio C 1994 Diradylglycerols stimulate phospholipase A2 and subsequent exocytosis in ram spermatozoa. Evidence that the effect is not mediated via protein kinase C. Biochemical Journal 297: 225–232.

Schieven GL, Kirihara JM, Burg DL, Geahlen RL & Ledbetter JA 1993 p72[syk] tyrosine kinase is activated by oxidizing conditions that induce lymphocyte tyrosine phosphorylation and Ca^{2+} signals. Journal of Biological Chemistry 268: 16688–16692.

Silvestroni L, Modesti A & Sartori C 1992 Insulin-sperm interaction: Effects on plasma membrane and binding to acrosome. Archives of Andrology 28: 201–211.

Spungin B & Breitbart H 1996 Calcium mobilization and influx during sperm exocytosis. Journal of Cell Science 109:1947–1955.

Spungin B, Margalit I & Breitbart H 1995 Sperm exocytosis reconstructed in a cell-free system. Evidence for the involvement of phospholipase C and actin filaments in membrane fusion. Journal of Cell Science 108: 2525–2535.

Tesarik K, Carreras A & Mendoza C 1996 Single cell analysis of tyrosine kinase dependent and independent Ca^{2+} fluxes in progesterone induced acrosome reaction. Molecular Human Reproduction 2: 225–232.

Tomes CN, McMaster CR & Saling PM 1996 Activation of mouse sperm phosphatidylinositol-4,5 bisphosphate-phospholipase C by zona pellucida is modulated by tyrosine phosphorylation. Molecular Reproduction and Development 43: 196–204.

Visconti PE, Bailey JL, Moore GD, Pan D, Olds-Clarke P. & Kopf GS. 1995a Capacitation of mouse spermatozoa, 1. Correlation between the capacitation state and protein tyrosine phosphorylation. Development 121: 1129–1137.

Visconti PE, Moore GD, Bailey JL, Leclerc P, Connors SA, Pan D, Olds-Clarke P, & Kopf GS 1995b Capacitation in mouse spermatozoa, II. Protein tyrosine phosphorylation and capacitation are regulated by a cAMP-dependent pathway. Development 121: 1139–1150.

Visconti PE, Muschietti JP, Flawia MM & Tezon JG 1990 Bicarbonate dependence of cAMP accumulation induced by phorbol esters in hamster spermatozoa. Biochimica et Biophysica Acta 1054: 231–236.

White DR & Aitken RJ 1989 Relationship between calcium, cAMP, ATP and intracellular pH and the capacity to express hyperactivated motility by hamster spermatozoa. Gamete Research 22: 63–178.

Wilden PA. & Broadway DC 1995 Combination of insulinomimetic agents H_2O_2 and vanadate enhances insulin receptor mediated tyrosine phosphorylation of IRS-1 leading to IRS-1 association with the phosphatidylinositol 3-kinase. Journal of Cell Biochemistry 58: 279–291.

Yanagimachi R, Huang TTF, Fleming AD, Kosower NS, & Nicolson GL 1983 Dithiothreitol, a disulphide-reducing agent, inhibits capacitation, the acrosome reaction, and interaction with eggs by guinea pig spermatozoa. Gamete Research 7: 145–154.

52

THE ROLE OF CARBOHYDRATES IN SPERM-EGG INTERACTION

E. Töpfer-Petersen, Z. Dostàlovà, and J. J. Calvete

Institut für Reproduktionsmedizin
Tierärztliche Hochschule Hannover
Bünteweg 15, 30559 Hannover
Germany

1. INTRODUCTION

Increasing evidence gained over the last 20 years support the concept that carbohydrates act as recognition determinants in a great variety of physiological and pathological biological processes (Varki, 1993; Sharon & Lis, 1997). This concept evolved with the realization that carbohydrates, by their multi-linkage monomers and branching structure, have an enormous potential for encoding biological information (Laine, 1994). The messages encoded in the structures of complex oligosaccharides ("glycocodes") are deciphered by complementary carbohydrate recognition domains (CRDs) of carbohydrate-binding proteins, i.e., glycosidases, glycosyltransferases, antibodies, and lectins (see below).

Fertilization is a complex multistep biological process, which guarantees species survival. An essential event in the mechanism of fertilization is gamete recognition through specific complementary molecules located on the external surfaces of the spermatozoon and the oocyte. The recognition and initial binding of spermatozoa and the glycoprotein network surrounding the investing egg, i.e., the marine invertebrate vitelline coat or the eutherian mammalian zona pellucida, are mediated by protein-carbohydrate interactions, and this basic mechanism appears to be conserved throughout the whole of evolution (Töpfer-Petersen & Calvete, 1996; Sinowatz et al., 1997).

2. THE MAMMALIAN ZONA PELLUCIDA

The mammalian zona pellucida (ZP) is a loose, 2–25 μm thick extracellular glycoprotein meshwork (Fig. 1), synthesized and secreted by growing mammalian oocytes and, in some species also granulosa cells. Mammalian zonae pellucidae typically consist of only 3 glycoproteins (Dunbar et al., 1994; Epifano & Dean, 1994) encoded by three different genes termed, according to their decreasing molecular size, ZPA, ZPB, and ZPC (Harris et al., 1994).

The Fate of the Male Germ Cell, edited by Ivell and Holstein
Plenum Press, New York, 1997

Besides its role in gamete interaction, the zona pellucida network protects the egg and the early embryo from physical damage, playing a vital role in preimplantation embryogenesis.

In the mouse, the gene organization, biosynthesis, ultrastructure, and function of the zona pellucida has been thoroughly investigated (Epifanio et al., 1995a; Wassarman et al., 1996; Liu et al., 1996; and references cited therein). The three mouse zona pellucida genes are single, copy located on chromosomes 19 (ZPB), 7 (ZPA), and 5 (ZPC), and are expressed in a coordinate, oocyte-specific manner during the growth phase of oogenesis (reviewed by Castle & Dean, 1996). Glycoproteins ZP2 (ZPA, 120 kDa) and ZP3 (ZPC, 83

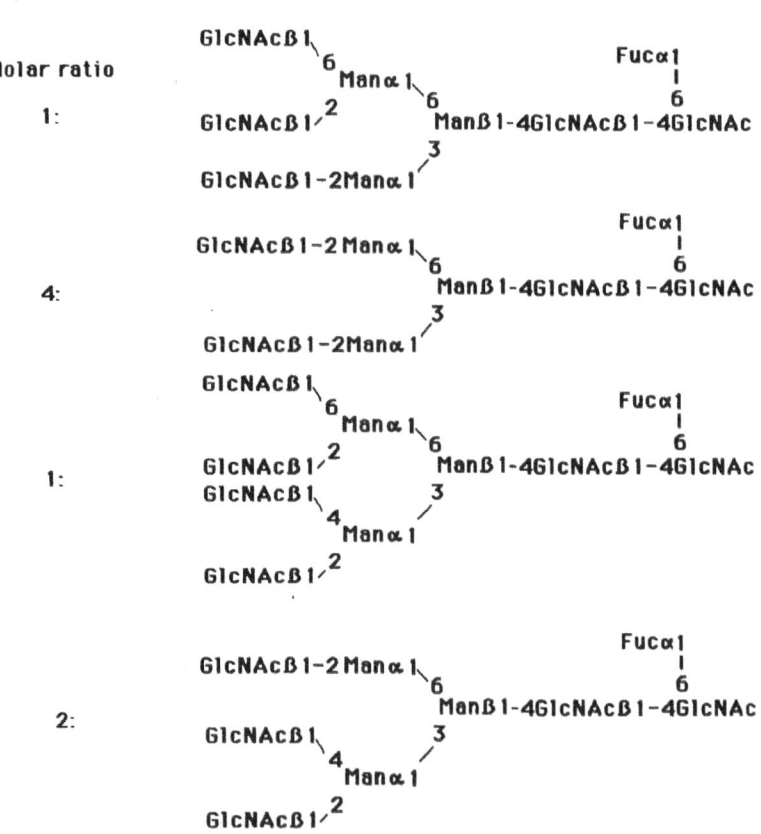

Figure 1. Electron micrography (x 1000) of a porcine zona pellucida-encased oocyte. The overall porous, fibrillar ultrastructure of the zona pellucida can be appreciated.

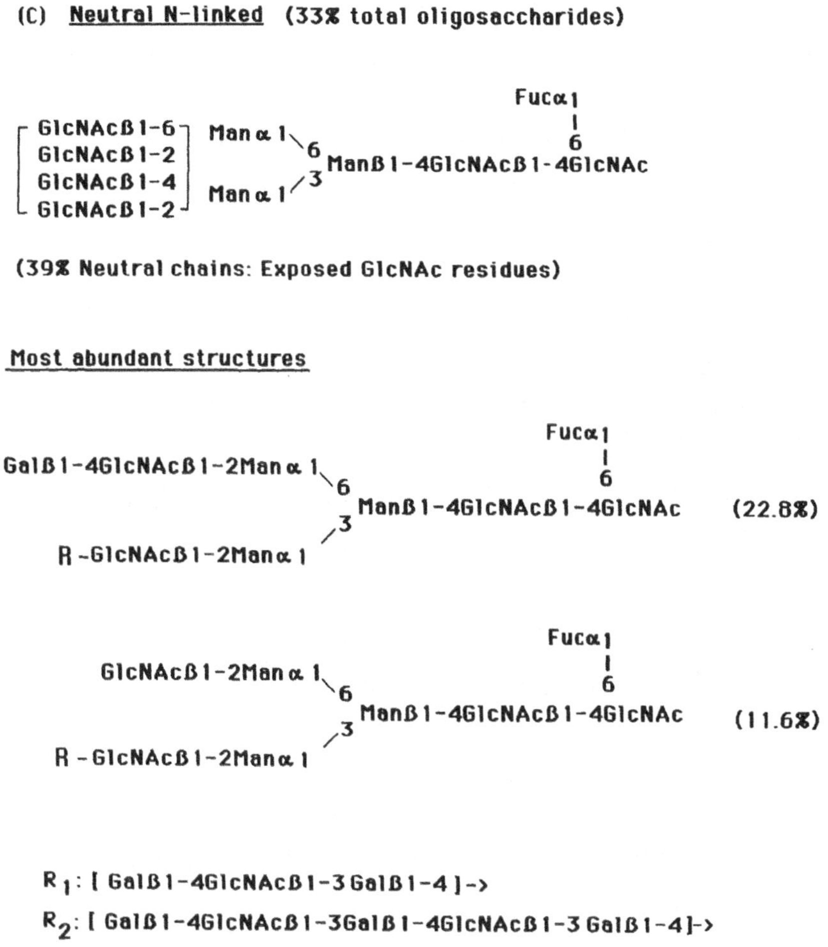

(C) **Neutral N-linked** (33% total oligosaccharides)

$$
\begin{bmatrix}
\text{GlcNAc}\beta 1\text{-6} \\
\text{GlcNAc}\beta 1\text{-2} \\
\text{GlcNAc}\beta 1\text{-4} \\
\text{GlcNAc}\beta 1\text{-2}
\end{bmatrix}
\begin{array}{l}
\text{Man}\,\alpha\,1 \\
\\
\text{Man}\,\alpha\,1
\end{array}
$$

Fucα1
|
6
Manα... ManΒ1-4GlcNAcΒ1-4GlcNAc

(39% Neutral chains: Exposed GlcNAc residues)

Most abundant structures

GalΒ1-4GlcNAcΒ1-2Manα1

Fucα1
|
6
ManΒ1-4GlcNAcΒ1-4GlcNAc   (22.8%)

R-GlcNAcΒ1-2Manα1

GlcNAcΒ1-2Manα1

Fucα1
|
6
ManΒ1-4GlcNAcΒ1-4GlcNAc   (11.6%)

R-GlcNAcΒ1-2Manα1

R₁: [GalΒ1-4GlcNAcΒ1-3GalΒ1-4]->
R₂: [GalΒ1-4GlcNAcΒ1-3GalΒ1-4GlcNAcΒ1-3 GalΒ1-4]->

Figure 1. (*Continued*)

(D)   0-linked neutral:

%

6   Gal ß(1-4)-GlcNAc ß(1-3)-GalNAc

14   GlcNAc ß(1-3)-Gal ß(1-3)-GalNAc

45   Gal ß(1-4)-GlcNAc ß(1-3)-Gal ß(1-3)-GalNAc

2   Gal α (1-3)-Gal ß(1-4)- GlcNAc ß(1-3)-Gal ß(1-3)-GalNAc

8   Gal ß(1-4)-GlcNAc ß(1-3)-Gal ß(1-4)- GlcNAc ß(1-3)-Gal ß(1-3)-GalNAc

1   [Gal ß(1-4)-GlcNAc ß(1-3)]$_3$ -Gal ß(1-3)-GalNAc

2   GlcNAc ß(1-3)-Gal ß(1-4)-GlcNAc ß(1-3)- Gal ß(1-3)-GalNAc

4   GalNAc ß(1-3/4)-GlcNAc ß(1-3)-Gal ß(1-3)-GalNAc

18   GalNAc

(E)   0-linked anionic:

```
NeuNGc ─┐
   or    ├ α (2-3)- Gal ß(1-3)-GalNAc
NeuNAc ─┘
```

```
NeuNGc ─┐
   or    ├ α (2-3)-Gal ß(1-4)-GlcNAc ß(1-3)-Gal ß(1-3)-GalNAc
NeuNAc ─┘
```

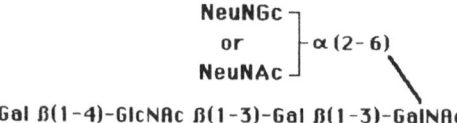

Gal ß(1-4)-GlcNAc ß(1-3)-Gal ß(1-3)-GalNAc

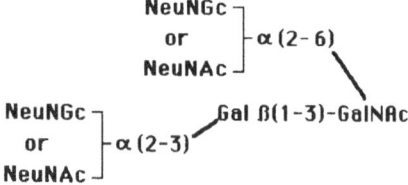

Figure 1. (*Continued*)

kDa) form a 2–3 µm long fibrillar structure of repeating heterodimers. These filaments are interconnected by dimers of disulphide-linked ZP1 (ZPB, 200 kDa). The three ZP proteins are held together by non-covalent bonds. Mouse glycoprotein ZPC plays a bifunctional role in sperm-egg interaction: it serves as the primary sperm receptor for capacitated acrosome-intact spermatozoa, and triggers the acrosome reaction upon interaction with bound sperm. mZPA seems to act as a secondary sperm receptor for acrosome reacted spermatozoa and plays a role in the prevention of post-fertilization polyspermy (Mortillo

& Wassarman,1991), and ZPB is thought to play mainly a structural role (Wassarman,1990; Litscher & Wassarman, 1996).

The sperm receptor activity of mouse ZP3 has been located with a restricted population of O-linked oligosaccharide chain(s) within a 55 kDa glycopeptide of the C-terminus (Litscher & Wassarman, 1996), a region of the protein which contains five serine residues and is encoded for exon 7 (Kinloch et al., 1995).

The amino acid sequences of the glycoproteins of the zona pellucida of several vertebrate species (mammals, fish, and anphibians) have been derived by cDNA cloning (Harris et al., 1994; Epifano et al., 1995b; Hedrick, 1996; and references therein). These studies show that all three ZP genes appear to be highly conserved during evolution. Thus, the primary structure of the Xenopus laevis egg vitelline coat denominated gp37, gp41/43, and gp69 correspond to mammalian ZPB, ZPC, and ZPA gene products, and show 30–45% amino acid sequence identity with the corresponding human, porcine, and murine zona pellucida proteins (Hedrick, 1996). Interestingly, a 260-amino-acid-spanning sequence found in both ZPA and ZPC from several mammalian species has a common characteristic amino acid pattern (and presumably tertiary structure) with the TGF-ß type III receptor, uromodulin, and the major zymogen granule membrane protein (GP-2). This ZP module defines a new family of mosaic proteins (Bork & Sander, 1992). In addition, a trefoil domain has been identified in ZPB, where it is located next to the ZP module (Bork, 1993; Otto & Wright, 1994). It appears, therefore, that ZPA, ZPB and ZPC display a modular structure, and that structural motifs have persisted through more than 650 million years of evolution.

The high amino acid sequence conservation would suggest that the biological role of each zona pellucida glycoprotein might have been also preserved in different species. However, the biosynthetic pathway and posttranslational processing of zona pellucida glycoproteins from different species is largely unknown, although these processes are likely to be important for both, the assembly of the macromolecular structure of the zona pellucida and the fertilization function of the glycoproteins. Thus, in contrast to what happens in the mouse, results from studies in rabbit, cynomolgus monkey, and pig indicate that each component of the zona pellucida is produced and secreted at specific stages of folliculogenesis, and that granulosa cells may be contributing to the synthesis of zona pellucida glycoproteins (Sinowatz et al., 1995; Martinez et al., 1996). Moreover, the mouse paradigm has not been tested in other mammals, and may not be representative of gamete interaction in other species. Indeed, it does not hold for the pig. Studies on porcine zona pellucida indicate that the sperm receptor activity is mainly associated with oligosaccharide chains attached to porcine 55 kDa α-protein, a member of the ZPB gene product family, although other zona pellucida glycoproteins have also a small but significant inhibitory effect on sperm-zona pellucida binding (Sinowatz et al., 1997). Similarly, rabbit RC55 glycoprotein, the homologue of porcine 55 kDa α-protein, appears to be involved in binding of sperm to the zona pellucida (Dunbar et al., 1994). Relevant to this point is the observation that variations in the distribution of sugar residues in the zona pellucida may contribute to the expression of species-specific determinants of mammalian oocytes (Skutelsky et al., 1994). Hence, acquisition of sperm receptor activity may be a consequence of species-specific, and perhaps tissue-specific, glycosylation and oligosaccharide processing of the distinct zona pellucida proteins.

3. CARBOHYDRATE STRUCTURES OF THE ZONA PELLUCIDA

Zona pellucida glycoproteins carry a complex pattern of both N- and O-linked oligosaccharides. Current knowledge on the structure of the oligosaccharides of mammalian zona pellucida glycoproteins is very limited. Recently, the ¹H-NMR solution structure of the major N- and O-linked carbohydrate chains of a mixture of porcine ZPB and ZPC glycoproteins, the N-linked sugar structures of a pool of mouse ZPA and ZPC, and the N-linked carbohydrate structures of unfractionated cow zonae pellucidae, have been reported (reviewed by Sinowatz et al., 1997; Katsuma et al., 1996). However, studies on the carbohydrate attachment sites; the occurrence of site heterogeneity (porcine ZPB and ZPC are each comprised of 30% (w/w) carbohydrate with three potential N-glycosylation sites each and three (ZPB) or six (ZPC) O-linked oligosaccharides, but more than 60 carbohydrate structures have been characterized!); the distribution of the carbohydrate chains in the three-dimensional structure of the intact zona pellucida network; and the biological relevance of particular oligosaccharides, are in their infancy.

The porcine ZPB and ZPC O-linked oligosaccharides consists of linear core structures of the type $[Gal\beta1,4\text{-}GlcNAc\beta1,3]_n\text{-}Gal\beta1,3\text{-}GalNAc\text{-}ZP$ which, in the case of the acidic carbohydrate chains are extended by polylactosamine structures of the type: $[NeuAc/NeuNGc]_{0-1}\text{-}a2,3\text{-}[Gal\beta1,4\text{-}GalNAc(6\text{-}OSO_3H)]_n$ $\beta1,3\text{-}Gal\beta1,4\text{-}core$. 16% of O-linked neutral chains contain non-reducing GlcNAc residues, 60% display terminal b1,4-linked galactose residues, 2% show the terminal Galα1,3-Gal sequence, and 22% end with GalNAc.

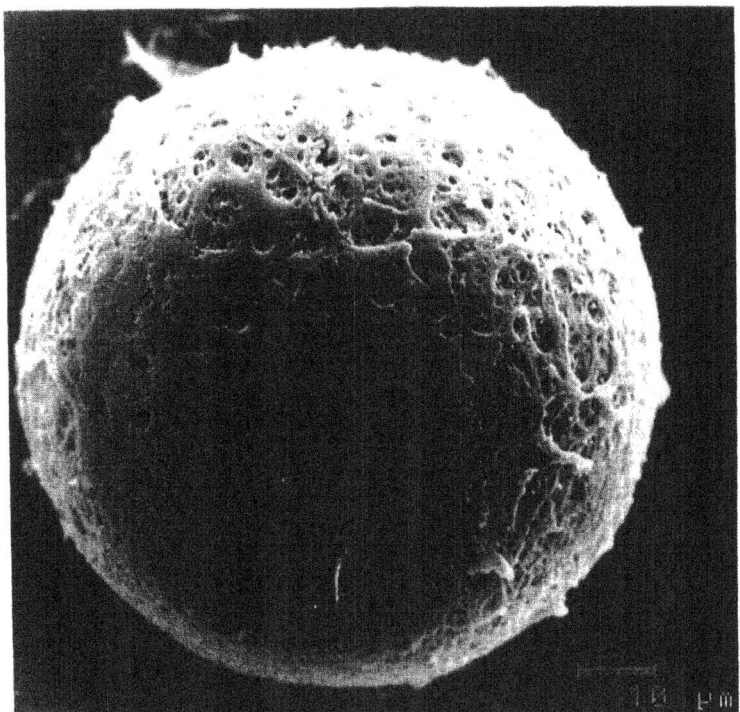

Figure 2. Schematic representation of carbohydrate structures isolated from a mixture of porcine ZPB and ZPC, and characterized by ¹H-NMR. The outer branches of acidic N-linked (A) and sulfated O-linked (B) oligosaccharides are made up by combinations of modules labeled (a), (b), and (c) in the upper carbohydrate structure. The structures of neutral N- and O-linked oligosaccharides are depicted in (C) and (D), respectively. (E) shows the structures of O-linked anionic carbohydrate chains. NeuNAc, N-acetyl neuraminic acid; NeuNGc, N-glycolyl neuraminic acid; Gal, galactose; GlcNAc, N-acetyl glucosamine; GalNAc, N-acetyl galactosamine; Fuc, fucose.

The N-linked glycans of porcine and murine zonae pellucidae possess basically the same structure, which consists of a fucosylated complex-type core. However, they have heterogeneous polylactosamine chains of the same type as the porcine O-linked oligosaccharides shown above, which in the case of the acidic glycans present N-acetyl neuraminic and N-acetyl glycolyl acid residues at the non-reducing ends and sulphate at the C6 position of N-acetylglucosamine residues (Fig. 2).

The ratio of di-/tri-/tetra-antennary chains, the amount of neutral chains, the sulfate and sialic acid contents, as well as the distribution of sialic acid differ between these species. In addition, mouse zona pellucida glycoproteins possess ß-GalNAc and α-Gal residues at the non-reducing ends of N-linked chains, which are absent from the porcine chains. On the other hand, in bovine, the major neutral chain is a high-mannose-type oligosaccharide and the acidic chains are di-, tri-, and tetra-antennary, fucosylated complex-type chains that have N-acetyllactosamine repeats in the non-reducing ends. High-mannose-type and hybrid type oligosaccharides are not present in porcine and mouse zona proteins. In addition, the amount of sulphate, which is high in the porcine zona pellucida glycoproteins, is very low in the bovine N-linked chains.

4. CARBOHYDRATE STRUCTURES DETERMINING GAMETE RECOGNITION

Terminal α-linked galactose (Bleil & Wassarman,1988) and terminal ß1,4-linked N-acetylglucosamine (Miller et al., 1992; Gong et al., 1995) residues have been suggested to be essential for the sperm receptor activity of mouse ZPC. This is a matter of debate, however. Hence, Thall et al. (1995) have demonstrated that genetic-engineered mice lacking Galα1,3Gal epitopes are fertile, indicating that terminal galactose residues are not strictly required for sperm adhesion to the zona pellucida. On the other hand, Litscher and colleagues (1995) have shown that oligosaccharide constructs bearing terminal (α- or ß-linked) galactose inhibit binding of mouse sperm to unfertilized eggs in vitro, while those containing terminal GlcNac were not active.

In the pig, apparently contradictory results have been reported. Yurewicz and co-workers have shown that the O-linked glycans attached to ZPB carry the bulk of the sperm-binding activity of intact zonae pellucidae (Yurewicz et al., 1991). The smallest of the O-linked-type oligosaccharide bearing sperm-zona pellucida binding inhibitory activity has been recently synthesized, and has the hexasaccharide structure: Galß1,4(HSO_3–6)GlcNAcß1,3-Galß1,4-GlcNAcß1,3-Galß1,3-GalNAc-ol (Spijker et al., 1996). On the other hand, Nakano and colleagues (Yonezawa, 1995; Nakano et al.,1996) have presented evidence that endo-ß-galactosidase-treated zona pellucida (EbG-ZP) retained sperm-binding inhibitory activity, and that the elimination of N-linked carbohydrate chains from EßG-ZPB markedly reduced its inhibitory effect, whereas release of O-linked carbohydrate chains hardly reduced the inhibitory activity. The authors concluded that the tri- or tetraantennary neutral N-linked carbohydrate chains of porcine EßG-ZPB may play a major role in mediating binding of sperm to the zona pellucida. Competition assays using endo-Lys-C-derived peptides of EßG-ZPB indicated that the active oligosaccharides are linked to Asn[67] and/or Asn[84] (Nakano et al., 1996).

Our own results show that glycopeptides of fetuin carrying O-linked oligosaccharides (binding in the 1–10 μM range) interfere with sperm-zona pellucida interactions, whereas those glycopeptides containing N-linked tri- and tetraantennary carbohydrate structures were effective in higher (>100 μM) concentrations (Blase, 1996).

These apparently contradictory results can be reconciled, assuming that the O-linked oligosaccharides account for the sperm receptor activity of intact porcine ZPB, and that this zona pellucida glycoprotein possesses, in addition, a cryptic sperm-binding site exposed after endo-β-glycosidase digestion.

5. SPERM-ASSOCIATED ZONA PELLUCIDA-BINDING LECTINS

A number of sperm proteins with affinity for zona pellucida glycoproteins have been identified in several mammalian species using different approaches, such as probing Western blots of sperm proteins with labelled zona pellucida proteins, identification of the epitopes for specific polyclonal or monoclonal antisera which inhibit sperm-zona binding, and crosslinking reagents. However, only few of the putative zona pellucida-binding proteins have been biochemically or structurally characterized, and the actual participation of any of these proteins in the in vivo process has not been demonstrated.

Sperm proteins known to have specific zona pellucida- and carbohydrate-binding sites include the mouse proteins β-1,4- galactosyltransferase - GalTase- and sp56; rabbit Sp17; a C-type human mannose-binding protein - MBP-; and the members of the porcine spermadhesin protein family known as AQN-1, AQN-3, and AWN (reviewed by Töpfer-Petersen & Calvete, 1996; Sinowatz et al., 1997). Mouse GalTase, rabbit Sp17, and human MBP are sperm type-I membrane proteins, whereas mouse sp56 and the porcine spermadhesins are peripherally associated proteins.

GalTase and sp56 bind GlcNAc and Gal residues, respectively, the sugar moieties that have been implicated in sperm-egg interaction in the mouse. On the other hand, boar spermadhesins AQN-1, AQN-3, and AWN display affinity for [± NeuAcα2,3]-Galβ1,3-GalNAc and, to a minor extent, for [± NeuAcα2,3]-Galβ1,4-GlcNAc-containing saccharides (Töpfer-Petersen & Calvete, 1996; Calvete et al., 1996). These sequences are abundant in the biologically active glycoconjugates of porcine zona pellucida. Moreover, mannose residues on the human zona pellucida have been suggested to be involved in sperm-egg recognition, and Benoff (1993) has provided evidence for the presence of a sperm membrane C-type mannose-binding lectin on human spermatozoa subjected to capacitating procedures. The sequence of a complete cDNA of the human sperm mannose-binding lectin which has been completed but not yet been published, encodes a protein with seven putative mannose-like-binding CRDs in a similar domain organization as the macrophage mannose receptor.

As a whole, these data support the concept formulated at the beginning of the present century that gamete recognition is mediated by specific complementary molecules located on the external surfaces of the spermatozoon and the oocyte (Lillie, 1913). The matter is rather more complex, however. Hence, apart from the sperm lectins cited above, a number of other zona pellucida binding proteins, with uncharacterized binding specificity, have been documented: (reviewed by Töpfer-Petersen et al., 1995; Töpfer-Petersen & Calvete, 1996). The wide variety of putative zona pellucida binding molecules identified, even in a single species, would indicate that fertilization is most likely mediated by multiple complementary receptor-ligand systems (Thaler & Cardullo, 1996). Different CRDs are able to discriminate different determinants among all the glycans presented. In addition, high-affinity protein-carbohydrate interaction is usually attained through oligomerization of CRDs (multivalency) (Rini, 1995). Thus, acting together, different sperm-associated zona pellucida-binding proteins may contribute in a complementary or hierarchical manner to the interactions, and cross-talking mechanisms between sperm and egg, which may collectively dictate the species-specificity of gamete recognition.

6. ACKNOWLEDGMENTS

The authors wish to thank Dr. Schwartz (Anatomisches Institut der Universität Göttingen) for the electron micrograph shown in Figure 1. The work of the authors has been financed by grants from the Bundesministerium für Bildung, Forschung und Technologie, Bonn, Germany (01KY9505) and from the Deutsche Forschungsgemeinschaft, Bonn, Germany (Tö 114/3-1 and Ca 209/1-1).

7. REFERENCES

Benoff S 1993 Preliminaries to fertilization. The role of cholesterol during capacitation of human spermatozoa. Human Reproduction 8: 2001–2008.

Blase N 1996 Doctoral thesis, Veterinary University Hannover, Germany.

Bleil JD & Wassarman PM (1988) Galactose at the nonreducing terminus of O-linked oligosaccharides of mouse egg zona pellucida glycoprotein ZP3 is essential for the glycoprotein's sperm receptor activity. Proceedings of the National Academy of Science of the USA 85: 6778–6782.

Bork P 1993 A trefoil domain in the major rabbit zona pellucida protein. Protein Science 2: 669–670.

Bork P & Sander C 1992 A large domain common to sperm receptors (Zp2 and Zp3) and TGF-β type III receptor. Federation of the European Biochemical Societies Letters 300: 237–240.

Calvete JJ, Carrera E, Sanz L & Töpfer-Petersen E 1996 Boar spermadhesins AQN-1 and AQN-3: oligosaccharide and zona pellucida binding characteristics. Biological Chemistry 377: 521–527.

Castle PE & Dean J 1996 Molecular genetics of the zona pellucida: implications for immunocontraceptive strategies. Journal of Reproduction and Fertility Supplement 50: 1–8.

Dunbar BS, Avery S, Lee V, Prasad S, Schwahn D, Schwoebel E, Skinner S & Wilkins B 1994 The mammalian zona pellucida: its biochemistry, immunochemistry, molecular biology, and developmental expression. Reproduction, Fertility and Development 6: 331–347.

Epifano O & Dean J 1994 Biology and structure of the zona pellucida: a target for immunocontraception. Reproduction, Fertility and Development 6: 319–330.

Epifano O, Liang L-F, Familari M, Moos MCjr & Dean J 1995a Coordinate expression of the three zona pellucida genes during mouse oogenesis. Development 121: 1947–1956.

Epifano O, Liang L-F & Dean J 1995b Mouse Zp1 encodes a zona pellucida protein homologous to egg envelope proteins in mammals and fish. Journal of Biological Chemistry 270: 27254–27258.

Gong X, Dubois DH, Miller DJ & Shur BD 1995 Activation of a G protein complex by aggregation of β-1,4-galactosyltransferase on the surface of sperm. Science 269: 1718–1721.

Harris JD, Hibler DW, Fonteno GK, Hsu KT, Yurewicz EC & Sacco AG 1994 Cloning and characterization of zona pellucida genes and cDNAs from a variety of mammalian species: the ZPA, ZPB and ZPC gene families. DNA Sequence 4: 361–393.

Hedrick JL 1996 Comparative structural and antigenic properties of zona pellucida glycoproteins. Journal of Reproduction and Fertility Supplement 50: 9–17.

Katsumata T, Noguchi S, Yonezawa N, Tanokura M, & Nakano M 1996 Structural characterization of the N-linked carbohydrate chains of the zona pellucida glycoproteins from bovine ovarian and fertilized eggs. European Journal of Biochemistry 240: 448–453.

Laine RA 1994 A calculation of all possible oligosaccharide isomers, both branched and linear yields 1.05 x 10^{12} structures or a reducing hexasaccharide: the isomer barrier to development of single method saccharide sequencing or synthesis systems. Glycobiology 4: 1–9.

Lillie F R 1913 The mechanism of fertilization. Science 38: 524–528.

Litscher ES, Juntunen K, Seppo A, Penttilä L, Renkonen O & Wassarman PM 1995 Oligosaccharide constructs with defined structures that inhibit binding of mouse sperm to unfertilized eggs in vitro. Biochemistry 34: 4662–4669.

Litscher ES & Wassarman PM 1996 Characterization of a mouse ZP3-derived glycopeptide, gp55, that exhibits sperm receptor and acrosome reaction-inducing activity in vivo. Biochemistry 35: 3980–3985.

Liu C, Litscher ES, Mortillo S, Sakai Y, Kinloch RA, Stewart CL & Wassarman PM 1996 Targeted disruption of the mZP3 gene results in production of eggs lacking a zona pellucida and infertility in female mice. Proceedings of the National Academy of Science of the USA 93: 5431–5436.

Martinez ML, Kontenot GK & Harris JD 1996 The expression and localization of zona pellucida glyoproteins and mRNA in cynomolgus monkeys (Macaca fascicularis). Journal of Reproduction and Fertility Supplement 50: 35–41.

Miller DJ, Macek MB & Shur BD 1992 Complementarity between sperm surface β-1,4-galactosyltransferase and egg-coat ZP3 mediates sperm-egg binding. Nature 357: 589–593.

Mortillo S & Wassarman PM 1991 Differential binding of gold-labeled zona pellucida glycoproteins mZP2 and mZP3 to mouse sperm membrane compartments. Development 113: 141–149.

Nakano M, Yonezawa N, Hatanaka Y & Noguchi S 1996 Structure and function of the N-linked carbohydrate chains of pig zona pellucida glycoproteins. Journal of Reproduction and Fertility Supplement 50: 25–34.

Otto B & Wright N 1994 Coming up clover. Current Biology 4: 835–838.

Rini JM 1995 Lectin structure. Annual Review of Biophysical and Biomolecular Structure 24: 551–577.

Sharon N & Lis H 1997 Glycoproteins: structure and function. In Glycosciences. Status and perspectives, pp 133–162. Eds H-J Gabius & S Gabius. Chapman & Hall, Weinheim.

Sinowatz F, Amselgruber W, Töpfer-Petersen E, Totzauer I, Calvete J & Plendl J 1995 Immunocytochemical characterization of porcine zona pellucida during follicular development. Anatomy and Embryology 191: 41–46.

Sinowatz F, Töpfer-Petersen E & Calvete JJ 1997 Glycobiology of Fertilization. In Glycosciences. Status and perspectives, pp 595–610. Eds H-J Gabius & S Gabius. Chapman & Hall, Weinheim.

Skutelsky E, Ranen E & Shalgi R 1994 Variations in the distribution of sugar residues in the zona pellucida as possible species-specific determinants of mammalian oocytes. Journal of Reproduction and Fertility 100: 35–41.

Spijker NM, Keuning CA, Hooglugt M, Veeneman GH & van Boeckel CAA 1996 Synthesis of a hexasaccharide corresponding to a porcine zona pellucida fragment that inhibits porcine sperm-oocyte interaction in vitro. Tetrahedron 52: 5945–5960.

Thaler CD & Cardullo RA 1996 The initial molecular interaction between mouse sperm and the zona pellucida is a complex binding event. Journal of Biological Chemistry 271: 23289–23297.

Thall AD, Maly P & Lowe JB 1995 Oocyte Galα1,3Gal epitopes implicated in sperm adhesion to the zona pellucida glycoprotein ZP3 are not required for fertilization in the mouse. Journal of Biological Chemistry 270: 21437–21440.

Töpfer-Petersen E, Calvete JJ, Sanz L & Sinowatz F 1995 Carbohydrate- and heparin-binding proteins in mammalian fertilization. Andrologia 27: 303–324.

Töpfer-Petersen E & Calvete JJ 1996 Sperm-associated protein candidates for primary zona pellucida-binding molecules: structure-function correlations of boar spermadhesins. Journal of Reproduction and Fertility Supplement 50: 55–61.

Varki A 1993 Biological roles of oligosaccharides: all of the theories are correct. Glycobiology 3: 97–130.

Wassarman PM 1990 Profile of a mammalian sperm receptor. Development 108:1–17.

Wassarman PM, Liu C & Litscher ES 1996 Constructing the mammalian egg zona pellucida: some new pieces of an old puzzle. Journal of Cell Science 109: 2001–2004.

Yonezawa N, Aoki H, Hatanaka Y & Nakano M 1995 Involvement of N-linked carbohydrate chains of pig zona pellucida in sperm-egg binding. European Journal of Biochemistry 233: 35–41.

Yurewicz EC, Pack BA & Sacco AG 1991 Isolation, composition, and biological activity of sugar chains of porcine oocyte zona pellucida 55K glycoproteins. Molecular Reproduction and Development 30: 126–134.

53

X-RAY CRYSTALLOGRAPHIC ANALYSIS OF BOAR PSP-I/PSP-II COMPLEX

A Zona Pellucida-Binding Protein

A. Romero,[1] P. F. Varela,[1] L. Sanz,[2] E. Töpfer-Petersen,[2] and J. J. Calvete[1,2]

[1]Departamento de Cristalografia
Instituto de Química-Física
CSIC, Madrid, Spain
[2]Institut für Reproduktionsmedizin
Tierärztliche Hochschule
Bünteweg 15, 30559 Hannover, Germany

PSP-I and PSP-II are major proteins isolated from porcine seminal plasma (PSP). Both, PSP-I (109 amino acids) and PSP-II (116 amino acids) are glycoproteins and display site heterogeneity, i.e. they contain a single N-glycosylated asparagine residue (PSP-I N^{50} and PSP-II N^{98}) which accomodates different glycan moieties (Calvete et al.,1993; Calvete et al., 1995a) PSP-II forms a non-covalent heterodimer with certain glycoforms of PSP-I (Calvete et al., 1995a) . Both subunits of the heterodimer belong to the spermadhesin protein family (Calvete et al., 1995b). In the pig, this group of proteins include sperm surface-associated lectins that are thought to mediate the initial binding of spermatozoa to carbohydrate structures of zona pellucida glycoproteins (Dostàlovà et al., 1995; Calvete et al., 1996). The PSP-I/PSP-III heterodimer contains a binding site for zona pellucida glycoproteins located around PSP-II N^{50} (Calvete et al., 1995a) The zona pellucida-binding characteristics and easiness of isolation of PSP-I/PSP-II (Calvete et al., 1995a), makes this spermadhesin a paradigm for studying protein-carbohydrate interactions involved in gamete recognition. Moreover, spermadhesins are composed of a single CUB domain architecture (Bork et al., 1993) whose three-dimensional structure remains to be determined.

Lectin mapping, carbohydrate analysis and electrospray mass spectrometry of boar seminal plasma PSP-II glycoforms show that its single N-glycosylation site displays a repertoire of carbohydrate structures consisting of the basic pentasaccharide core Manα1–6[Manα1–3]Manβ1–4GlcNAcb1–4GlcNAc with a fucosyl residue α1–6-linked to the innermost N-acetylglucosamine residue. Other glycoforms display fucosylated hybrid-type or monoantennary complex-type chains, some of which contain α2–6-linked sialic acid. N-acetylgalactosamine, possibly in Galβ1–3GalNAc sequence, appear to be present in most of the PSP-II glycoforms (Solís et al., 1997). Initial analysis indicate that PSP-II-associated PSP-I glycoforms contain similar glycan chains, while those PSP-I glycoforms which do not form heterodimers with PSP-II are glycosylated with mannose-rich oligosaccharides (Calvete et al., 1993). This raises the possibility that PSP-II reclutes PSP-I glycoforms into a heterodimer through a carbohydrate recognition mechanism. If this hypothesis turns out to be correct, structural analysis of PSP-I/PSP-II may not only reveal the structure of the CUB domain, but also may shed light on the carbohydrate-binding mechanism used by spermadhesin PSP-II to bind zona pellucida glycoconjugates.

We have recently crystallized the glycoprotein PSP-I/PSP-II heterodimer (Romero et al., 1996). Hexagonal (space group $P6_{1,5}22$) and trigonal (space group $P3_{1,2}21$) crystals diffract beyond 2.9 and 2.5 Å resolution, respectively. The three-dimensional crystal structure of PSP-I/PSP-II, solved by the heavy atom derivative Multiple Isomorphous Replacement (MIR) method, shows that each spermadhesin molecule possesses a ß-sandwich fold with each sheet consisting of both antiparallel and parallel ß-strands. The PSP-I/PSP-II heterodimer surface involves homologous faces of the subunits.

The arrangement of amino acid side chains around PSP-II Asn[50] resemble the topology of the carbohydrate-recognition domains (CRDs) found in a number of animal and legume lectins. In addition, there is high positive electron density in the vicinity of PSP-II Asn[50]. This might indicate that the carbohydrate-binding site of PSP-II is actually filled with a sugar residue(s). At present, however, we can determine neither the identity of this putative carbohydrate, nor whether this moiety is part of an oligosaccharide chain of PSP-I (of the same heterodimer or of a neighbor molecule) or forms part of a PSP-II glycan chain. Studies are underway to assess this question, which is relevant for establishing the carbohydrate binding specificity of PSP-II and, by analogy, a model for the spermadhesin-zona pellucida interaction. Nevertheless, our data confirm the relative location of the carbohydrate binding site within the spermadhesin (CUB) fold.

ACKNOWLEDGMENTS

This work has been supported by grant from the Deutsche Forschungsgemeinschaft, Bonn, Germany (Tö 114/3-1 and Ca 209/1-1) and the Dirección General de Investigación Científica y Técnica, Madrid, Spain (PB93-0120 and PB92-0096).

REFERENCES

Bork P & Beckmann G 1993 The CUB domain. A widespread module in developmentally regulated proteins. Journal of Molecular Biology 231: 539–545.

Calvete JJ, Solís D, Sanz L, Díaz-Mauriño T, Schäfer W, Mann K & Töpfer-Petersen E 1993 Characterization of two glycosylated boar spermadhesins. European Journal of Biochemistry 218: 719–725.

Calvete JJ, Mann K, Schäfer W, Raida M, Sanz L & Töpfer-Petersen E 1995a Boar spermadhesin PSP-II: location of posttranslational modifications, heterodimer formation with PSP-I glycoforms and effect of dimerization on the ligand-binding capabilities of the subunits. FEBS Letters 365: 179–182.

Calvete JJ, Sanz L, Dostàlovà Z & Töpfer-Petersen E 1995b Spermadhesins: sperm-coating proteins involved in capacitation and zona pellucida binding. Fertilität 11: 35–40.

Calvete JJ, Carrera E, Sanz L & Töpfer-Petersen E 1996 Boar spermadhesins AQN-1 and AQN-3: oligosaccharide and zona pellucida binding characteristics. Biological Chemistry 377: 521–527.

Dostàlovà Z, Calvete JJ, Sanz L & Töpfer-Petersen E 1995 Boar spermadhesin AWN-1. Oligosaccharide and zona pellucida binding characteristics. European Journal of Biochemistry 230: 329–336.

Romero A, Varela PF, Sanz L, Töpfer-Petersen E & Calvete JJ 1996 Crystallization and preliminary X-ray diffraction analysis of boar seminal plasma spermadhesin PSP-I/PSP-II, a heterodimer of two CUB domains. FEBS Letters 382: 15–17.

Solís D, Calvete JJ, Sanz L, Hettel, Raida M, Díaz-Mauriño T & Töpfer-Petersen E 1997 Fractionation and characterization of boar seminal plasma spermadhesin PSP-II glycoforms reveal the presence of uncommon N-acetylgalactosamine-containing N-linked oligosaccharides. Glycoconjugate Journal 14: 57–66.

54

THE ZONA PELLUCIDA "RECEPTORS"

E. Hinsch, W. Hägele, W.-B. Schill, and K.-D. Hinsch

Center of Dermatology and Andrology
Justus-Liebig-Universität Giessen
Gaffkystr. 14, 35385 Giessen
Germany

1. SUMMARY

Binding of mammalian sperm to the zona pellucida and the induction of the acrosome reaction are prerequisites for successful oocyte fertilization. In the mouse model, the zona pellucida consists of three sulfated glycoproteins, ZP1, ZP2, and ZP3. Zona pellucida proteins are secreted to form a filamentous zona matrix in which ZP2 and ZP3 complex into co-polymers cross-linked by ZP1. ZP3 is the ligand for primary sperm binding and important for the induction of the acrosome reaction. The zona pellucida glycoprotein ZP2 is also crucially involved in the process of fertilization. Previous reports suggest that ZP2 mediates secondary binding of spermatozoa and that cleavage of ZP2 by proteases released through cortical granule reaction causes zona "hardening" and thus prevents polyspermy. Human and mouse ZP2 proteins differ in the primary structure as derived from cDNA clones. We designed an immunological approach to search for ZP2 domains with functional relevance. Antisera were generated against synthetic peptides derived (a) from ZP2 amino acid sequences that are homologous in human and mouse ZP2 amino acid sequences (AS ZP2–20) or (b) from human ZP2 amino acid sequences that differ from the mouse ZP2 sequence (AS ZP2–26). Immunochemical studies with microbisected bovine zonae pellucidae demonstrated that both antisera, AS ZP2–20 and AS ZP2–26, specifically detected ZP2 protein. Using the competition-hemizona-assay, sperm binding to antibody treated bovine hemizonae pellucidae were compared with control hemizonae (given as hemizona index). Antiserum AS ZP2–20 significantly inhibited binding of spermatozoa to test hemizonae ($p<0.0001$), whereas treatment of hemizonae with AS ZP2–26 did not influence sperm-egg interaction. Our results show that antibodies against ZP2 peptides react with bovine zonae pellucidae and can be used as markers for ZP2. Furthermore, AS ZP2–20 identifies a ZP2 epitope that is possibly of functional relevance for sperm-egg interaction.

The Fate of the Male Germ Cell, edited by Ivell and Holstein
Plenum Press, New York, 1997

2. INTRODUCTION

2.1. The Sperm-Zona Pellucida Interaction

The male component necessary for successful reproduction depends on a large variety of biological processes working in concert (Yanagimachi, 1994). Following ejaculation, mammalian spermatozoa acquire full fertilizing ability during residence in the female reproductive tract. The physiological changes that spermatozoa undergo until they are able to perform the acrosome reaction are collectively termed as capacitation. At the endpoint of this process, capacitated spermatozoa bind to the zona pellucida and undergo acrosomal exocytosis. The acrosome reaction is an irreversible exocytotic event and consists of fusion of the outer acrosomal membrane with the plasma membrane and the dispersion of its enzymatic components. Subsequently, acrosome-reacted spermatozoa penetrate the zona pellucida and fertilize the oocyte by fusing with the oolemma and entering the oocyte (Figure 1). As a consequence of fertilization, extensive release of cortical granules occurs. The contents of cortical granules are proposed to modify the zona pellucida (zona "hardening") resulting in a block for polyspermy (Wassarman, 1987).

Sperm-zona pellucida interaction is a complex phenomenon, and its molecular mechanisms are not yet well understood. At least three steps can be distinguished. First, binding of acrosome intact spermatozoa with the zona pellucida; second, binding of acrosome-reacting and acrosome-reacted spermatozoa with the zona; and third, penetration of acrosome-reacted spermatozoa through the zona. The first two processes are designated as the primary and secondary zona binding, respectively. However, most studies in the past did not distinguish between the first two steps, and it is not clear whether the

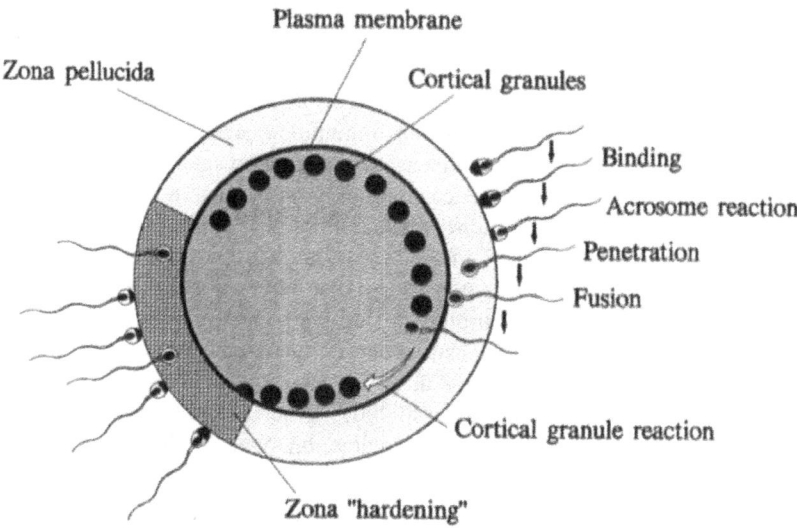

Figure 1. Diagram of the fertilization pathway in mice. The sequence of events includes attachment of sperm to the zona pellucida, followed by binding of sperm to the zona, completion of acrosome reaction, penetration through the zona pellucida, sperm-egg fusion, cortical granule reaction, and zona "hardening" (modified from Wassarman, 1987).

many molecules that were proposed to be involved in sperm-zona interaction are of functional importance for primary and/or for secondary zona binding (Yanagimachi, 1994).

2.2. The Mammalian Zona Pellucida

In mammals, the egg is protected by the zona pellucida, a relatively thick, translucent, acellular coat surrounding the plasma membrane of the oocyte. The zona pellucida appears during oocyte growth and increases in width as oocytes increase in diameter (Wassarman & Mortillo, 1991). The size and rigidity of the zona pellucida varies from one species to another, as does the thickness (e.g. 1 - 2 μm in marsupials and about 16 μm in pigs) (Dunbar, 1983). In general, the outer zona pellucida surface of a mature oocyte has a fenestrated sponge-like structure, whereas the inner surface shows a fine granular (Phillips & Shalgi, 1982) or microtubular appearance (Familiari et al., 1992). The spongy structure of the outer zona may reflect the maturity of the sperm penetrability, because it has been shown that human zonae pellucidae with a more compact and smooth outer surface are less penetrable than those with a spongy appearance (Familiari et al., 1988).

In the mouse, the best studied species so far, the zona pellucida of fully grown oocytes is about 5 μm thick (Dunbar, 1991), contains about 3 ng of protein, and is permeable to large macromolecules. In early stages of oocyte growth, zona pellucida material appears as patches of fine filaments between the oocyte and follicle cells. As growth continues, the zona pellucida becomes a denser and thicker meshwork of interconnected filaments completely surrounding the oocyte and largely separating it from the follicle cells. The zona pellucida contains sperm "receptors" that mediate sperm-egg interaction as a prelude to fertilization; it also participates in a secondary block for polyspermy following fertilization of eggs (Wassarman, 1988).

Depending on species, the mammalian zona pellucida comprises different polypeptides (Dunbar, 1991). These protein components have a polypeptide backbone varying in glycosylation and thus yielding extensive heterogeneity of charge. The mouse zona pellucida consists of three sulfated glycoproteins, named ZP1 (185 - 200 kDa), ZP2 (120 - 140 kDa), and ZP3 (83 kDa) (Wassarman, 1988). Likewise, the zonae pellucidae of hamster, human, rabbit, rat, and horse are each composed of three families of glycoproteins; the porcine zona consists of three families of acidic glycoproteins and five different components, whereas the cat has only two families of glycoproteins (Yanagimachi, 1994). However, there is confusion in the literature as to the nomenclature of zona pellucida proteins in different species. The terms ZP1, ZP2, and ZP3 usually refer to molecular masses. With the advances in molecular biology, it became apparent that a more direct approach to answer the questions about the composition and functions of the zona pellucida in different species was to clone the genes that code for the zona pellucida proteins. Recently, full length cDNA clones of ZP1, ZP2 or ZP3 proteins for different mammalian species have been isolated (Harris et al., 1994; Epifano et al., 1995). Genes encoding ZP2 and ZP3 have been shown to be conserved among mammals. DNA sequences of ZP3 cDNA-coding regions show extensive homology between species studied so far. Because of the confusion in nomenclature of particular zona pellucida proteins with distinct apparent molecular masses in different species, the terms ZPA, ZPB, and ZPC (encoding for mouse ZP2, ZP1, and ZP3, respectively) have been coined to express functions and respective genes coding for these proteins (Harris et al., 1994).

2.2.1. Zona Pellucida "Hardening". Following fertilization, the zona pellucida is modified in the "hardening reaction" to prevent polyspermy (Fig. 1). Cortical granule exocytosis is the likely cause for this modification. Partial hydrolysis of zona proteins following fertilization was reported in various animals, including rat, mouse, pig and human.

In the mouse, both ZP2 and ZP3 are converted to a form called $ZP2_f$ and $ZP3_f$, respectively. Not all molecules are converted to $ZP2_f$ or $ZP3_f$, but it is sufficient to render the zona impermeable by spermatozoa. $ZP3_f$ does not induce acrosomal exocytosis, and it does not bind to acrosome intact spermatozoa anymore. In contrast to $ZP2_f$, no apparent change of electrophoretic mobility of $ZP3_f$ was observed (Bleil & Wassarman, 1980a,b, 1983). Because O-linked ZP3 carbohydrates play an important role for sperm-ZP3 binding, the loss of biological activity has been proposed to be caused by the release of cortical granule-derived N-acetylglucosaminidase (Miller et al., 1993).

Enzymatically converted ZP2 protein, $ZP2_f$, no longer interacts with acrosome-reacted spermatozoa (Bleil & Wassarman, 1986). It arises by a proteolytic cleavage that is mediated by a cortical granule derived non-trypsin-like protease (Moller & Wassarman, 1989) and can be visualized by a shift in electrophoretic mobility under reducing conditions. In the mouse, it becomes a 90 kDa glycoprotein which remains as a part of the zona due to persisting disulfide bonds.

It should be noted that species-specific differences can be observed in the mechanisms by which eggs of a species prevent polyspermy. This can be explained by a further mechanism that takes place at the oolemma, the plasma membrane block. Austin (1961), who made those observations, consequently divided mammals into three groups: group I, depending primarily on the zona reaction, group II depends primarily on the plasma membrane block, and group III depending on both mechanisms. Importantly, in all mammals the zona pellucida and the egg plasma membrane work synergistically to reduce the chance of polyspermy. In other words, eggs of all species have at least two safeguards against polyspermy (Yanagimachi, 1994).

2.2.2. The ZP1 (ZPB) Protein. ZP1 provides the structural integrity of the zona pellucida network. The molar ratios of mouse ZP1, ZP2, and ZP3 are 1:10:10 (Moos et al., 1995). Recently, arrangements of glycoproteins in mouse zona filaments based on electron microscopy, immunological, and chemical cross-linking experiments were proposed (Wassarman, 1988). As shown in Figure 2, the zona pellucida network consists of ZP2/ZP3 heterodimers that are stabilized by ZP1 through intermolecular disulfide bonds. Negatively stained zona pellucida filaments resemble "beads on a string", with each bead (9.5 nm in diameter) located about every 17 nm (center to center distance) along the axis of the filament. The filaments are interconnected by ZP1, giving rise to a three-dimensional matrix. Proteolysis of ZP1 as well as reduction of intermolecular disulfides lead to solubilization of zonae pellucidae and disruption of interconnections between individual zona pellucida filaments (Greve & Wassarman, 1985). In the hamster, ZP1 has an apparent molecular mass of about 200 kDa under non-reducing conditions and migrates in the presence of reducing conditions on SDS-PAGE with an apparent molecular mass of 103 kDa (Moller et al., 1990). These data indicate that hamster ZP1, like mouse ZP1, is composed of two polypeptides held together by intermolecular disulfides. Full length ZP1 (ZPB) cDNA clones and genes were cloned and characterized from a variety of mammalian species (Harris et al., 1994; Epifano et al., 1995). The single copy mouse ZP1 gene at the carboxy-terminal half is significantly similar to a corresponding region of mouse ZP2. Furthermore, it has been reported that mouse ZP1 encodes a zona pellucida protein homologous to egg envelope proteins in mammals and fish. The conservation of this region in fish egg envelope indicates that this protein domain has been conserved in species that diverged 650 million years ago (Epifano et al., 1995).

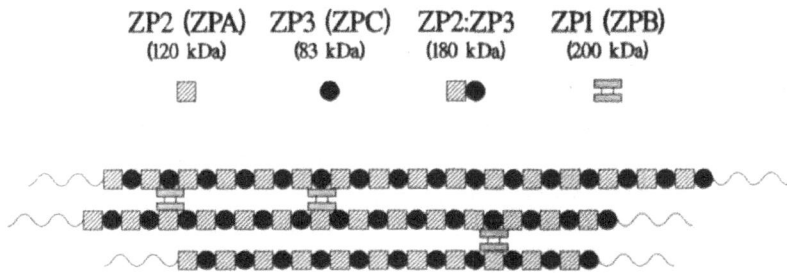

Figure 2. Diagram of the arrangement of glycoproteins in mouse zona pellucida filaments. This model features polymerization of ZP2-ZP3 dimers into long zona pellucida filaments, with filament interconnections attributable to ZP1. (From Wassarman et al., 1989.)

2.2.3. The ZP2 (ZPA) Receptor. The mouse ZP2 (ZPA) protein is a sulfated zona pellucida glycoprotein with an apparent molecular mass of 120 kDa. ZPA cDNA clones and genes were cloned and characterized from various mammals, including human, mouse, cat, dog, pig, and rabbit (Harris et al., 1994). ZP2 is the molecule responsible for a mechanism called "secondary binding" of sperm to the egg's glycocalix.

Followed by the ZP3-induced acrosome reaction, the spermatozoon, which can no longer interact with ZP3, is then postulated to bind to ZP2 (Bleil et al., 1988). This interaction is thought to be required for the second step of sperm-zona pellucida interaction by maintaining the association of the spermatozoon with the zona as it progresses through this extracellular coat. ZP2 thus functions as a secondary "receptor". The most probable complementary sperm protein(s) for ZP2 are, thus far, proacrosin/acrosin. These enzymes are believed to be located at the inner acrosomal membrane, which displays sites that recognize the zona (Yanagimachi, 1981) and bind to ZP2 (Bleil et al., 1988; Mortillo & Wassarman, 1991). Barros et al. (1992) detected acrosin on the inner acrosomal membrane by immunological methods and found that acrosin of hamster spermatozoa disappeared rather quickly after the acrosome reaction, whereas that of guinea pig spermatozoa did not. In earlier reports a fucoidin-binding compound was localized on the inner acrosomal membrane and on or in the equatorial segment of guinea pig spermatozoa that underwent acrosomal exocytosis (Huang & Yanagimachi, 1984). The fucoidin-binding substance could be proacrosin/acrosin, because in boar spermatozoa this fucoidin-binding protein has been identified as acrosin (Töpfer-Petersen & Henschen, 1987). However, Richardson and co-workers (1991) showed that, after zona pellucida-induced acrosomal exocytosis in rabbit spermatozoa, acrosin is located in the anterior region of the postacrosomal area and not in the inner acrosomal membrane. Tesarik et al. (1990) reported that in acrosome-reacted human spermatozoa acrosin is mainly found in the equatorial segment and, to a lesser extent, also in the inner acrosomal membrane. In the golden hamster it has been shown that, after sperm had undergone acrosomal exocytosis, the less diffusable portion of contents (including pieces of membranes) is firmly attached to the zona pellucida (Yanagimachi & Phillips, 1984). This remnant or acrosomal ghost displays strong acrosin activity and has been discussed to facilitate tight sperm adhesion to and penetration through the zona pellucida (Tesarik et al., 1988; Yanagimachi & Teichmann, 1972). PH-20, a 64 kDa protein with hyaluronidase activity that migrates from the postacrosomal region to the inner acrosomal membrane after the acrosome reaction (Cowan et al. 1991), has been proposed as a possible primary and/or secondary zona receptor for guinea pig and cynomologus macaque spermatozoa (Overstreet et al., 1995).

The different views and uncertainties in regard to chemical nature and site of the sperm counterpart(s) for the secondary zona receptor reflect the ignorance in this field of research and thus demands intensive efforts to bring light into this "black box" of science.

2.2.4. The ZP3 (ZPC) Receptor. The mouse ZP3 (ZPC) protein is a sulfated zona pellucida glycoprotein with an apparent molecular mass of 83 kDa. ZPC cDNA clones and genes were cloned and characterized from various mammals, including human, hamster, cow, mouse, cat, dog, pig, rabbit (Harris et al., 1994) and marmoset (Thillai-Koothan et al., 1993). In the mouse, the ZP3 protein is responsible for species-specific binding of spermatozoa to the zona pellucida and for the induction of the acrosome reaction (Wassarman, 1990a,b; Saling, 1991).

It is not known whether ZP3 binding and acrosome reaction activities are mediated via interaction with one or more sperm receptors. Extensive work in various species has resulted in the identification and isolation of this primary "receptor" for sperm (Bleil & Wassarman, 1980a,b; Sacco et al., 1981; Shabanowitz & O'Rand, 1988). Initial binding of sperm to the zona pellucida is supported by ZP3 via O-linked oligosaccharide side chains and complementary sperm binding proteins present in the sperm plasma membrane (Florman & Wassarman, 1985; Saling, 1989; Bleil & Wassarman, 1990; Wassarman 1990b; Saling, 1991). Pronase digestion of ZP3 results in small oligopeptides (1,500–6,000 Da) that bind to sperm and are as effective as native ZP3 in competitively inhibiting sperm binding to intact ZP3 (Florman & Wassarman, 1985). Acrosome reaction-triggering activity, in contrast, is lost when ZP3 is digested into small glycopeptides and seems to depend upon the integrity of the peptide backbone (Florman et al., 1984). It has been proposed that ZP3 binding and ZP3-induced acrosomal exocytosis can be dissociated from each other because sperm binding and acrosome reaction seem to represent two independent processes (Lee et al., 1987; Kopf, 1990). There are differences in the concentration dependence of ZP3 to express sperm binding activity and acrosome reaction-inducing activity. Induction of the acrosome reaction appears to be dependent on the correct spatial orientation of carbohydrate side chains of the ZP3 glycoprotein (Florman et al., 1984; Leyton & Saling, 1989a,b), which are thought to be responsible for aggregating sperm surface macromolecules that induce the acrosome reaction.

The molecular identity of the sperm surface receptor (or receptors) for ZP3 is beginning to be unveiled. A 95 kDa mouse (and human) protein with characteristics of a protein tyrosine kinase has been identified and proposed as a receptor for ZP3 (Leyton & Saling, 1989a,b; Leyton et al., 1992; Burks et al., 1995). On the other hand, it has been demonstrated that sp56 is the mouse sperm protein responsible for recognition of ZP3 (Bleil & Wassarman, 1990; Bookbinder et al., 1995). According to this model, multiple interactions of ZP3 at the sperm surface would lead to aggregation (patching) of sp56, which in turn would generate the signal to trigger membrane fusion (Bleil & Wassarman, 1990).

Different views also exist in regard to the signaling cascades that follow the interaction between ZP3 and its receptor(s) leading to acrosomal exocytosis. It has been proposed that murine (and human) sperm-zona pellucida interaction leads to tyrosine phosphorylation of p95 (referred to as zona receptor kinase - ZRK); ZRK, therefore, contains intrinsic transmembrane signalling potential (Leyton & Saling, 1989a; Burks et al., 1995). Other investigators have proposed that activation of GTP-binding proteins (G_i class) by ZP3 functions as a signal transducing element distal to ZP3-mediated interactions. This results in coupling of receptor occupancy to changes in ionic conductance and/or a variety of intracellular second messenger cascade systems (Kopf et al., 1986; Kopf, 1990).

2.3. The Aim of the Study

Because ZP proteins play a major role in the fertilization process, the molecular mechanisms of sperm-egg interaction and the involvement of individual zona pellucida proteins are subjects of intense investigations. Thus far, however, work has mainly been focused on ZP3 as the primary sperm receptor protein and the agonist that induces acrosomal exocytosis. For comparative studies and for the evaluation of the molecular mechanisms of spermatozoa-zona pellucida interaction in general, efforts have been made to elucidate the structures and physiological properties of mammalian ZP3 protein. These studies were not only important for the identification of functional domains of the zona pellucida; they were also important for identifying specific antigenic determinants of the zona that would be of use for the development of contraceptive vaccines. Thus, the immunological cross-reactivity of antisera against mammalian zona pellucida proteins has generated interest in the use of these proteins as antigens for the development of contraceptive agents (Sacco et al., 1981; Skinner et al., 1984; Henderson et al., 1988; Millar et al., 1989; Epifano & Dean, 1994; Naz et al., 1995).

We have used an immunological approach to identify and localize ZP2 protein and to evaluate possible functional roles of selected ZP2 domains. We show that anti-ZP2 peptide antibodies are sensitive and specific markers for ZP2 in mammalian oocytes. We also demonstrate that the antibodies are useful tools for the assessment of defined ZP2 epitopes that reveal physiological function in the sperm-zona pellucida interaction process.

3. MATERIALS AND METHODS

3.1. Generation of Anti-ZP2 Peptide Antisera

The sequences of the synthetic peptides were deduced from cDNA clones coding for mammalian ZP2 (Liang et al., 1990; Liang & Dean, 1993; Harris et al., 1994). As depicted in Figure 3, two amino acid sequences were chosen: (a) a ZP2 amino acid sequence that is homologous in the human and mouse ZP2 amino acid sequence (ZP2-common sequence) and (b) a human ZP2 amino acid sequence that differs from the mouse ZP2 sequence

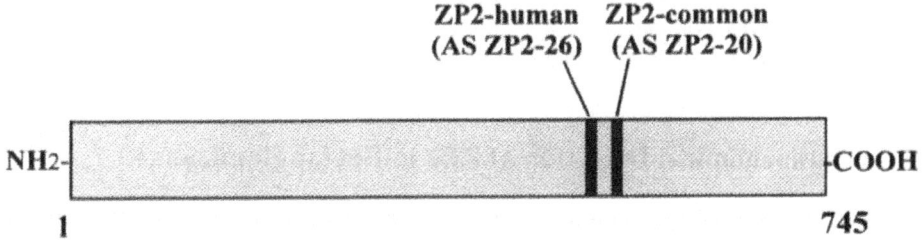

Figure 3. Diagram of the human ZP2 (ZPA) amino acid sequence (Liang & Dean, 1993). The positions of amino acid sequences chosen for generation of anti-ZP2 synthetic peptide antisera AS ZP2–20 and AS ZP2–26 are indicated.

(ZP2-human sequence). Peptides synthesized by the Merrifield solid-phase method (Barany & Merrifield, 1979) were kindly donated by Dr. Krause, Marburg, Germany. Cross-linking to keyhole limpet hemocyanin was performed as described (Fitzpatrick & Bundy, 1978). Generation of antisera and evaluation of titers of specific antibodies by ELISA were performed as described (Hinsch et al., 1989).

3.2. Semen Collection and Preparation

Frozen/thawed bovine semen, kindly provided by Dr. F. Müller-Schlösser (Besamungsstation Giessen der ZBH, Giessen Germany), was used. Ejaculates from mature Holstein Frisian/black and white bulls were collected with the aid of an artificial vagina. High quality semen complying with the official standards of motility and morphology as determined by the Arbeitsgemeinschaft Deutscher Rinderzüchter (ADR) was employed. Ejaculates were diluted at room temperature with glycerol-TRIS extender supplemented with egg-yolk. Subsequently, semen was processed at room temperature followed by a 3 h incubation at 4°C. Finally, semen was cryopreserved in a nitrogen atmosphere under appropriate conditions by using a freezing processor (Heede-Nielsson, Danmark). Thawing was performed by carefully moving the straws into a waterbath at 38°C for 25 sec.

3.3. Preparation of Oocytes

Bovine ovaries were obtained from the local slaughterhouse. Ovaries were removed after sacrificing and instantly cooled on ice. About two hours after the removal of ovaries, oocytes were prepared at room temperature and used for subsequent treatment.

Bovine mature oocytes were obtained by puncturing follicles of excised ovaries with fine steel needles and rinsing the aspirates. Oocytes were collected under a dissecting microscope and harvested free from cumulus cells using mouth-operated micropipets.

3.4. Competition-Hemizona-Assay

The competition-hemizona-assay was performed as shown in Figure 4. Denuded bovine oocytes were equally microbisected using a micromanipulator. Hemizonae were separated from oocyte particles by micropipetting. Matching hemizonae were incubated with test sera or pre-immune sera (deluted 1:30) for 2h at 38.5 °C, while bovine spermatozoa were prepared and capacitated for 4h at 38.5 °C. Thoroughly washed hemizonae were placed in a sperm suspension containing 0.5×10^6 motile spermatozoa/ml for 4h. Following sperm hemizona co-incubation, each hemizona was rinsed to remove loosely attached spermatozoa. The number of spermatozoa tightly bound to the outer surface of each hemizona was counted, and the hemizona index (HZI) was calculated (HZI = [number of sperm bound for test hemizona/number of sperm bound for control hemizona] x 100).

3.5. Immunochemical Detection of ZP2 in Bovine Hemizonae

Detection of anti-ZP2 antibodies bound to hemizonae was essentially performed as previously described (Hinsch et al., 1994; Oehninger et al., 1996). Hemizonae were prepared and mounted to glass slides and air dried over night. They can be stored exsiccated at room temperature for several weeks. For immunochemical studies, hemizonae were treated with antisera or pre-immune sera. Specific binding was visualized with anti-rabbit-

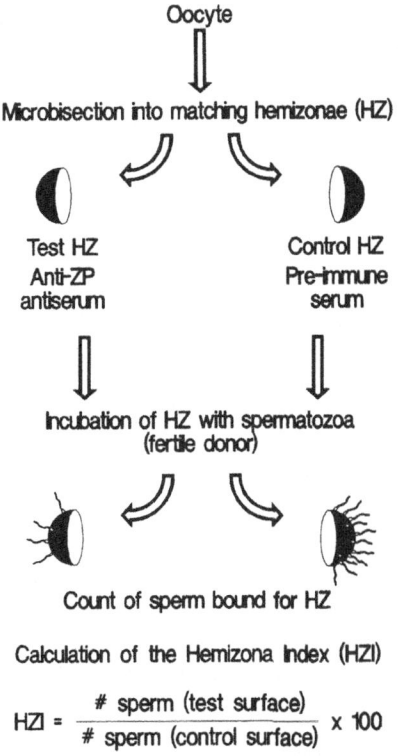

Oocyte

Microbisection into matching hemizonae (HZ)

Test HZ
Anti-ZP
antiserum

Control HZ
Pre-immune
serum

Incubation of HZ with spermatozoa
(fertile donor)

Count of sperm bound for HZ

Calculation of the Hemizona Index (HZI)

$$HZI = \frac{\text{\# sperm (test surface)}}{\text{\# sperm (control surface)}} \times 100$$

Figure 4. Diagram of the experimental design of the bovine competition-hemizona-assay.

IgG-antibodies (DAKO, Hamburg, Germany), the peroxidase-antiperoxidase method (DAKO, Hamburg, Germany) and 3,3'-diaminobenzidine (DAB, Sigma, Deisenhofen, Germany) as color reagent. Briefly, hemizonae were rehydrated with phosphate buffered saline (PBS). This step and all following steps were performed in a moist chamber at room temperature. After rehydration, PBS was removed and substituted with PBS supplemented with 3% (w/v) bovine serum albumin (PBS/BSA 3%) and incubated for 2 h. Thereafter, hemizonae were incubated with antisera AS ZP2–20 or AS ZP2–26 (diluted 1:30) for 1 h. The matching hemizonae were incubated with pre-immune sera (diluted 1:30) for 1 h. After removal of antisera and thoroughly washing, hemizonae were incubated with anti-rabbit-IgG-antibodies (diluted 1:100) for 30 min. Washed hemizonae were then incubated with peroxidase-antiperoxidase-immune-complex (diluted 1:100). Finally, hemizonae were rinsed with PBS, and the color reagent DAB was added for 20 min. Thereafter, hemizonae were washed and fixed in graded alcohols. Slides were mounted with Corbit balsam (Hecht, Kiel, Germany) and a cover slip. Results of immunostaining were evaluated and photographed using a Zeiss Axioskop microscope.

3.6. Statistical Analysis

Results are presented as means ± SEM. Statistical analysis was carried out using the one sample t-test.

Table 1. Competition-hemizona-assay with anti-ZP2 peptide antisera

Antiserum	# Sperm (test HZ)	# Sperm (control HZ)	HZI	p-value (HZI)	n
AS ZP2-20	28.60(± 3.10)	55.70(± 4.28)	51.26(± 4.38)	< 0.0001	10
AS ZP2-26	47.00(± 5.94)	45.40(± 5.28)	103.12(± 2.97)	0.3212	10

\# = values are mean numbers of sperm bound for outer HZ surface ± SEM.

4. RESULTS

Antisera were generated against synthetic peptides corresponding to defined ZP2 epitopes (Fig. 3). Selection of the synthetic peptides was based on their probability to be antigenic sequences as predicted by analysis of the primary structure (Krchnak et al., 1987). As determined by ELISA, the antisera specifically recognized their respective peptide employed as immunogen and did not cross-react with control peptides (not shown).

The competition-hemizona-assay (Fig. 4) was developed to assess the capacity of anti-ZP2 peptide antisera to interfere with the process of sperm-zona pellucida binding. As depicted in Table 1, test hemizonae preincubated with AS ZP2–20 revealed an average of about 29 spermatozoa bound to the outer surface, whereas the corresponding control hemizonae yielded a mean value of 56 tightly bound spermatozoa. The calculated hemizona index (about 51 on average) as well as the statistical analysis revealed a significant inhibition of sperm binding to the zona pellucida by AS ZP2–20 as compared to controls (p<0.0001). On the other hand, treatment of hemizonae with AS ZP2–26 did not influence sperm-zona interaction. Hemizonae treated with AS ZP2–26 bound 47 spermatozoa on average, and the number of sperm attached to control hemizonae was about 45 spermatozoa. The hemizona index (103 on average) suggests no influence of AS ZP2–26 antibodies on sperm-zona binding. No significant changes in sperm binding were calculated (p=0.3212).

Figure 5 presents phase contrast micrographs of hemizonae that were photographed after completion of the competition-hemizona-assay. Representative experiments with

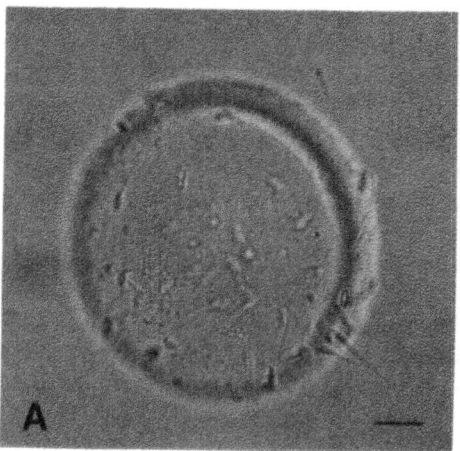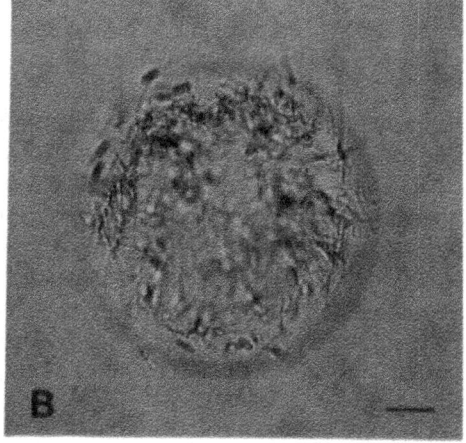

Figure 5. Phase contrast micrographs of bovine hemizonae after completion of the competition-hemizona-assay. Matching hemizonae were pre-treated with appropriate dilutions of AS ZP2–20 (A) and pre-immune serum (B), respectively. Bar: 20 μm.

hemizonae that were pre-treated either with antiserum AS ZP2–20 (Fig. 5A) or with the respective pre-immune serum (Fig. 5B) are shown. If a hemizona was incubated with AS ZP2–20, only few spermatozoa were tightly bound to its surface, whereas the matching control hemizona that was incubated with the same dilution of the respective pre-immune serum carried considerably more spermatozoa.

We further sought to evaluate, whether anti-ZP2 antibodies in fact bind to bovine hemizonae. As shown in Figure 6, No. 1, AS ZP2–20 treated hemizonae exhibited a homogeneous staining pattern throughout the zona. The intensity of color reaction resulted in a bright ring at the edge of the bisected zona pellucida; the amount of substrate deposition in the central part of the hemizona was about equal. Staining of the hemizona incubated with AS ZP2–26 appeared as a dark ring of substrate deposition close to the edge, whereas staining faded out towards the center of the zona pellucida (Fig. 6, No. 3). The periphery of the hemizona revealed a ragged, fading staining pattern towards the outer border, whereas the central portion of the hemizona exhibits a more smooth and homogeneous staining. Fern-like structures are visible in almost all hemizonae; the origin of this phenomenon is not clear, but it most probably represents artifacts due to mechanical and chemical processing of the zonae. Almost no staining was observed when matching hemizonae were treated with corresponding pre-immune sera (Fig. 6, No. 2 and No. 4).

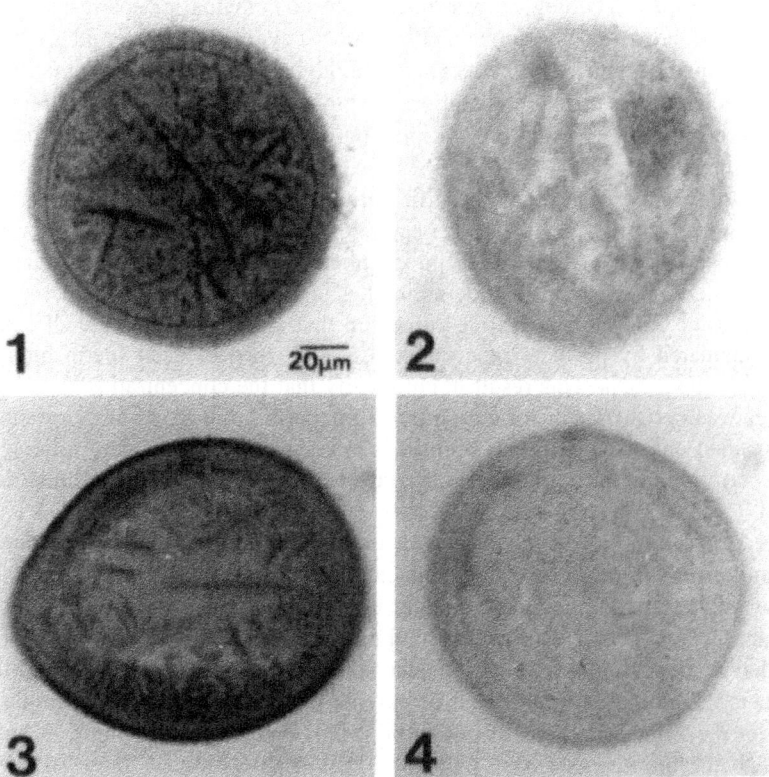

Figure 6. Immunochemical localization of ZP2 protein in bovine hemizonae. Air-dried hemizonae were incubated with antiserum (1:30 dilution) AS ZP2–20 (No. 1) or with antiserum AS ZP2–26 (No. 3). The incubation of the matching hemizona with the respective pre-immune serum (1:30 dilution) was used as control (No. 2 and No. 4).

5. DISCUSSION

We have generated and characterized anti-peptide antibodies against defined human ZP2 (ZPA) epitopes based on published ZP2 (ZPA) cDNA sequences.

The use of synthetic ZP2 peptides as immunogen provides several advantages over antibodies against purified ZP2 protein. One of the major problems of generating highly specific antibodies against zona pellucida proteins is the paucity of biological material. Synthetic peptides can be synthesized in sufficient quantities and with a high homogeneity. No contaminating proteins may interfere in the outcome of the immunization and generate non-specific antibodies. In a recent study, East & Dean (1984) generated monoclonal antibodies against purified mouse ZP2. Two of five antibodies cross-reacted with mouse ZP3. At that time, the authors discussed the possibility of the presence of common structural elements. Because Greve et al. (1982) produced a polyclonal antiserum to murine ZP2 that was unable to immunoprecipitate ZP3 and comparison of ZP2 and ZP3 cDNA reveales no amino acid sequence similarities, so the monoclonal antibodies (e.g. those generated by East & Dean, 1984) most probably do not detect a linear ZP2 amino acid epitope. The results obtained by East & Dean (1984) show that monoclonal antibodies against purified zona pellucida proteins do not guarantee antigen specificity.

Antisera against synthetic ZP2 peptides allow the characterization of defined linear epitopes of discrete ZP2 protein antigens. They can be used as tools for the determination of possible epitopes of biological relevance, such as domains for sperm-zona binding. It has to be noted, however, that antisera raised against synthetic ZP2 peptides react with the protein backbone of the ZP2 protein. Therefore, glycosylation as well as modifications of zona pellucida proteins during preparation and processing of zona pellucida (e.g. proteolysis, protein denaturation) might prevent binding of antibodies to their antigen and lead to inconsistent results.

Our study shows that highly specific antibodies and bovine gametes can be used as an in vitro model for the study of sperm-egg interaction. The results demonstrate that antibodies against a defined linear ZP2 epitope interfere with sperm-egg interaction. In the bovine system, ZP2-common peptide antibodies significantly inhibited sperm-zona binding, whereas the ZP2-human peptide antiserum that also detects bovine ZP2 protein did not. As calculated by the hemizona index, AS ZP2–20 decreased sperm binding to the hemizona to about 50%. This result might signify that the concentration or the affinity of antibodies was not sufficient for a total block of sperm binding. Another reason could be that more than one zona pellucida domain is responsible for secondary binding and thus further sperm binding sites that are independent of the ZP2-common peptide domain are provided.

AS ZP2–20 antigen is identical with the respective mouse and pig amino acid sequence and differs in one amino acid from cat, dog, and rabbit ZP2 amino acid sequences. Inhibition of sperm-zona pellucida binding indicates that this common ZP2 sequence resembles a species-independent sperm binding site. Recently, Koyoma et al. (1991) reported that a monoclonal antibody against porcine zona pellucida protein (MAb-5-H4) that detects the aminoterminal region of porcine pZP1 (which is a heterodimer and consists under reducing conditions of pZP2 and pZP4) could inhibit binding of human spermatozoa to the homologous zona pellucida. Epitope mapping of MAb-5-H4 revealed a particular amino acid sequence of the aminoterminal region of the porcine zona pellucida glycoprotein pZP4 as reactive epitope. This domain shows sequence homology to human ZP2 (ZPA) (8 out of 10 amino acids) but differs in 4 out of 10 amino acids with mouse ZP2 (ZPA); this epitope, in contrast to AS ZP2–20 and AS ZP2–26 that detect epitopes more towards the C-terminal region, is located at the aminoterminal of the ZPA sequence. Antibodies generated against the corresponding synthetic peptide recognized zona pellucida from pigs, humans and rabbits but not from mouse (Hasegawa et al., 1995). The antibodies inhibited porcine fertilization in vitro and had no effect on sperm-zona binding immediately after insemination. Applying the bovine competition-hemizona-assay, we found, after a 4h period of sperm-zona pellucida co-incubation, a significant decrease of tight sperm attachment to AS ZP2–20 treated hemizonae; however, initial loose binding was not affected by anti-ZP2 antibodies. The data obtained by Hasegawa et al. (1995) together with the results presented in this study suggest that antibodies against two different ZP2 epitopes, a species independent and a more species specific domain, inhibit secondary binding of spermatozoa to the zona pellucida. It has to be noted that proper sperm-zona binding and fertilization success could also be effected by steric hinderance of zona conformation or indirectly by inhibition of sperm enzyme activity through antibodies bound to a domain where ZP proteins are cleaved. However, since AS ZP2–26 antibodies strongly bind to hemizonae but do not affect sperm binding capacity, steric hindrance of AS ZP2–20 antibodies seems to be less probable.

We conclude that the ZP2-common peptide sequence reflects a ZP2 domain that is of physiological importance for sperm-zona pellucida binding. Our results may contribute to the understanding of crucial events that occur during sperm-zona pellucida interaction leading to fertilization. Moreover, the molecular mechanisms involved in this interaction and their regulation by physiological and pharmacological interventions may be unveiled. We estimate that our model may become very useful for the investigation of the molecular mechanisms of sperm-egg interaction. Further studies based on those results may also lead to the acquisition of knowledge with potential impact on human reproductive medicine (i.e. diagnosis and treatment of infertility and development of improved contraceptive strategies).

6. ACKNOWLEDGMENTS

The authors wish to thank Mrs. S. Gröger for excellent technical assistance.

7. REFERENCES

Austin CR 1961 The mammalian egg. Charles C Thomas, Springfield, Illinois.

Barany G & Merrifield RB 1979 Solid-phase peptide synthesis. In The Peptide Vol. 2A, pp 1–284. Eds E Gross & J Meienhofer. Academic press, New York.

Barros C, Capote C, Perez C, Crosby JA, Becker M & Deloannes A 1992 Immunodetection of acrosin during the acrosome reaction of hamster, guinea pig, and human spermatozoa. Biological Research (Chile) 25: 31–40.

Bleil JD & Wassarman PM 1980a Mammalian sperm-egg interaction: identification of a glycoprotein in mouse egg zonae pellucidae possessing receptor activity for sperm. Cell 20: 873–882.

Bleil JD & Wassarman PM 1980b Structure and function of the zona pellucida: identification and characterization of the proteins of the mouse oocyte's zona pellucida. Developmental Biology 76: 185–202.

Bleil JD & Wassarman PM 1983 Sperm-egg interactions in the mouse: sequence of events and induction of the acrosome reaction by a zona pellucida glycoprotein. Developmental Biology 95: 317–324.

Bleil JD & Wassarman PM 1986 Autoradiographic visualization of the mouse egg's sperm receptor bound to sperm. Journal of Cell Biology 102: 1363–1371.

Bleil JD, Greve JM & Wassarman PM 1988 Identification of a secondary sperm receptor in the mouse egg zona pellucida: role in maintenance of binding of acrosome-reacted sperm to eggs. Developmental Biology 128: 376–385.

Bleil JD & Wassarman PM 1990 Identification of a ZP3-binding protein acrosome-intact mouse sperm by photoaffinity cross-linking. Proceedings of the National Academy of Sciences USA 87: 5563–5567.

Bookbinder LH, Cheng A & Bleil JD 1995 Tissue- and species-specific expression of sp56, a mouse sperm fertilization protein. Science 269: 86–89.

Burks DJ, Carballada R, Moore HD & Saling PM 1995 The interaction of a tyrosine kinase from human sperm with the zona pellucida at fertilization. Science 269: 83–86.

Cowan AE, Myles DG & Koppel DE 1991 Migration of the guinea pig sperm membrane protein pH-20 from one localized surface domain to another does not occur by a simple diffusion-trapping mechanism. Developmental Biology 144: 189–198.

Dunbar BS 1983 Morphological, biochemical and immunochemical characterization of the mammalian zona pellucida. In Mechanism and Control of Animal Fertilization, pp 139–175. Ed. JF Hartman. Academic press, New York.

Dunbar BS, Prasad SV & Timmons TM 1991 Comparative Structure and Function of Mammalian Zonae Pellucidae. In A Comparative Overview of Mammalian Fertilization, pp 97–113. Eds BS Dunbar & MG O'Rand. Plenum Press, New York.

East IJ & Dean J 1984 Monoclonal antibodies as probes of the distribution of ZP-2, the major sulfated glycoprotein of the murine zona pellucida. Journal of Cell Biology 98: 795–800.

Epifano O & Dean J 1994 Biology and structure of the zona pellucida: a target for immuncontraception. Reproduction, Fertility and Development 6: 319–330.

Epifano O, Liang LF & Dean J 1995 Mouse ZP1 encodes a zona pellucida protein homologous to egg envilope proteins in mammals and fish. Journal of Biological Chemistry 270(45): 27254–27258.

Familiari G, Nottola SA, Micara G, Aragona C & Motta PM 1988 Is the sperm-binding capacity of the zona pellucida linked to its surface structure? Journal of in vitro Fertilization and Embryo Transfer 6: 134–143.

Familiari G, Nottola SA, Macchiarelli G, Micara G, Aragona C & Motta PM 1992 Human zona pellucida during in vitro fertilization. Molecular Reproduction and Development 32: 51–61.

Fitzpatrick FA & Bundy GL 1978 Hapten mimic elicits antibodies recognizing prostaglandin E2. Proceedings of the National Academy of Sciences USA 75: 2689–2693.

Florman HM, Bechtol KB & Wassarman PM 1984 Enzymatic dissection of the functions of the mouse egg's receptor for sperm. Developmental Biology 106: 243–255.

Florman HM & Wassarman PM 1985 O-linked oligosaccharides of mouse egg ZP3 account for its sperm receptor activity. Cell 41: 313–324.

Greve JM, Salzmann GS, Roller RJ & Wassarman PM 1982 Biosynthesis of the major zona pellucida glycoprotein secreted by oocytes during mammalian oogenesis. Cell 31: 749–759

Greve JM & Wassarman PM 1985 Mouse egg extracellular coat is a matrix of interconnected filaments possessing a structural repeat. Journal of Molecular Biology 181: 253–264.

Harris JD, Hibler DW, Fontenot GK, Hsu KT, Yurewicz EC & Sacco AG 1994 Cloning and characterization of zona pellucida genes and cDNAs from a variety of mammalian species: The ZPA, ZPB and ZPC gene families. DNA sequence. Journal of Sequencing and Mapping 4: 361–393.

Hasegawa A, Yamasaki N, Inoue M, Koyama K & Isojima S 1995 Analysis of an epitope sequence recognized by a monoclonal antibody MAb-5H4 against a porcine zona pellucida glycoprotein (pZP4) that blocks fertilization. Journal of Reproduction and Fertility 105: 295–302.

Henderson CJ, Hulme MJ & Aitken RJ 1988 Contraceptive potential of antibodies to the zona pellucida. Journal of Reproduction and Fertility 83: 325–343.

Hinsch KD, Tychowiecka I, Gausepohl H, Frank R, Rosenthal W & Schultz G 1989 Tissue distribution of ß$_1$- and ß$_2$-subunits of regulatory guanine nucleotide-binding proteins. Biochimica et Biophysica Acta 1013: 60–67.

Hinsch KD, Hinsch E, Meinecke B, Töpfer-Petersen E, Pfisterer S & Schill WB 1994 Identification of mouse ZP3 protein in mammalian oocytes with antisera against synthetic ZP3 peptides. Biology of Reproduction 51: 193–204.

Huang TTF & Yanagimachi R 1984 Fucoidin inhibits attachment of guinea pig spermatozoa to the zona pellucida through binding to the inner acrosomal membrane and equatorial domains. Experimental Cell Research 153: 363–373.

Kopf GS, Woolkalis MJ & Gerton GL 1986 Evidence of a guanine nucleotide-binding regulatory protein in invertebrate and mammalian sperm: identification by islet-activating protein-catalyzed ADP-ribosylation and immunochemical methods. Journal of Biological Chemistry 261: 7327–7331.

Kopf GS 1990 Zona pellucida-mediated signal transduction in mammalian spermatozoa. Journal of Reproduction and Fertility 42: 33–49.

Koyama K, Hasegawa A, Yamasaki N, Inoue M & Isojima S 1991 Blocking of human sperm-zona interaction by monoclonal antibodies to a glycoprotein family (ZP4) of porcine zona pellucida. Biology of Reproduction 45: 727–735.

Krchnak V, Mach O & Maly A 1987 Computer prediction of potential immunogenic determinants from protein amino acid sequences. Analytical Biochemistry 165: 200–207.

Lee MA, Kopf GS & Storey BT 1987 Effects of phorbol esters and a diacylglycerol on the mouse sperm acrosome reaction induced by the zona pellucida. Biology of Reproduction 36: 617–627.

Leyton L & Saling P 1989a 95 kD sperm proteins bind ZP3 and serve as tyrosine kinase substrates in response to zona binding. Cell 57: 1123–1130.

Leyton L & Saling P 1989b Evidence that aggregation of mouse sperm receptors by ZP3 triggers the acrosome reaction. Journal of Cell Biology 108: 2163–2168.

Leyton L, Legum P, Bunch D & Saling P 1992 Regulation of mouse gamete interaction by a sperm tyrosine kinase. Proceedings of the National Academy of Sciences USA 89: 11692–11695.

Liang LF, Chamow SM & Dean J 1990 Oocyte-specific expression of mouse ZP-2: developmental regulation of the zona pellucida genes. Molecular and Cellular Biology 10: 1507–1515.

Liang LF & Dean J 1993 Conservation of mammalian secondary sperm receptor genes enables the promoter of the human gene to function in mouse oocytes. Developmental Biology 156: 399–408.

Millar SE, Chamow SM, Baur AW, Oliver C, Robey F & Dean J 1989 Vaccination with a synthetic zona pellucida peptide procedures long-term contraception in female mice. Science 246: 935–938.

Miller DJ, Gong X, Decker G & Shur BD 1993 Egg corticle granule N-acetylglucosaminidase is required for the mouse zona block to polyspermy. Journal of Cell Biology 123: 1431–1440.

Moller CC & Wassarman PM 1989 Characterization of a proteinase that cleaves zona pellucida glycoprotein ZP2 following activation of mouse eggs. Developmental Biology 132: 103–112.

Moller CC, Bleil JD, Kinloch RA & Wasserman PM 1990 Structural and functional relationships between mouse and hamster zona pellucida glycoproteins. Developmental Biology 137: 276–286.

Moos J, Faundes D, Kopf GS & Shultz RM 1995 Composition of the human zona pellucida and modifications following fertilization. Mol Hum Reprod 1, Human Reproduction 10: 2467–2471.

Mortillo S & Wassarman PM 1991 Differential binding of gold-labeled zona pellucida glycoproteins mZP2 and mZP3 to mouse sperm membrane compartments. Development 113: 141–149.

Naz RK, Sacco A, Singh O, Pal R & Talwar GP 1995 Development of contraceptive vaccines for humans using antigens derived from gametes (spermatozoa and zona pellucida) and hormones (human chorionic gonadotrophin): current status. Human Reproduction Update (1): 1–18.

Oehninger S, Hinsch E, Kolm P, Schill WB, Hodgen GD & Hinsch KD 1996 Use of specific zona pellucida protein 3 (ZP3) antiserum as a clinical marker for human zona pellucida integrity and function. Fertility and Sterility 65: 139–145.

Overstreet JW, Lin J, Judin AI, Meyers SA, Primakoff P, Myles DG, Katz DF & Vandevoort CA 1995 Location of the pH-20 protein on acrosome-intact and acrosome-reacted spermatozoa of cynomolgus macaques. Biology of Reproduction 52: 105–114.

Phillips DM & Shalgi R 1982 Surface architecture in the mouse and hamster zona pellucida and oocyte. Journal of Ultrastructure Research 72: 1–12.

Richardson RT, Nikolajczyk BS, Abdullah LH, Beavers JC & O'Rand MG 1991 Localization of rabbit sperm acrosin during the acrosome reaction induced by immobilized zona matrix. Biology of Reproduction 45: 20–26.

Sacco AG, Yurewicz EC, Subramanin MG & DeMayo FJ 1981 Zona pellucida composition: Species crossreactivity and contraceptive potential of antiserum to a purified pig zona antigen (PPZA). Biology of Reproduction 25: 997–1008.

Saling PM 1989 Mammalian sperm interaction with extracellular matrices of the egg. Oxford Reviews of Reproductive Biology 11: 339–88.

Saling PM 1991 How the egg regulates sperm function during gamete interaction: Facts and fantasies. Biology of Reproduction 44: 246–251.

Shabanowitz RB & O'Rand MG 1988 Molecular changes in the human zona pellucida associated with fertilization and human sperm receptor. Annals of the New York Academy of Sciences: 621–632.

Skinner SM, Mills T, Kirchick HJ & Dunbar BS 1984 Immunisation with zona pellucida proteins results in abnormal ovarian follicular differentiation and inhibition of gonadotropin-induced steroid secretion. Endocrinology 115: 2418–2432.

Tesarik J, Drahorad J & Peknicova J 1988 Subcellular immunochemical localization of acrosin in human spermatozoa during the acrosome reaction and zona pellucida penetration. Fertility and Sterility 50: 133–141.

Tesarik J, Drahorad J, Testart J & Mendoza C 1990 Acrosin activation follows its surface exposure and precedes membrane fusion in human sperm acrosome reaction. Development 110: 391–400.

Thillai-Koothan P, van Duin M & Aitken RJ 1993 Cloning, sequencing, and oocyte-specific expression of the marmoset sperm receptor protein ZP3. Zygote 2: 1–9.

Töpfer-Petersen E & Henschen A 1987 Acrosin shows zona and fucose binding, novel properties for a serine proteinase. FEBS Letters 226: 38–42.

Wassarman PM 1987 Early events in mammalian fertilization. Annual Review of Cell Biology 3: 109–142.

Wassarman PM 1988 Zona pellucida glycoproteins. Annual Reviews of Biochemistry 57: 415–442.

Wassarman PM, Bleil JD, Fimiani C, Florman HM, Greve JM, Kinloch R, Moller CC, Mortillo S, Roller R, Salzmann G & Vasquez M 1989 In The Mammalian Egg Coat: Structure and Function, pp 18–37. Ed J Dietl. Springer-Verlag, Berlin and New York.

Wassarman PM 1990a Profile of a mammalian sperm receptor. Development 108: 1–17.

Wassarman PM 1990b Regulation of mammalian fertilization by zona pellucida glycoproteins. Journal of Reproduction and Fertility 42: 79–87.

Wassarman P & Mortillo S 1991 Structure of the mouse egg extracellular coat. International Review of Cytology 130: 85–110.

Yanagimachi R & Teichman RJ 1972 Cytochemical demonstration of acrosomal proteinase in mammalian and avian spermatozoa by a silver proteinate method. Biology of Reproduction 6: 87–97.

Yanagimachi R 1981 Mechanisms of fertilization in mammals. In Fertilization and embryonic development in vitro, pp 81–187. Eds L Mastroianni & JD Biggers. Plenum Press, New York.

Yanagimachi R & Phillips DM 1984 The status of acrosomal caps of hamster spermatozoa immediately before fertilization in vitro. Gamete Research 9: 1–19.

Yanagimachi R 1994 Mammalian fertilization. In The Physiology of Reproduction, Second Edition, pp 189–317. Eds E Knobil & JD Neill. Raven Press, Ltd., New York.

INDEX

The manufacturer's authorised representative in the EU is Springer
Nature Customer Service Centre GmbH, Europaplatz 3, 69115 Heidelberg,
Germany. If you have any concerns regarding our products, please
contact ProductSafety@springernature.com

Printed and bound by CPI Group (UK) Ltd, Croydon, CR0 4YY
23/04/2026
02095607-0008